Tribological Properties, Performance, and Applications of Biocomposites

Tribological Properties, Performance, and Applications of Biocomposites

Edited by Chandrasekar Muthukumar, Senthilkumar Krishnasamy, Senthil Muthu Kumar Thiagamani, and Ganesan Chinnachamy

WILEY-VCH

The Editors

Dr. Chandrasekar Muthukumar
Hindustan Institute of Technology and Science
Department of Aeronautical Engineering
603103 Chennai, Tamil Nadu
India

Dr. Senthilkumar Krishnasamy
PSG Institute of Technology and Applied Research
Department of Mechanical Engineering
641062 Coimbatore, Tamilnadu, India

Dr. Senthil Muthu Kumar Thiagamani
Kalasalingam Academy of Research and Education
Department of Mechanical Engineering
626126 Krishnankoil, Tamil Nadu
India

Dr. Ganesan Chinnachamy
Hindustan Institute of Technology and Science
Department of Aeronautical Engineering
Padur, Kelambakkam
603103 Chennai, Tamil Nadu
India

Cover Image: © Vink Fan/Shutterstock

Library of Congress Card No.: applied for

British Library Cataloguing-in-Publication Data
A catalogue record for this book is available from the British Library.

Bibliographic information published by the Deutsche Nationalbibliothek
The Deutsche Nationalbibliothek lists this publication in the Deutsche Nationalbibliografie; detailed bibliographic data are available on the Internet at <http://dnb.d-nb.de>.

Print ISBN: 978-3-527-35053-7
ePDF ISBN: 978-3-527-83806-6
ePub ISBN: 978-3-527-83807-3
oBook ISBN: 978-3-527-83808-0

Typesetting Straive, Chennai, India
Printing and Binding CPI Group (UK) Ltd, Croydon, CR0 4YY

C9783527350537_290923

Contents

Preface *xiii*

1 **Tribological Characterization of Biocomposites: An Overview** *1*
Manickam Ramesh, Thangamani Vinitha, and Manickam Tamil Selvan
1.1 Introduction *1*
1.2 Tribological Characterization *2*
1.2.1 Flax Reinforcement *3*
1.2.2 Coconut Coir Reinforcement *4*
1.2.3 Banana Reinforcement *4*
1.2.4 Hemp Reinforcement *4*
1.2.5 Ramie Reinforcement *5*
1.2.6 *Calotropis gigantea* Reinforcement *5*
1.2.7 Kenaf Reinforcement *6*
1.2.8 Betel Nut Fibers *7*
1.3 Parameters Influencing the Tribological Characteristics *8*
1.3.1 Impact of Reinforcement Orientation on Wear Behavior *8*
1.3.2 Effect of Reinforcement Volume Fraction on Wear Behavior *9*
1.3.3 Effect of Fillers on Wear Behavior *11*
1.3.4 Influence of Surface Modification on Wear Behavior *11*
1.4 Morphology Analysis of Tribological Characteristics *12*
1.5 Conclusion *14*
References *15*

2 **Tribological Properties of the Natural Fiber-Reinforced Epoxy Composites** *19*
Lin Feng Ng and Mohd Yazid Yahya
2.1 Introduction *19*
2.2 Fiber-Reinforced Composites *20*
2.3 Cellulosic Natural Fibers *22*
2.4 Impact of Tribology on the Environment and Industry *23*

2.5 Tribological Properties of FRPs *25*
2.5.1 Tribological Properties of Natural Fiber-Reinforced Epoxy Composites *25*
2.5.2 Tribological Properties of Natural Fiber-Reinforced Epoxy Hybrid Composites *30*
2.6 Conclusion *33*
References *34*

3 Wear Properties of Flax/Epoxy-Based Composites With Different Machining Parameters *39*
K.R. Sumesh, Petr Spatenka, and G. Rajeshkumar
3.1 Introduction *39*
3.2 Materials and Methods *40*
3.2.1 Method *40*
3.2.2 Wear Testing *40*
3.3 Results and Discussion *41*
3.3.1 Wear Results *41*
3.4 Conclusions *43*
References *43*

4 Polyester-Based Biocomposites for Tribological Applications *47*
Anand Gobiraman, Santhosh Nagaraja, and Vishvanathperumal Sathiyamoorthi
4.1 Introduction: Background and Driving Forces *47*
4.2 Materials and Methods *49*
4.2.1 Natural Fibers *49*
4.2.2 Polyester–Natural Fiber Composites *49*
4.2.3 Hybrid Polyester—Composites *50*
4.2.4 Methods of Production of Biocomposites *51*
4.2.4.1 Stratification *51*
4.2.4.2 Hand Lay-Up Method *52*
4.2.4.3 Vacuum Bagging Technique *52*
4.2.4.4 Tribological Tests on Natural Fiber-Reinforced Polyester-Based Biocomposites *52*
4.3 Tribological Characteristics of Polyester-Based Biocomposites *53*
4.4 Polyester-Based Biocomposites for Tribological Applications *59*
4.5 Conclusions *60*
References *61*

5 Tribological Properties of the Natural Fiber-Reinforced Vinyl Ester Composites *65*
Krushna Gouda, Muthukumar Chandrasekar, Vellaichamy Parthasarathy, Senthilkumar Krishnasamy, and Senthil Muthu Kumar Thiagamani
5.1 Introduction *65*
5.2 Natural Fiber-Based VE Composite *70*

5.3 Problems Associated with Natural Fiber-Based Composite *71*
5.4 Conclusion *71*
References *71*

6 Friction and Sliding Wear Properties of the Natural Fiber-Reinforced Polypropylene Composites *75*
Emel Kuram
6.1 Introduction *75*
6.2 Polypropylene *76*
6.3 Natural Fibers *76*
6.4 Natural Fiber-Reinforced PP Composites *80*
6.5 Tribological Properties of Natural Fiber-Reinforced PP Composites *83*
6.5.1 Friction Coefficient of Natural Fiber-Reinforced PP Composites *86*
6.5.2 Wear Behavior of Natural Fiber-Reinforced PP Composites *90*
6.6 Conclusions *94*
Acknowledgments *94*
References *95*

7 Wear Behavior of the Natural Fiber-Reinforced Thermoplastic Composites *105*
Ramu Sundaramoorthy, Vellaichamy Parthasarathy, Jeyanthi Subramanian, Lin Feng Ng, and Naveen Jesuarockiam
7.1 Introduction *105*
7.2 Wear Testing Methods *105*
7.3 Factors Affecting Wear Behavior of the Composite *107*
7.4 Motion Type *107*
7.5 Load *107*
7.6 Velocity *107*
7.7 Temperature *108*
7.8 Test Duration *108*
7.9 Performance Metrics From the Wear Test *108*
7.10 Wear Studies on Natural Fiber-Reinforced Thermoplastic Composites *109*
7.11 Conclusion *113*
References *113*

8 Tribological Characterization of the Natural Fiber-Reinforced Polyimide Composites *115*
Anand Gobiraman, Santhosh Nagaraja, and Vishvanathperumal Sathiyamoorthi
8.1 Introduction: Background and Driving Forces *115*
8.2 Materials and Methods *117*
8.2.1 Natural Fibers *118*
8.2.2 Methods of Production of Natural Fiber-Reinforced Polymer Composites *119*

8.2.2.1 Stratification *120*
8.2.2.2 Hand Layup *120*
8.2.2.3 Vacuum Bagging Technique *121*
8.3 Polyimides *121*
8.4 Natural Fibers/Polyimides Composites *121*
8.5 Tribological Applications of Natural Fibers/Polyimides Composites *122*
8.6 Conclusions *124*
References *125*

9 Investigations of the Friction and Wear Resistance of the Natural Fiber-Reinforced Polyamide Composites *129*
İbrahim Can Kaymaz, Alperen Doğru, Miray Batıkan Kandemir, and Mehmet Özgür Seydibeyoğlu
9.1 Introduction *129*
9.1.1 Thermosetting *130*
9.1.2 Thermoplastics *130*
9.1.3 Thermoplastic Composites *131*
9.1.4 Thermoplastic Polymer Matrix *132*
9.1.5 Fibers *133*
9.2 Natural Fiber-Reinforcement Polyamide *134*
9.2.1 Polyamide *134*
9.2.2 Natural Fibers *135*
9.2.2.1 Animal Fiber *135*
9.2.2.2 Plant Fiber *135*
9.2.3 Mineral Fiber *141*
9.2.4 Production *141*
9.3 Friction and Wear Resistance at Natural Fiber-Reinforcement Polyamide *142*
9.3.1 Friction *142*
9.3.2 Wear *145*
9.3.3 Testing and Measurement *148*
9.3.3.1 Friction Test Methodologies *148*
9.3.3.2 Wear Test Methodologies *148*
9.3.4 Applications *149*
References *150*

10 Friction and Wear Resistance of the Natural Fiber-Reinforced Polymer Composites With Metal Oxide Fillers *159*
Niket Suresh Powar and Mariyappan Shanmugam
10.1 Introduction *159*
10.2 Oil Palm Fiber *160*
10.3 Jute Fiber *161*
10.4 Bamboo Fiber *162*

10.5 Coconut Fiber *164*
10.6 Conclusion *164*
References *165*

11 Investigation of Sliding Wear Properties of Nanofiller-Based Biocomposites *167*
Ajish Babu, Anusree Thilak, Harikrishnan Pulikkalparambil, Sandhya Alice Varghese, Sanjay Mavinkere Rangappa, Kuruvilla Joseph, and Suchart Siengchin
11.1 Introduction *167*
11.2 Wear General Aspects *168*
11.3 Methods to Measure Wear *170*
11.4 Sliding Wear in Polymer Composites *171*
11.5 Sliding Wear in Biocomposites, General *173*
11.5.1 Sliding Wear Property of Biofiller Incorporated Biopolymer Composite *173*
11.5.2 Sliding Wear Property of Synthetic/Inorganic Filler Incorporated Biopolymer Composite *175*
11.6 Conclusion *177*
Acknowledgment *177*
References *177*

12 Friction and Wear Properties of Biocomposites for Dental, Orthopedic, and Biomedical Applications *185*
Piyush Gaur, Chandrasekar Muthukumar, and V. Parthasarathy
12.1 Introduction *185*
12.2 Desired Properties and Classification of Biomaterials *188*
12.2.1 Desired Properties of Biomaterials *188*
12.2.2 Classification of Biomaterials *189*
12.2.2.1 Metallic Biomaterials *189*
12.2.2.2 Ceramic Biomaterials *193*
12.2.2.3 Composite Biomaterials *193*
12.3 Wear of Biomaterials *194*
12.3.1 Wear Testing Methods *195*
12.3.2 Friction and Wear Characterization Techniques for Biomaterials *196*
12.4 Friction and Wear Properties of Biocomposites Used in Different Biomedical Applications *197*
12.4.1 Dental Applications *197*
12.4.1.1 Friction and Wear of Dental Resins *199*
12.4.2 Orthopedic Applications *200*
12.4.2.1 Friction and Wear of Biocomposites in Orthopedics Applications *203*
12.5 Conclusion *207*
References *207*

13 Wear and Friction Behavior of Biocomposites Fabricated Through Additive Manufacturing *219*
Manickam Ramesh, Kanagaraj Niranjana, and Manickam Tamil Selvan
13.1 Introduction *219*
13.2 Additive Manufacturing of Biocomposites *220*
13.3 Fabrication of Biocomposites Using AM *222*
13.4 Types of Wear Behavior Based on Its Processes, Effects, and Environment *222*
13.4.1 Adhesion Wear *223*
13.4.2 Abrasive Wear *224*
13.4.3 Erosive Wear *225*
13.4.4 Fatigue Wear *226*
13.4.5 Corrosive or Oxidative Wear *226*
13.4.6 Fretting Wear *226*
13.5 Determining the Level of Specimen Deterioration *227*
13.6 Wear and Frictional Characteristics of AM Products *228*
13.7 Method of Testing the Wear and Friction in the AM Parts *228*
13.7.1 Pin-on-Disk or Tribometer *239*
13.7.2 Pin-on-Drum *239*
13.7.3 Repeated Impact Wear Test *240*
13.7.4 Acoustic Emission Monitoring Test *241*
13.7.5 Rubbing Test *241*
13.8 Conclusion *241*
References *242*

14 Influence of Fiber Treatment on the Wear Properties of Biocomposites *247*
Anthony Chidi Ezika, Emmanuel Rotimi Sadiku, Raphael Stone Odera, Uzoma Ebenezer Enwerem, Victor Ugochukwu Okpechi, Martin Emeka Ibenta, and Shadrack Chukwuebuka Ugwu
14.1 Introduction *247*
14.2 Fibers *248*
14.2.1 NF Reinforcement *249*
14.2.2 Treatments of NFs *250*
14.2.2.1 Types of NF Treatment *250*
14.3 Biocomposites *254*
14.3.1 Classification of Biocomposites *254*
14.3.2 Natural Fiber-Polymer Composites *254*
14.3.3 Tribological Properties of NF-Reinforced Composites *255*
14.4 Influence of Fiber Treatment on the Wear Properties of NF-Filled Polymer *256*
14.4.1 Influence of Fiber Treatment on the Wear Properties of NF-Reinforced Epoxy Composites *257*
14.4.2 Influence of Fiber Treatment on the Wear Properties of NF- Reinforced Polyester Composites *260*

14.4.3 Influence of Fiber Treatment on the Wear Behavior of NF Reinforced Vinyl Ester Composite *262*
14.4.4 Influence of Fiber Treatment on the Wear Properties of NF- Reinforced Polypropylene Composites *263*
14.4.5 Influence of Fiber Treatment on the Wear Properties of NF-Reinforced Polylactic Acid Composites *265*
14.4.6 Influence of Fiber Treatment on the Wear Properties of NF-Reinforced High-Density Polyethylene Composites *267*
14.4.7 Influence of Fiber Treatment on the Wear Properties of NF- Reinforced Low-Density Polyethylene Composites *267*
14.4.8 Influence of Fiber Treatment on the Wear Properties of NF- Reinforced PET Composites *269*
14.4.9 Influence of Fiber Treatment on the Wear Properties of NF- Reinforced Polyamide Composites *269*
14.4.10 Influence of Fiber Treatment on the Wear Properties of NF-Reinforced Hybrid Biocomposites (Fiber Blending + Polymer Blending) Composites *270*
14.5 Conclusion *273*
References *273*

Index *285*

Preface

Biocomposites are touted as a promising substitute to conventional composites in various applications due to their biodegradable characteristics, low density, and moderate mechanical properties. As these materials are intended for applications in the automotive, aerospace, biomedical industries, etc., it is essential to explore their tribological behavior.

The chapters in this book cover a broad range of topics, including the tribological properties of biocomposite based on thermoset and thermoplastic resins, the influence of fiber pretreatment techniques, and nanofillers for the enhanced tribological performance. In addition to this, the characterization methods employed to assess wear and friction characteristics of the biocomposites were discussed.

This book aims to provide researchers, engineers, and students with a comprehensive knowledge of tribology in the context of fundamental principles, experimental techniques, and practical applications. This book will serve as a valuable resource to understand the factors influencing the tribological performance of the biocomposite materials and their suitability for various applications. We sincerely hope that this book will inspire further research and innovation in the field of tribology and can unlock the full potential of biocomposites.

The contributors to this book are distinguished researchers and experts in the field of tribology and biocomposites. Their collective expertise enables us to gain deeper insights into the tribological behavior of biocomposites, making this book a valuable reference for both academia and industry. So, we express our sincere appreciation to the contributors.

1

Tribological Characterization of Biocomposites: An Overview

Manickam Ramesh[1], Thangamani Vinitha[2], and Manickam Tamil Selvan[1]

[1] *KIT-Kalaignarkarunanidhi Institute of Technology, Department of Mechanical Engineering, Coimbatore, 641402, Tamil Nadu, India*
[2] *KIT-Kalaignarkarunanidhi Institute of Technology, Department of Aeronautical Engineering, Coimbatore, 641402, Tamil Nadu, India*

1.1 Introduction

The desire for environmentally acceptable materials fueled the establishment of a slew of pollution control laws, and the engineering imperative for cost-effectiveness in all areas pushed us to seek low-cost alternatives. Fossil-emerged sources are limited; hence, researchers and experts are now looking for alternative sources of conventional sources. Green technology, housing, solutions, energy, lifestyle, and materials are all part of the green environment [1, 2]. Natural fiber polymer composites (NFPCs) are a versatile material with a wide range of applications due to their capabilities and unique characteristics. Sisal, coconut coir, jute, *Calotropis gigantea*, kenaf, palm, banana, bamboo, bagasse, flax, and hemp are among the most often used natural fibers. Natural fibers provide several advantages, including low price, low mass per volume, minimal energy inputs, and superior mechanical qualities [3, 4]. Plant fibers, on the other hand, have some limitations. They can absorb moisture from the environment, resulting in a weak connection between the resin and the reinforcement. To overcome these conditions, fibers require some chemical treatment to modify their surfaces [5]. NFPCs are used for a wide range of engineering applications, like structural/nonstructural and tribological applications, because of their significant qualities. Biocomposites are used in the automobile industry to produce different parts like window linings, bike mudguard headliners, package trays, cupboards, and other vehicle internal spare parts. Other applications such as sliding panels, linkage, bearings, and bushings are fast-growing. Bio-composites are occasionally subjected to a variety of tribological loading environments, exposing the component to various forms of wear mechanisms such as adhesive, erosion, corrosion wear, and two-and three-body abrasive sliding wear. To improve the usefulness of composites in various technical sectors, it is essential to examine and investigate their tribological performance [6].

Tribological Properties, Performance, and Applications of Biocomposites, First Edition.
Edited by Chandrasekar Muthukumar, Senthilkumar Krishnasamy,
Senthil Muthu Kumar Thiagamani, and Ganesan Chinnachamy.

1.2 Tribological Characterization

The term "tribology" derives from the Greek word tribos, which means "rubbing." Tribology is the study of lubrication, wear, and friction between surfaces in relative motion. Friction is the force that prevents two bodies from moving in the same direction. The friction coefficient is the proportion of the frictional force to the perpendicular force acting on the outer layer. The sources of tribological effects are presented in Figure 1.1 [7].

It is a dimensionless number that indicates how much friction exists between two surfaces. Tribometers are devices that are used to measure the friction coefficient. The friction coefficient is independent of the contact area and sliding speed and is determined by the surface roughness and nature of the material. When two surfaces come into contact, they wear away at each other, removing material. Adhesive wear, abrasive wear, surface fatigue wear, and corrosive wear are the four types of wear that can occur as a result of friction. When two adjacent surfaces slide against each other, adhesive wear occurs. Large values of friction coefficients occur from adhesion wear. Abrasive wear is the wear that occurs when a rough surface moves over a soft surface [7].

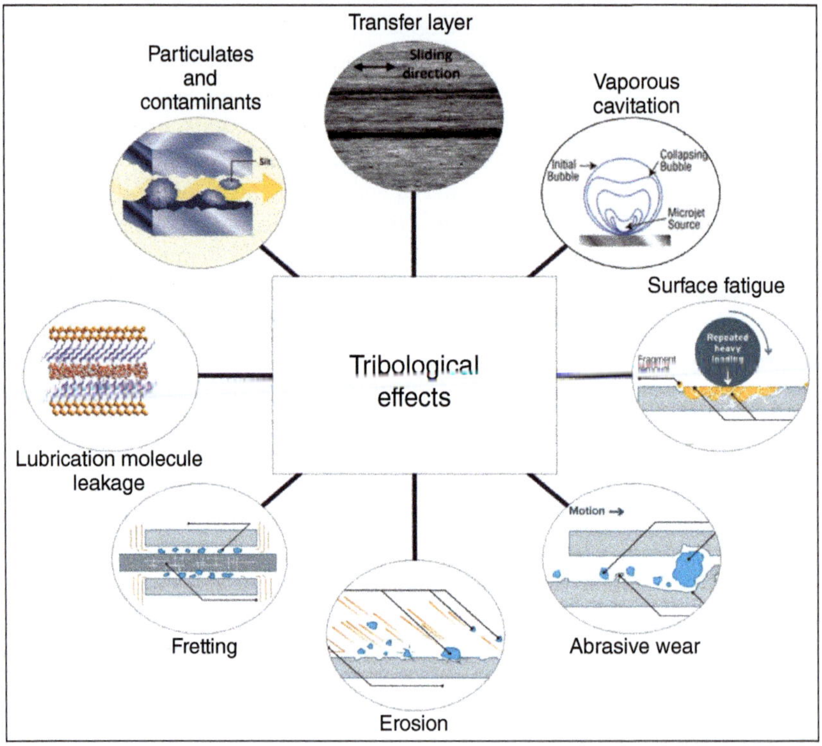

Figure 1.1 Various sources of tribological effects. Source: Karthikeyan et al. [7]/with permission of Sage Publications, Inc.

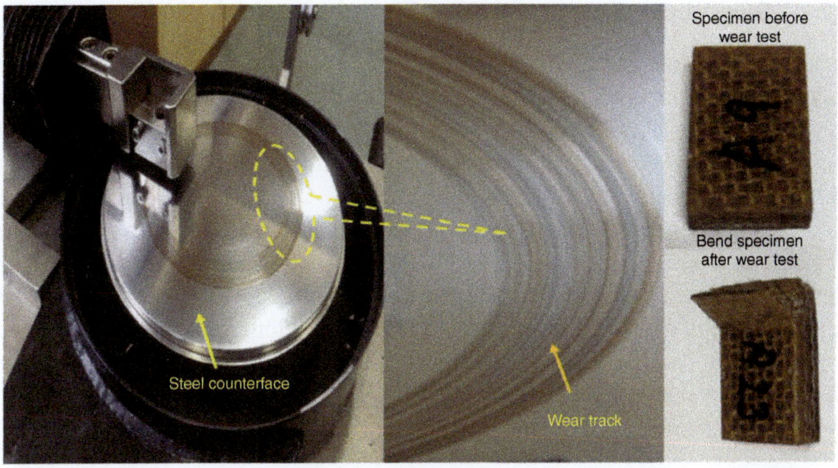

Figure 1.2 Wear experiment for jute/hemp/sisal epoxy laminates. Source: Chaudhary et al. [6]/with permission of Springer Nature.

1.2.1 Flax Reinforcement

The tribological performance of low-cost and regionally available jute/hemp/flax fibers was checked by using the hand-layup method. The tribological performance of the novel bio-composites was evaluated in terms of frictional characteristics and sliding wear in dry contact, employing a range of process parameters such as applied load, sliding speed, and sliding distance. Natural fibers with epoxy polymer improved the wear resistance rate, and the effect of speed is insignificant on the coefficient of friction (CoF) at higher speeds [6]. Natural fibers (jute, hemp, and flax) combined with epoxy polymers improved the tribological performance of all the laminates studied. At higher speeds, the effect of speed on CoF is minimal, while the impact of applied load on CoF is insignificant. Each form of specimen has a distinct average CoF. As a result, the friction conditions on the created composites are influenced by the type of natural fiber used. Epoxy's wear performance has been significantly improved by using various combinations of natural fibers as reinforcement. Sliding speed, in addition to the applied stresses, has a significant influence on the wear performance of the composites generated [6, 7]. Wear experimental setup is depicted in Figure 1.2 [6].

An investigation has been made on the friction and tribo properties of natural fiber 3D braided yarn polylactic acid (PLA) composites with woven fabric reinforcement exposed to dry sliding. For various weights and sliding velocities, the impact of different fiber weight fractions is examined. The natural-fiber-reinforced PLA composites are found to have a high CoF and wear rate. The addition of natural fiber braided yarn to PLA, aliphatic polyester, increases the frictional force and decreases the composite specimen height loss. With increasing normal loads, the wear rate and specific wear rate of pure PLA and natural fiber/PLA composites increase. The specific wear rate of PLA is reduced by roughly 95% when natural fiber reinforcement is used at 35 wt% [8].

1.2.2 Coconut Coir Reinforcement

Coir, obtained from the fibrous middle layer of coconut fruits, is a hard and stiff biodegradable lignocellulosic fiber. It has a high lignin content that makes it weather resistant and strong, and it can be chemically modified [9]. It is found that coir fibers, when chemically treated, improve their interfacial interaction with PLA resin and are thermally stable up to 265 °C. Coir-fiber-reinforced polyester laminates were made and evaluated for wear and frictional behavior using a block-on-disc (BOD) machine. The worn-out surfaces are examined by scanning electron microscopy (SEM). By adjusting the applied stresses, the particular wear rate and friction coefficient are investigated as a function of sliding distance. It is found that the composites have better wear performance than the neat polyester. SEM observation showed that there was no pull-out, tear, or breakage of fibers, but there was deformation and micro-plowing in the resinous regions [10].

1.2.3 Banana Reinforcement

Banana fibers, known as Musa fiber, are extracted from the outermost layer of the banana trees by the retting process, which improves the quality of the fiber. Banana-reinforced epoxy composites were fabricated by hand-layup, and their mechanical and tribological behavior was tested at different orientations using a pin-on-disc tribometer. It is found that at 0°, the wear rate is less than at 90°, as the fiber area is at its maximum at 90° orientation [11].

It is observed in the tribological behavior of banana/coir composites reinforced with glass fibers that were made using a compression molding procedure. The fibers are silane treated, and the composites' wear resistance is tested on a pin-on-disc tribometer by altering the force on the pin, disc speed, and fiber weight percentage. According to the wear characterization performed using SEM, the dry body abrasive wear test demonstrated that integrating natural fiber with synthetic fibers enhances the wear resistance capability and increases the wear life of the composites. According to Taguchi's design of experiments and analysis of variance methods, the most influencing factor is the type of composite fiber (96.11%), followed by the speed of disc rotation (1.85%), and the load on the pin (1.85%) [12].

1.2.4 Hemp Reinforcement

Hemp fiber is extracted from the stem of the plant by various processing techniques such as retting, decortications, softening, combing, and spinning, and it is the oldest fiber with a high heat capacity. More research work has been done to understand the mechanical behavior of hemp fiber [7]. The tribological behavior of hemp-phenolic resin-based fiber was studied by few researchers. Modifiers such as alumina, vermiculite, and graphite are used to improve the frictional properties. Phenolic (C_6H_6O) resin as constant 20, mineral reinforcement as constant value of 10, and alumina (Al_2O_3), graphite, and vermiculite as constant value of 5. The tribological properties were assessed by the IS2742–4 standard chase machine. The nomenclature is presented in Table 1.1. The result revealed that with 5% weight hemp fiber, it exhibits

Table 1.1 Composite nomenclature.

Composition (wt%)	1	2	3	4
C_6H_6O matrix	Constant	Constant	Constant	Constant
Mineral reinforcement	Constant	Constant	Constant	Constant
Al_2O_3	Constant	Constant	Constant	Constant
Graphite	Constant	Constant	Constant	Constant
Vermiculite	Constant	Constant	Constant	Constant
Hemp reinforcement	5	10	15	20
$BaSO_4$	50	45	40	35

Source: Adapted from Kumar et al. [13].

the highest friction performance, the highest friction stability, and the lowest wear and least friction variability compared to 15–20% weight hemp fiber [13].

1.2.5 Ramie Reinforcement

Ramie fiber, also known as *Boehmeria nivea*, comes from the stem of the nettle family. The water absorption behavior of ramie/epoxy composites and their effects on mechanical and tribological behavior fabricated by hand layup method with 30% fiber loading are investigated. Friction and wear behavior were also tested against the dry specimen. The worn-out surfaces and fractured spots were examined by SEM. The ASTM D3039 standard is used for the mechanical characterization.

Friction and wear behavior changed after the water absorption test was done on Ducom TR-20 LE-PHM400 pin-on-disc tribometer. The tribological performance is done by varying the sliding speeds (1, 1.5, and $2\,m\,s^{-1}$) with different normal loads (10, 15, 20, 25, and 30 N) at a constant sliding distance of 1000 m. Due to the hydrophilic nature of natural fibers, the fiber loading increases, which increases the weight gain percentage. With the increase in sliding speed, the wear loss increases significantly from 1 to $1.5\,m\,s^{-1}$ and reduces from 1.5 to $2\,m\,s^{-1}$ sliding speed [14]. The wear rate of ramie fiber is presented in Figure 1.3 [14].

1.2.6 *Calotropis gigantea* Reinforcement

Calotropis gigantea, popularly known as mudar or milkweed, is a wasteland plant that has yet to be commercially exploited. *Calotropis* is a member of the Asclepiadaceae family, which has 280 genera and 2000 species with a global distribution. It is most common in the subtropics and tropics and rare in colder climates. This shrub is evergreen, perennial, and soft-wooded. Mudar is a medium-sized shrub with a stem diameter of 25 cm and a height of 2–3 m. *C. gigantea* fibers are preferred as they are abundant and biodegradable [15] It is produced and tested in composites with various amounts of multi-walled carbon nanotube (MWCNT) in phenolic resin reinforced with *C. gigantea* fiber for tribological performance (0.25, 0.5, 1, and 2 wt%).

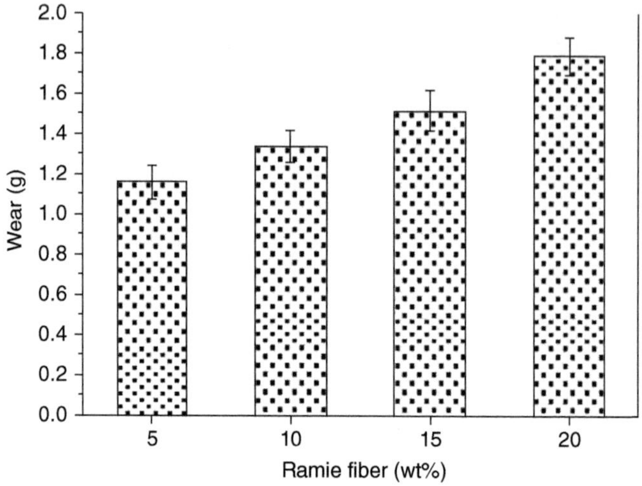

Figure 1.3 Ramie fiber (wt%) vs. wear (g). Source: Adapted from Kumar et al. [14].

The manufactured composites were tested on a conventional pin-on-disc tribometer for adhesive wear and frictional behavior against an EN-31 steel disc under varied applied loads (up to 100 N). The results revealed that adding MWCNT to a phenolic resin composite reinforced with *C. gigantea* fibers considerably enhanced the pressure velocity limit and lowered the friction coefficient value. The friction coefficient decreases as the applied load increases. The amount of MWCNT in a product has a significant influence on the rate of wear. The contact temperature of sliding surfaces rises as applied tension increases. The worn surfaces of the samples were examined using a field-emission scanning electron microscope. According to the tribological study, *C. gigantea* fiber has a lot of promise as a tribomaterial. MWCNT has a significant impact on reducing composite wear. The value of pure phenolic resin indicated the maximum friction coefficient at varied applied loads [16].

Ramesh et al. [17] studied the tribological behavior of surface-treated *C. gigantea* with sodium hydroxide and potassium permanganate solutions. The composites were prepared by compression molding by reinforcing them with untreated, sodium hydroxide and potassium permanganate treated fibers. Friction force and CoF were found to vary with the time of wear. It was also found that the frictional force value and rate of wear are the lowest for sodium hydroxide-treated *C. gigantea* fibers compared to other treated fibers.

1.2.7 Kenaf Reinforcement

Kenaf fibers come from the family Malvaceae, which has a pithy stem surrounded by fibers. It can grow up to a height of 5 m, resulting in a high production of fibers. The most attractive point is that kenaf plants contain 20–25% fiber. The adhesive wear property of thermoplastic treated kenaf fiber reinforced with polyurethane is investigated. The wear property is found by using the BOD apparatus. The composite performance was studied by varying the sliding distance, orientation of fibers, and applied loads. It was found that the adhesive wear property is better in the antiparallel direction compared to normal and parallel directions. The friction coefficient is less in thermoplastic kenaf fibers reinforced with polyurethane [18]. A newly built BOD machine is used to investigate the adhesive wear and friction characteristics of kenaf fiber composites made by closed mold process against clean epoxy at 50 N applied load, sliding lengths (0–4.2 km), and sliding velocity of 2.8 m s^{-1}. The wear performance of the epoxy was improved by around 70% when kenaf fiber-reinforced composites were used. The debonding of the fibers was thus shown to dominate the wear processes [19].

1.2.8 Betel Nut Fibers

The wear and frictional performance of betel nut fiber as reinforcement for polyester composites were investigated. The fibers were removed from the betel nut fruit after it had been steeped in water for 48 hours [20]. The composites were made with 13 layers of randomly dispersed betel nut fibers and 15 layers of polyester composites using a manual lay-up approach. The wear behavior was tested by BOD machine. The wear behavior was tested at different conditions, i.e. by varying the applied loads in the range of 5–30 N, sliding distance from 0 to 6.72 m, and sliding velocity of 2.8 m s^{-1} with different fiber orientations.

The betel nut fiber in parallel orientation concerning sliding speed has better wear and frictional performance compared to other orientations. The performance is enhanced by about 98% to 73%. Debonding of fibers is the main mechanism to be found in parallel orientation. Micro and macrocracks, plastic deformation, debonding, and fiber pullout dominated the wear mechanism for AP—O, whereas microcracks, plastic deformation, debonding, and fiber pullout dominated the wear mechanism for P—O. Meanwhile, SEM revealed micro-and micro-cracks, polyester fracture, delamination, detachment, pullout, and fiber breaking in N—O [20]. The surface roughness with respect to fiber orientation is given in Figure 1.4 [20]. A comparative study on the performance of chopped-strand mat glass-fiber-reinforced polyester as an alternative to treated betelnut-fiber-reinforced polyester composites was made. The tribological performance check was carried out both in dry and wet conditions, and it was found that treated betelnut-fiber-reinforced polyester composites revealed an increase in wear of 98% and 90.8% with a reduction in friction coefficient of about 9.4% and 80%, respectively. The presence of trichomes on the outer surface improves the interfacial adhesion strength (i.e. high fiber surface roughness), thereby minimizing fiber pullout and debonding during sliding [21]. The mechanical and tribological properties of walnut shell powder reinforced with polypropylene resin with different weight fractions, namely 30%, 40%, and 50%, are studied. A pin-on-disc tribometer is used to calculate the material loss, which shows that increasing the walnut shell powder fraction decreases the specific wear rate, thereby increasing the wear resistance [22]. The physical, chemical, tensile, and surface roughness properties of *Caryota urens* fibers, often called "toddy palm, jaggery palm, or wine palm," were investigated. It may reach a height of 20 m and is branched with an expansion at the internodes, similar to a palm. It was found that the combination of fibers with barites improved the tribological performance [23].

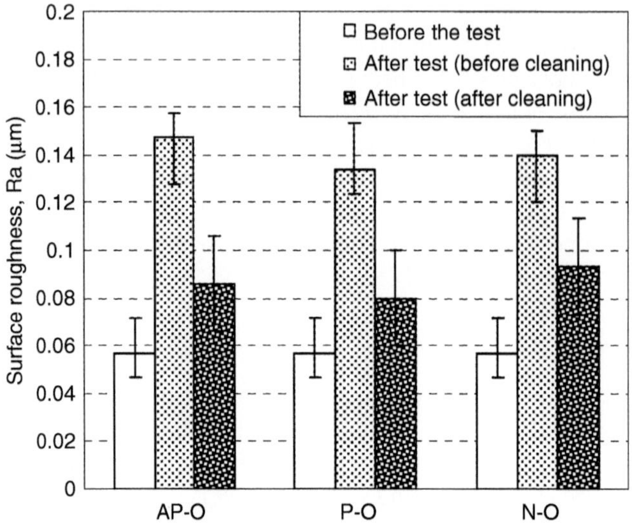

Figure 1.4 Surface roughness vs. betel nut fiber orientation. Source: Gill and Yousif [20]/with permission of Sage Publications.

1.3 Parameters Influencing the Tribological Characteristics

Parameters such as fiber orientation, fiber volume fraction, and filler percentage play a vital role in influencing the tribological characteristics, which impact wear loss, frictional force, and CoF. With the exception of biodegradable polymers, biocomposite has been studied in a variety of natural polymers. Furthermore, research on the effects of plant fiber reinforcement in biomatrix on wear properties has been found to be exceedingly limited. Furthermore, it was discovered that relatively little study has been undertaken on the wear features of plant fiber-based hybrid laminates in various matrix systems [7]. Natural fiber composites with optimal fiber alignment provide the best tensile, flexural, and wear characteristics [24].

1.3.1 Impact of Reinforcement Orientation on Wear Behavior

The reinforcement orientation plays a predominant role in influencing the wearer's behavior. Sisal fiber reinforced with epoxy, where fiber is oriented as parallel or antiparallel as shown in Figure 1.5 with respect to the wear sliding direction [25].

The experimental outcome revealed that specific wear vs. load, where parallel fiber orientation has the maximum specific wear rate corresponding to the load condition, antiparallel fiber orientation initially has 1.2×10^{-3} m^3 Nm^{-1} then normal fiber orientation has the minimum value of specific wear rate at 0.8×10^{-3} m^3 Nm^{-1} as shown in Figure 1.6 [26].

Parallel orientation, random orientation, and antiparallel orientation are needed for fabricating banana-fiber-reinforced epoxy composites in different orientations.

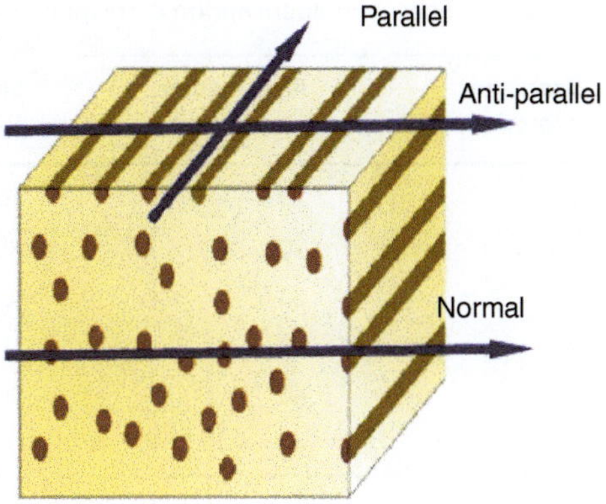

Figure 1.5 Reinforcement alignment at parallel, antiparallel, and normal directions. Source: Adapted from Ruggiero et al. [25].

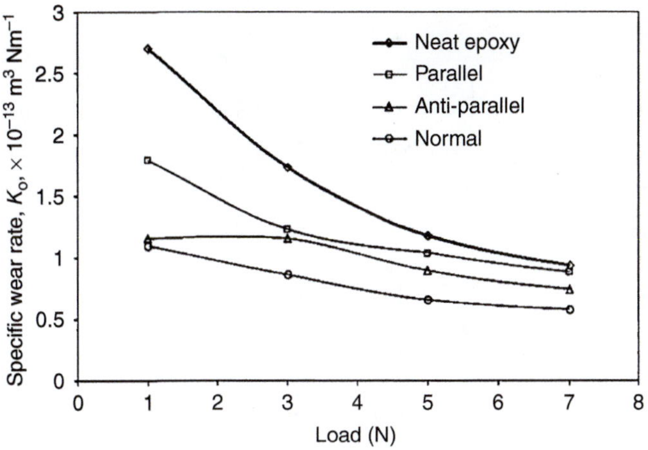

Figure 1.6 Specific wear rate for fiber orientation. Source: Adapted from Chaudhary and Ahmad [26].

Results revealed that tribological behavior of antiparallel orientation > parallel orientation > random orientation. Frictional performance is higher for antiparallel orientation with 44% at minimum value of sliding velocity, and random orientation has lower frictional performance of 9.89% at maximum sliding velocity [27]. Then fiber orientation has conflict with impact of single fiber arrangement, impact of multilayer fiber orientation, impact of continuous fiber alignment, and effect of discontinuous reinforcement orientation [28].

1.3.2 Effect of Reinforcement Volume Fraction on Wear Behavior

An investigation of sisal-fiber- and glass-fiber-reinforced polymer laminates showed the reinforcement volume% is to be discussed. In an experimental study, sisal fiber was taken at 100% and 50%, and the composites were fabricated by compression molding machines (Table 1.2). From the results, it appears that the CoF increased as fiber volume percentage increased. The presence of more sisal strands in the resin

Table 1.2 Specimen abbreviation corresponding to fiber volume fraction.

Specimen abbreviation	Fiber volume fraction (%)	Sisal fiber (% in vol)	Glass fiber (% in vol)
1	20	100	0
2	20	50	50
3	20	0	100
4	40	100	0
5	40	50	50
6	40	0	100

Source: Adapted from Ramesh et al. [29].

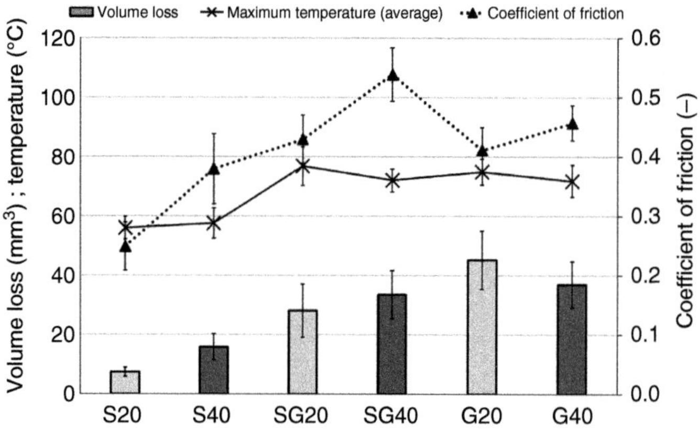

Figure 1.7 Influence of fiber volume fraction. Source: Adapted from Ramesh et al. [29].

of the laminate resulted in more discontinuous phases in the resin, which resulted in maximum friction value.

The minimum temperatures recorded for S20 and S40 samples may help elucidate how virgin sisal laminates have the maximum wear resistance. The polymeric resin of the materials softens at high temperatures, allowing for more layer interaction between the sample and the anti-side, leading to improved actual contact zone and wear. The tribological characteristics of *Ipomoea carnea* fiber-reinforced laminates were investigated in a research work. The fiber volume percentage was varied using 10%, 20%, 30%, and 40% of fibers at different ranges of sliding velocity in the tests. Abrasive wear loss may be considerably reduced by incorporating *I. carnea* into epoxy. At a particle level of 30% fiber volume percentage, the best frictional resistance characteristic was attained. However, due to weak interfacial adhesion, an excessive amount of fiber (40 wt%) causes the particles to tear away from the matrix resin throughout the test (Figure 1.7).

1.3.3 Effect of Fillers on Wear Behavior

The influence of date palm seed fillers on the wear characteristics of polymer laminates in a variety of experiment conditions is studied. The outcomes suggest that improving the date seed filler from 15% to 25% reduces the friction coefficient for polyester laminates filled with date seed filler under maximum contact pressure, and that under reduced friction force and reduced sliding velocity, date seed filler can increase by up to 10%. In practice, adding supplementary fillers to a natural-fiber-reinforced material improves wear performance by improving the interfacial strength of the reinforcement and resin [30, 31]. Natural-fiber-reinforced laminates containing filler particles such as MWCNTs, silicon dioxide, graphite, titanium dioxide, graphene, and others are a growing field where major studies may be done. Despite this, significant progress has been made in the wear analysis of composite materials [32]. Jute-reinforced polypropylene laminate tribological performance revealed that the CoF reduces as the load rises. The CoF begins to rise as the filler loading increments, but it progressively declines. The minimum CoF was discovered under high loads and 0% fiber loading. At 4 kg of weight, the CoF was 0.3854, while at 8 kg of load, it was 0.3325. This might be due to a strong binding between resin and reinforcement, resulting in a lower composite CoF [33].

The influence of nanomaterials oil palm empty fruit bundle additives on fiber-reinforced kenaf in nonwoven matting form with laminates is studied. Due to the nanotechnology fillers' capacity to eliminate free spaces, the research finding implies that utilizing nano-oil palm empty fruit bundle filler with kenaf-reinforced composites enhances mechanical strength when compared to kenaf epoxy composites. A nanofiller hybridization composite's fracture toughness has been increased by 28.3% [34]. The influence of fillers on the tribological characteristics of glass-coir-reinforced polymer laminates with various proportions of molybdenum disulfide and titanium dioxide fillers is used to make different compositions. When evaluated against unfilled and titanium dioxide filler-filled glass-coir-reinforced epoxy composites, molybdenum disulfide-integrated filler demonstrated improved tribological performance. The inclusion of molybdenum disulfide as a filler will improve the CoF. This is due to molybdenum disulfide's structure, which generates a thin black layer that offers cushioning [35].

1.3.4 Influence of Surface Modification on Wear Behavior

The influence of surface treatment on *C. gigantea* fiber-reinforced epoxy laminates is that the rate of wear of raw *C. gigantea* fiber laminates is notably higher than that of alkaline and potassium permanganate-treated materials, as shown in Figure 1.8. The alkaline-treated *C. gigantea* fiber composites wore out less than the potassium permanganate-treated *C. gigantea* fiber composites. The rate of wear was shown to be enhanced due to the improved coupling between alkali-treated *C. gigantea* fiber and resin. Overall, the wear behavior of alkaline-treated *C. gigantea* fiber composites was superior to that of the other materials, and the wear behavior of these laminates

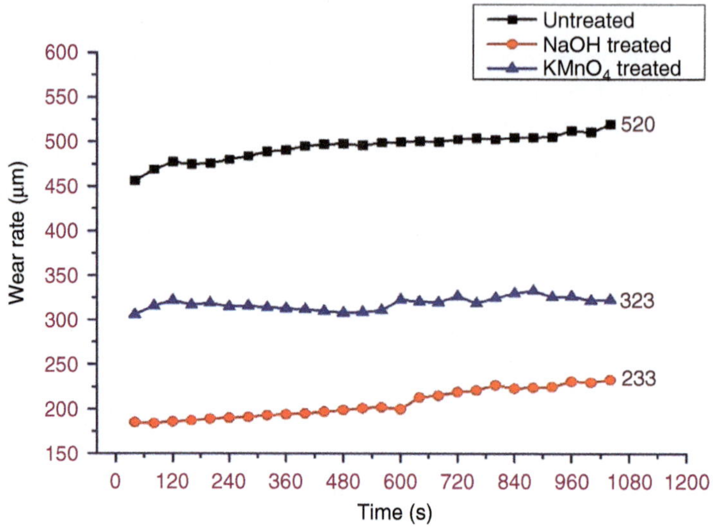

Figure 1.8 Wear rate analysis of *Calotropis gigantea*-fiber-reinforced laminates. Source: Ramesh et al. [17]/John Wiley & Sons.

was found to be superior between 400 and 600 s. These composites are used in a variety of technical features, including brake lines and discs.

The impact of the silane solution concentration on the wear properties of maize-stem-fiber-reinforced polymer laminates was investigated. According to the observations, the silane-treated maize-stem-fiber-reinforced polymer laminates exhibited a possible low-density feature. The silane-treated corn stalk fiber did not considerably enhance the frictional behavior of the polymer composites, but it did greatly increase the abrasive wear. When compared to the raw maize-stem-fiber-reinforced laminates, the cumulative wear rate of the silane-treated maize-stem-fiber-reinforced laminates fell by 22.8% from 4.792×10^{-7} cm^3 (Nm)$^{-1}$ to 3.699×10^{-7} cm^3 (Nm)$^{-1}$, indicating the best wear-resistant behavior [36, 37].

A research using coccinea indica fiber as a reinforcement to explore and assess the possibilities for improving the wear resistance of epoxy laminates. Raw coccinea indica fiber reinforced with epoxy laminates was made, and two distinct surface treatments, 5 wt% alkaline treated and 2 wt% silane treated, were conducted. The laminate was made utilizing a manual lamination and open mold method for all specimens, with fiber lengths of 30 mm and a fiber proportion of 35 wt%. The weight loss of alkaline and silane-processed laminates was decreased by 14% and 38%, respectively, whereas the CoF was decreased by 40% and 38%, according to the results (Figure 1.9) [38, 39].

1.4 Morphology Analysis of Tribological Characteristics

Morphological study plays a crucial role in tribological characterization, while SEM is used to examine the worn-out region after the experiments. They clearly guide the analysis of surface layer and show the fiber interaction between matrix in detail.

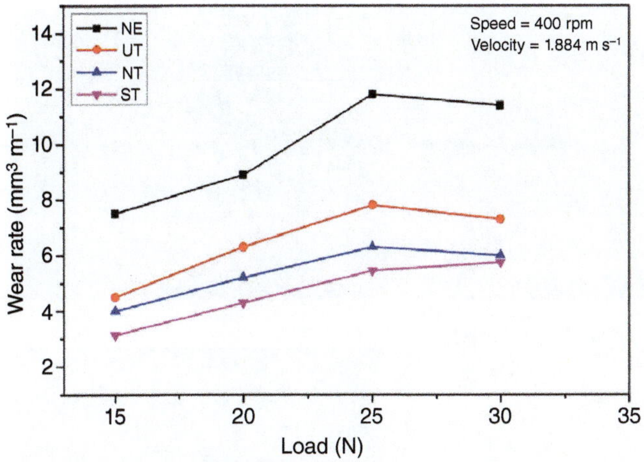

Figure 1.9 Load vs. wear rate with respect to surface treatment. Source: Mylsamy et al. [38]/with permission of Elsevier.

The impact of silane-treated maize-stem-fiber-reinforced composites is analyzed. The corn stalk fibers are silane treated at different weight fractions of untreated, 1% silane treated, 5% silane treated, 9% silane treated, and 13% silane treated. The worn surface morphology was analyzed using a TESCAN VEGA3 model SEM with an EDX-ray spectroscopy at 15 kV. The examined laminates were spray coated with gold dust using a model graphite coating technique to increase the conduction of the sample surfaces. SEM study of surface roughness demonstrated that silane-treated fibers were ideal for the formation of secondary peaks on the various polymeric layers, which might enhance wear and morphological qualities significantly (Figure 1.10) [36, 40].

The wear layer for topological analysis of sodium-hydroxide-surface-modified chopped abaca-fiber-reinforced laminates was researched. SEM morphological characterization provided visual evidence that abaca fiber epoxy with 25 wt% abaca content had the lowest abrasive wear rate and remarkable wear resistance. Fiber fracture, matrix breaking, microcracking, large fractures, and pit generation were notable wear causes on the worn surfaces. Composites with no or low fiber content wore down quicker, according to SEM micrographs. In 25% weights fiber-epoxy samples, fiber pullouts produced by applied load were found to be minimal (Figure 1.11) [41].

The structural, thermal, and wear performance of untreated, NaOH, $C_5H_9NO_4$, and combinations of NaOH + $C_5H_9NO_4$-treated sisal reinforcement polymer laminates were studied. On the surface of the UT fibers, wax, lignin, hemicellulose, silica, pectin, and other impurities can be seen, and they were primarily responsible for the poor reinforcement–resin interaction. Alkali treatment was responsible for the elimination of wax and various contaminants from the reinforced surface. As a result, the waviness of the fiber increases and the diameter of the fiber decreases. This could explain the improvement in fiber–matrix bonding [42].

The natural-waste-reinforced epoxy multilayer laminates and their mechanical and wear behavior are investigated . Composites with varying weight proportions of

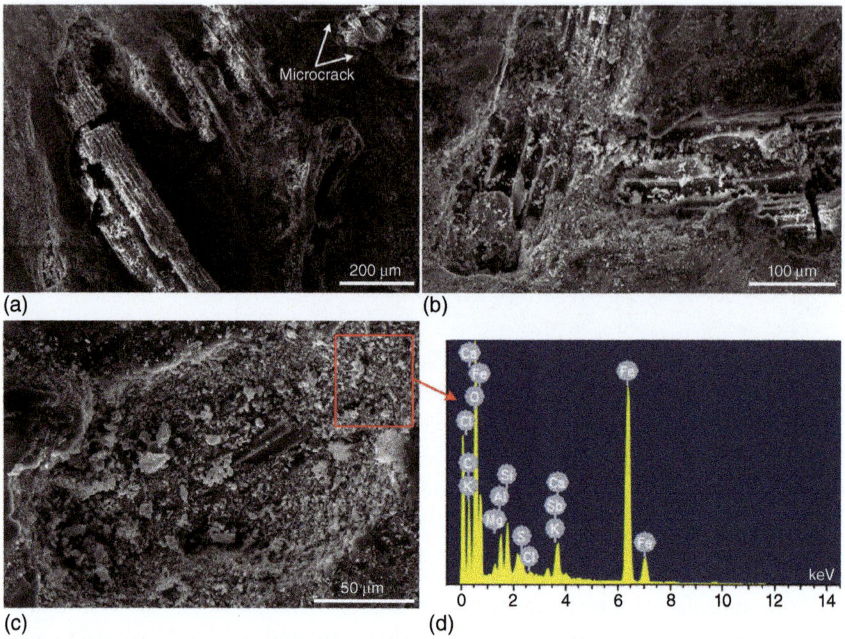

Figure 1.10 SEM image for CMU; (a) SEM image of 200 μm; (b) 100 μm; (c) 50 μm; (d) EDX analysis. Source: Liu et al. [36]/with permission of Elsevier.

fiber content were fabricated with varying lengths of reinforcement and resin. They used a Carl Zeiss Sigma 300 microscope to study the microstructural and structural appearance. The uniform dispersion of banana and hemp fiber within the vinyl ester bio-hybrid composites was investigated using SEM. It was discovered that the resin and the fibers in hemp and banana fiber reinforced with vinyl ester had a significant interaction. It was also observed that the banana and hemp fiber (20% each) diffused nicely. The hemp fiber is shedding within the vinyl ester resin, according to SEM photographs. This is due to the fiber's consistent distribution. The inclusion of hemp and banana fibers increases material characteristics as well as surface roughness, which is self-evident. As an outcome, the vinyl ester and 20% (banana + hemp) bio-hybrid composite had a smoother texture than the other materials [43].

1.5 Conclusion

Recent research in the green tribology industry includes the quest for energy-saving substances that also have sustainable and environmental properties, as well as adapting to worldwide demand through a cost-effective approach. As a result, plant-based fiber-reinforced and self-lubricated epoxy laminates are gaining popularity as a viable tribomaterial option with cost savings. One of the most essential criteria in determining the usefulness of biocomposites in appropriate tribo-applications is reinforcement selection. With the use of brief literary classics,

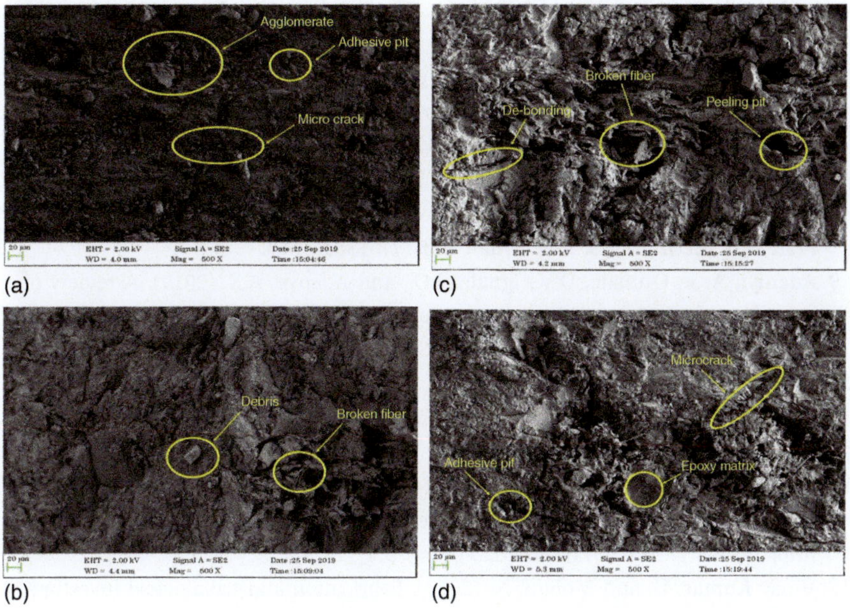

Figure 1.11 Morphological characterization of abaca fiber–epoxy laminates. Source: Kurien et al. [41]/with permission of Springer Nature.

this chapter has studied numerous aspects of natural fibers and the impact associated with wear performance in order to build effective tribo-composites.

References

1 Ramesh, M. (2016). Kenaf (*Hibiscus cannabinus* L.) fibre based bio-materials: a review on processing and properties. *Prog. Mater. Sci.* 78: 1–92.

2 Ramesh, M. (2018). Flax (*Linum usitatissimum* L.) fibre reinforced polymer composite materials: a review on preparation, properties and prospects. *Prog. Mater. Sci.* 102: 109–166.

3 Paul, R., Gouda, K., and Bhowmik, S. (2021). Effect of different constraint on tribological behaviour of natural fibre/filler reinforced polymeric composites: a review. *Silicon* 13 (8): 2785–2807. https://doi.org/10.1007/s12633-020-00613-z.

4 Ramesh, M., Palanikumar, K., and Reddy, K.H. (2017). Plant fibre based bio-composites: sustainable and renewable green materials. *Renew. Sustain Energy Rev.* 79: 558–584.

5 Ramesh, M., Deepa, C., Kumar, L.R. et al. (2020). Life-cycle and environmental impact assessments on processing of plant fibres and its bio-composites: a critical review. *J. Ind. Text.* https://doi.org/10.1177/1528083720924730.

6 Chaudhary, V., Bajpai, P.K., and Maheshwari, S. (2018). An investigation on wear and dynamic mechanical behavior of jute/hemp/flax reinforced composites and its hybrids for tribological applications. *Fibers Polym.* 19 (2): 403–415. https://doi.org/10.1007/s12221-018-7759-6.

7 Karthikeyan, S., Rajini, N., Jawaid, M. et al. (2017). A review on tribological properties of natural fiber based sustainable hybrid composite. *Proc. Inst. Mech. Eng. J. Eng. Tribol.* 231 (12): 1616–1634.

8 Kanakannavar, S., Pitchaimani, J., and Ramesh, M. (2021). Tribological behaviour of natural fibre 3D braided woven fabric reinforced PLA composites. *Proc. Inst. Mech. Eng. J. Eng. Tribol.* 235 (7): 1353–1364.

9 Adeniyi, A.G., Onifade, D.V., Ighalo, J.O., and Adeoye, A.S. (2019). A review of coir fiber reinforced polymer composites. *Composites, Part B* 176: 107305. https://doi.org/10.1016/j.compositesb.2019.107305.

10 Yousif, B.F. (2008). Frictional and wear performance of polyester composites based on coir fibres. *Proc. Inst. Mech. Eng. J. Eng. Tribol.* 223: 51–59. https://doi.org/10.1243/13506501JET455.

11 Chavali, P.J. and Taru, G.B. (2020). Effect of fiber orientation on mechanical and tribological properties of Banana-reinforced composites. *J. Fail. Anal. Prev.* https://doi.org/10.1007/s11668-020-01048-9.

12 Vinay Kumar, D. and Mohan, N. (2019). Fabrication and tribological investigation of coconut coir/banana fiber/glass fiber reinforced hybrid polymer matrix composites—a Taguchi's approach. *Mater. Res. Exp.* https://doi.org/10.1088/2053-1591/ab3d4a.

13 Kumar, N., Singh, T., Grewal, J.S., and Patnaik, A. (2019). Experimental investigation on the physical, mechanical and tribological properties of hemp fiber-based non-asbestos organic brake friction composites. *Mat. Res. Exp.* 6 (8): 085710.

14 Kumar, R., Haq, M.I.U., Sharma, S.M., and Raina, A. (2021). Effect of water absorption on mechanical and tribological properties of Indian ramie/epoxy composites. *Proc. Inst. Mech. Eng. J. Eng. Tribol.* https://doi.org/10.1177/13506501211005635.

15 Ramesh, G., Subramanian, K., Sathiyamurthy, S., and Jayabal, S. (2020). Study on influence of fiber length and fiber weight proportion in mechanical behavior of *Calotropis gigantea* fiber-epoxy composites. *Polym. Compos.* 41 (11): 4899–4906. https://doi.org/10.1002/pc.25761.

16 Dwivedi, U.K., Singh, S., Shukla, S.C., and Hashmi, S.A.R. (2022). Investigation on tribo-characteristics of *Calotropis gigantea* fiber-reinforced-CNT modified polymer composites. *Polym. Bull.* 79 (2): 697–707. https://doi.org/10.1007/s00289-020-03532-6.

17 Ramesh, M., Deepa, C., Rajeshkumar, L. et al. (2021). Influence of fiber surface treatment on the tribological properties of *Calotropis gigantea* plant fiber reinforced polymer composites. *Polym. Compos.* 42 (9): 4308–4317. https://doi.org/10.1002/pc.26149.

18 Narish, S., Yousif, B.F., and Rilling, D. (2010). Adhesive wear of thermoplastic composite based on kenaf fibres. *Proc. Inst. Mech. Eng. J. Eng. Tribol.* 225 (2): 101–109. https://doi.org/10.1177/2041305X10394053.

19 Chin, C.W., Yousif, B.F., Ayer, J., and Lama, K. (2010). Tribological behaviour of KFRE composite. *Int. J. Mod. Phys. B* 24 (28): 5589–5599. https://doi.org/10.1142/S0217979210056232.

20 Gill, N.S. and Yousif, B.F. (2009). Wear and frictional performance of betelnut fibre-reinforced polyester composite. *Proc. Inst. Mech. Eng. J. Eng. Tribol.* 223 (2): 183–194. https://doi.org/10.1243/13506501JET516.

21 Nirmal, U., Hashim, J., Lau, S.T. et al. (2012). Betelnut fibres as an alternative to glass fibres to reinforce thermoset composites: a comparative study. *Text. Res. J.* 82 (11): 1107–1120. https://doi.org/10.1177/0040517512439945.

22 Ali, A. and Ogaili, F. (2020). Mechanical and tribological properties of walnut/polypropylene natural composites. *J. Mech. Eng. Res. Dev.* 43 (4): 372–380.

23 Krishnan, G.S., Babu, L.G., Kumaran, P. et al. (2020). Investigation of *Caryota urens* fibers on physical, chemical, mechanical and tribological properties for brake pad applications. *Mat. Res. Exp.* 7 (1): 015310.

24 Parbin, S., Waghmare, N.K., Singh, S.K., and Khan, S. (2019). Mechanical properties of natural fiber reinforced epoxy composites: a review. *Proc. Comput. Sci.* 152: 375–379.

25 Ruggiero, A., Valasek, P., and Muller, M. (2016). Exploitation of waste date seeds of *Phoenix dactylifera* in form of polymeric particle biocomposite: investigation on adhesion, cohesion and wear. *Comp. Eng.* 104: 9–16.

26 Chaudhary, V. and Ahmad, F. (2020). A review on plant fiber reinforced thermoset polymers for structural and frictional composites. *Polym. Test.* 91: 106792.

27 Sumithra, H. and Sidda Reddy, B. (2018). A review on tribological behaviour of natural reinforced composites. *J. Reinforced Plast. Compos.* 37 (5): 349–353.

28 Ramesh, M., Rajeshkumar, L., Deepa, C. et al. (2021). Impact of silane treatment on characterization of *Ipomoea staphylina* plant fiber reinforced epoxy composites. *J. Nat. Fibers* 19: 1–12.

29 Ramesh, M., Balaji, D., Rajeshkumar, L. et al. (2021). Tribological behavior of glass/sisal fiber reinforced polyester composites. In: *Vegetable Fiber Composites and Their Technological Applications*, 445–459. Singapore: Springer.

30 Basumatary, K.K., Mohanta, N., and Acharya, S.K. (2014). Effect of fiber loading on abrasive wear behaviour of *Ipomoea carnea* reinforced epoxy composite. *Int. J. Plast. Technol.* 18 (1): 64–74.

31 Ibrahem, R.A. (2015). Effect of date palm seeds on the tribological behaviour of polyester composites under different testing conditions. *J. Mater. Sci. Eng.ss* 4 (6).

32 Kumar, S. and Singh, K.K. (2020). Tribological behaviour of fibre-reinforced thermoset polymer composites: a review. *Proc. Inst. Mech. Eng. J. Mater. Des. Appl.* 234 (11): 1439–1449.

33 Ramesh, M., Tamil Selvan, M., and Niranjana, K. (2022). Hygrothermal aging, kinetics of moisture absorption, degradation mechanism and their influence on performance of the natural fibre reinforced composites. In: *Aging Effects on Natural Fiber-Reinforced Polymer Composites*, 257–277. Singapore: Springer.

34 Ramesh, M., Deepa, C., Tamil Selvan, M., and Reddy, K.H. (2020). Effect of alkalization on characterization of ripe bulrush (*Typha domingensis*) grass fiber reinforced epoxy composites. *J. Nat. Fibers* 1–12.

35 Vijay, B.R. and Srikantappa, A.S. (2020). Influence of fillers on tribological behaviour of glass-coir reinforced epoxy composites–an ANN approach. *IOP Conf. Ser. Mater. Sci. Eng.* 852 (1): 012065.

36 Liu, Y., Xie, J., Wu, N. et al. (2019). Influence of silane treatment on the mechanical, tribological and morphological properties of corn stalk fiber reinforced polymer composites. *Tribol. Int.* 131: 398–405.

37 Mohammed, L., Ansari, M.N., Pua, G. et al. (2015). A review on natural fiber reinforced polymer composite and its applications. *Int. J. Polym. Sci.* 2015: https://doi.org/10.1155/2015/243947.

38 Mylsamy, B., Chinnasamy, V., Palaniappan, S.K. et al. (2020). Effect of surface treatment on the tribological properties of *Coccinia indica* cellulosic fiber reinforced polymer composites. *J. Mater. Res. Technol.* 9 (6): 16423–16434.

39 Vigneshwaran, S., Sundarakannan, R., John, K.M. et al. (2020). Recent advancement in the natural fiber polymer composites: a comprehensive review. *J. Clean. Prod.* 277: 124109.

40 Ramesh, M., Deepa, C., Selvan, M.T. et al. (2021). Mechanical and water absorption properties of *Calotropis gigantea* plant fibers reinforced polymer composites. *Mater. Today Proc.* 46: 3367–3372.

41 Kurien, R.A., Selvaraj, D.P., and Koshy, C.P. (2021). Worn surface morphological characterization of NaOH-treated chopped abaca fiber reinforced epoxy composites. *J. Bio- Tribo-Corrosion* 7 (1): 1–8.

42 Behera, S., Gautam, R.K., Mohan, S., and Chattopadhyay, A. (2021). Dry sliding wear behavior of chemically treated sisal fiber reinforced epoxy composites. *J. Nat. Fibers* 1–14.

43 Suresh, S., Sudhakara, D., and Vinod, B. (2019). Investigation of bio-waste natural fiber–reinforced polymer hybrid composite: effect on mechanical and tribological characteristics of biodegradable composites. *Mech. Soft Mater.* 1 (1): 1–16.

2

Tribological Properties of the Natural Fiber-Reinforced Epoxy Composites

Lin Feng Ng[1,2] and Mohd Yazid Yahya[1,2]

[1] *Centre for Advanced Composite Materials (CACM), Universiti Teknologi Malaysia, 81310 Johor Bahru, Malaysia*
[2] *Faculty of Mechanical Engineering, Universiti Teknologi Malaysia, 81310 Johor Bahru, Malaysia*

2.1 Introduction

Over the past decades, material evolution has successfully aroused the development of advanced and promising composite materials for a broad range of engineering applications. Fiber-reinforced polymers (FRPs) are created by merging multiple separate phases, such as reinforcement and matrix, each with its own set of properties. The combination of different phases endows the composite materials with improved functional properties compared to each individual constituent. They have been well-known candidates to replace metallic alloys in many engineering sectors such as aerospace and automotive. The lightweight and excellent fatigue performances of composite materials are among the driving factors that escalate the use of such materials. Additionally, composite materials offer more flexibility to the designer to tailor the material properties in accordance with the standards of particular industrial applications [1]. However, the mechanical, thermal, morphological, and tribological performances are greatly influenced by several factors. The kind of reinforcement and polymer matrix, as well as fiber-matrix compatibility, filler content, and fiber orientation, all have their role in the overall performance of composite materials for long-term applications. Therefore, comprehensive research is essential to exploring the optimum performance of composite materials.

Wear and friction are two common phenomena that could lead to the damage and failure of the entire structure. These events occur due to the relative motion of two solid surfaces in contact [2]. While developing advanced materials with high mechanical performance, attention should also be given to tribological properties, as wear and friction are involved in almost all industrial applications. At present, developing eco-friendly and high-wear performance materials to reduce pollutant emissions is still regarded as a fundamental topic in research communities. When scrutinizing the industry trend, most composite materials are incorporated with synthetic fibers, particularly glass fiber, due to their high mechanical

Tribological Properties, Performance, and Applications of Biocomposites, First Edition.
Edited by Chandrasekar Muthukumar, Senthilkumar Krishnasamy,
Senthil Muthu Kumar Thiagamani, and Ganesan Chinnachamy.

strength and low-cost characteristics. The ever-growing concern related to the waste accumulation brought on by the inordinate use of synthetic fibers is alerting researchers and scientists to search for alternative materials to reduce the reliance on synthetic fibers. Natural fibers can be used as a possible reinforcement for composite materials, which is thought to be a more environmentally friendly option. The availability of natural fibers and simplicity of manufacture have prompted researchers to investigate their practicality to meet the requirements of tribological applications [3].

2.2 Fiber-Reinforced Composites

The initial intention of incorporating fiber in composite materials is to reduce the use of plastics. Later, it was discovered that adding fiber to a composite could increase its mechanical performance. Due to their improved mechanical performance along with lightweight characteristics, FRPs have found applications in a wide range of engineering sectors. Apart from their considerable mechanical performance and lightweight characteristics, FRPs have shown excellent fatigue resistance to sustain cyclic loadings compared to metallic alloys. Moreover, FRPs displayed outstanding specific properties compared to metallic alloys [4]. In spite of several remarkable properties of FRPs, they are not demerit-free materials. In comparison with metallic alloys, FRPs have relatively weak impact resistance, and the internal damages resulting from the initial stage of the impact event are not visible. These internal damages could drastically undermine the mechanical performance of FRPs. The difficulty in inspecting the internal damages of FRPs due to an impact event has become an eminent challenge to the engineering sectors.

Several factors have been identified that could significantly influence the miscellaneous properties of composite materials. Fiber length is one of the factors that has a decisive effect on the functional properties of composite materials. The aspect ratio (l/d), which is the ratio of length to diameter, may be used to define the fiber in general. The fiber with a high aspect ratio could offer greater mechanical properties to the FRPs. In particular, increasing the fiber length greatly improves the mechanical performance of FRPs. The improvement is mainly attributed to the increase in load-carrying capacity of the FRPs. When an external load is exerted on the FRPs, the stress is transmitted from the polymer matrix to the fibers via the interfacial shear. The stress at the fiber tips is zero, but it increases along the fiber length. Therefore, the fiber length should be higher than the critical fiber length so as to carry the maximum amount of stress. The critical fiber length can be expressed in Eq. (2.1).

$$\frac{L_c}{D} = \frac{\sigma_t}{2\tau_i} \tag{2.1}$$

where D is the fibre diameter; σ_t is the fibre strength, and τ_i is the interfacial shear strength.

In accordance with Eq. (2.1), by increasing the interfacial shear strength, the critical fiber length drops. This demonstrates a strong relationship between critical fiber length and interfacial shear strength. It is worth noting that the critical fiber length may vary with respect to fiber types and fiber surface treatment. However, when it comes to natural fiber-reinforced composites, long fiber tends to entangle due to the formation of hydrogen bonding, leading to weak mechanical properties.

Aside from the length of the fibers, the composition of the fibers is also important in influencing the mechanical performance of FRPs. Since fiber is the primary load-bearing constituent in FRPs, increasing the fiber content will undoubtedly augment the mechanical strength and stiffness of FRPs. Nonetheless, it is worth mentioning that increasing the fiber content over a certain limit could deteriorate the mechanical performance of FRPs. The main role of the polymer matrix is to hold the fiber and provide excellent structural integrity to the FRPs. Once the fiber content exceeds a critical limit, insufficient polymer matrix to hold the fiber may result in premature failure, thus weakening the mechanical properties of FRPs. Generally, the optimum fiber content is within the range of 30–40 wt%. For short fiber-reinforced composites, incorporating a small amount of fiber could have a detrimental effect on the FRPs, implying that the mechanical properties of such FRPs could be lower than those of neat polymers. The fiber tips are often the stress concentration points that may cause crack initiation. In comparison with long fiber-reinforced composites, short fiber-reinforced composites have more fiber tips and stress concentration points, and hence the mechanical properties of such composites are weaker than those of long fiber-reinforced composites. To overcome this problem, a higher amount of fiber should be incorporated in order to optimize the reinforcing effect of FRPs.

To attain the exemplary mechanical properties of FRPs, the fiber-matrix interfacial adhesion should not be neglected. A weak fiber-matrix interfacial adhesion retards the load-transferring efficiency from the matrix to the reinforcement. Given the hydrophilic nature of natural fibers, their compatibility with hydrophobic polymer matrices is poor. Thus, fiber surface pre-treatment is the single most essential stage prior to the fabrication of FRPs. A few surface treatment methods are available to strengthen the bonding level between fiber and matrix, including chemical, physical, and biological. Chemical treatment can undermine the hydrophilicity of the natural fibers and improve fiber-matrix adhesion [5, 6]. The amounts of lignin, hemicellulose, and impurities are apparently dropped upon the chemical treatment. The enhanced surface roughness of the fiber resulting from the partial elimination of superfluous chemical elements allows effective bonding with the polymer matrix. Besides improving the fiber-matrix adhesion, chemical treatment can lower the hydrophilicity of the fiber and expose more reactive hydroxyl groups of cellulose to be linked with the polymer matrix. Because cellulose is the principal chemical ingredient that gives natural fibers their mechanical strength, exposing more of its reactive hydroxyl groups is good for their mechanical properties. Physical treatment is simpler and cleaner than the chemical method. It tends to alter the surface structure and properties without significantly modifying the chemical contents of the fibers. The fiber surface is roughened, allowing mechanical anchoring between the fiber and the matrix. Corona, cold plasmas, stretching, and thermal treatments are examples of physical methods. One of the most common physical treatments for cellulosic natural fibers is the Corona treatment. It modifies the fiber surface energy and increases the number of aldehyde groups, and therefore the fiber's wettability is enhanced. Plasma treatment is regarded as an alternative method to restrict the use of chemicals for fiber surface treatment owing to the growing concern about environmental pollution. Nonetheless, it has been shown that plasma treatment might cause natural fibers to degrade, so it is not recommended to be used for fiber surface treatment [7]. Another method to improve fiber-matrix adhesion is through biological surface treatment. Biological surface treatment is performed with the aid of fungi, enzymes, or bacteria to alter the surface morphology and properties. However, biological treatment using fungi is relatively cheaper compared to enzymes. The surface roughness of the fiber is increased by the fungus, which promotes mechanical interaction between the fiber and the matrix.

2.3 Cellulosic Natural Fibers

Environmental pollution has caused massive devastation around the world. Waste accumulation is one of the most challenging issues nowadays, as many products are nonbiodegradable and nonrecyclable. The extensive use of man-made fibers such as carbon, aramid, and glass fibers has been recognized as one of the driving factors for waste accumulation. The ever-growing concern for environmental pollution is motivating researchers and scientists to search for alternative candidates to replace synthetic fibers. Over the years, natural fibers have gained the attention of researchers and scientists worldwide. Natural fibers are increasingly being used as potential reinforcement in the development of sustainable and environmentally friendly FRPs on a global scale. The high popularity of natural fibers in recent years is due to their low-cost, recyclability, nonabrasive, nontoxic, lightweight, biodegradable, carbon dioxide neutral and high availability [8]. Intensive research studies focusing on composites based on natural fibers have been conducted to explore their feasibility as a replacement for those synthetic fiber-based composites. Despite innumerable attractive features offered by natural fibers, the complete replacement of synthetic fibers with natural fibers is not realistic, as the mechanical strength and stiffness of natural fibers are lower than those of synthetic fibers. The specific mechanical properties, on the other hand, are equivalent to those of synthetic fibers due to their lightweight nature. Their high specific properties make them a good candidate, particularly for transportation sectors, as they could reduce energy consumption and greenhouse gas emissions.

Natural fibers can be grouped into several categories based on their origins: bast, seed, and leaf fibers. In other words, natural fibers are extracted from their origins, and their chemical composition may vary. Natural fibers mainly comprise cellulose, hemicellulose, pectin, lignin, and wax. The chemical composition could have a profound impact on their mechanical properties. Among the chemical compositions of natural fibers, cellulose is the key load-bearing component that provides strength to the fibers [9]. Generally, natural fibers extracted from the bast of the plant have high cellulose content, giving them higher mechanical properties than those extracted from seed and leaf fibers. In contrast, natural fibers with significant amount of lignin have a lesser affinity for water molecules as the lignin acts as a protective layer against moisture penetration. Feng et al. [10] disclosed that the addition of kenaf fiber in the kenaf/pineapple leaf fiber-reinforced hybrid composites tends to undermine the moisture absorption due to the higher lignin content in kenaf fiber. High affinity for water molecules is one of the most critical weaknesses of natural fibers, as moisture uptake could reduce the load capacity of natural fiber-based composites. There are several straightforward methods to reduce the moisture uptake of natural fiber-based composites. Hybridization with synthetic fiber is among the most popular techniques to resolve the moisture absorption of natural fiber-based composites. Ng et al. [11] revealed that the addition of glass fiber in the pineapple leaf/glass fiber-reinforced hybrid composites could drop the moisture uptake percentage and diffusion coefficient tremendously.

2.4 Impact of Tribology on the Environment and Industry

In recent years, there has been a growing demand for lightweight and high-strength materials with eco-friendly characteristics. In the aerospace and automotive sectors, innovative engineering design to develop materials with reduced weight without apparently deteriorating their mechanical performance is particularly desirable. Apart from their mechanical performance and lightweight characteristics, these materials should accommodate excellent tribological properties by means of higher wear resistance and coefficients of friction. Friction and wear have been recognized as one of the main factors that cause material failure and equipment damage [12]. In addition, they were also recognized to have a profound effect on the efficiency and service life of machinery in the industry. The wear and corrosion of materials could impose eminent damages and losses on the industry if measures are not taken to avoid wear and corrosion. Since material surfaces are often susceptible to wear and corrosion, understanding tribology is vital in the field of surface technology. Tribology can be defined as the science involving friction, wear, and lubrication. Considering these properties, FRPs are regarded as excellent candidates to reach the mileage of advanced materials in a judicious manner. Enhancing the wear performance of the materials is advantageous to the service life of the components and the economic value of the entire structure as the maintenance cost can be reduced and the inspection interval can be extended.

In tribology, wear and friction are two important phenomena that occur when two solid surfaces move relative to one another. This relative motion eventually leads to energy dissipation and material degradation. When looking into the total energy consumption at a worldwide level, tribological contact accounts for 23%, where 20% of this energy consumption is used to resist friction, and the remaining 3% is used for the reproduction of the damaged equipment because of the tribological phenomena [13]. In the automobile industry, it has been discovered that around 33%

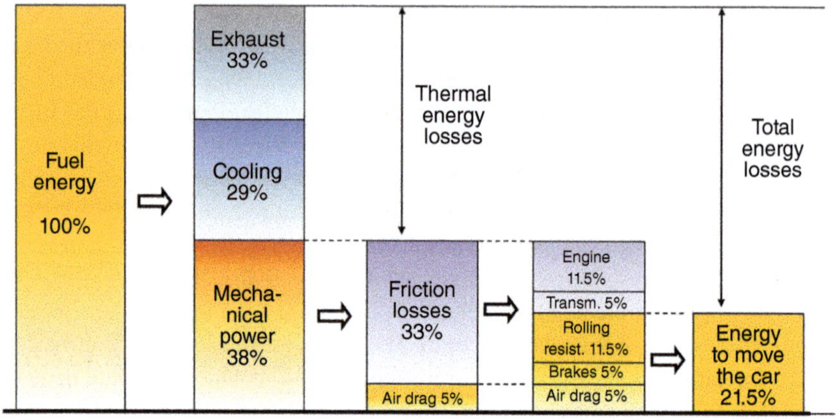

Figure 2.1 Energy consumption percentage of a vehicle. Source: Holmberg et al. [14]/with permission of Elsevier.

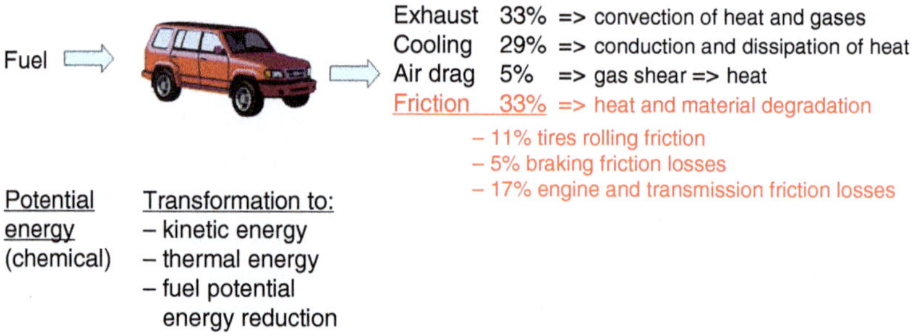

Figure 2.2 Fuel energy consumption of vehicle. Source: Holmberg et al. [14]/with permission of Elsevier.

of energy consumption is used to overcome friction in various vehicle components. The energy consumption percentage of a vehicle is shown in Figure 2.1. Figure 2.2 depicts the fuel energy consumption of a vehicle traveling at a velocity of 60 km/h. Today, a significant amount of energy produced is still based on the burning of fossil fuels. In 2017, up to 64.5% of global electricity was generated through the burning of fossil fuels. Similar to worldwide electricity, the burning of fossil fuels is the main energy source to propel a vehicle. Nonetheless, the burning of fossil fuels produces a large amount of greenhouse gases, which are one of the driving factors of climate change and air pollution. It is worth noting that transportation sectors account for 20% of global energy consumption and 18% of total anthropogenic greenhouse gas emissions [14]. Environmental pollution eventually causes an increasing number of deaths around the world. Therefore, improving the tribological properties is essential not only to extend the service life of the materials but also to reduce greenhouse gas emissions and preserve the natural environment. Efforts have been devoted to the material design to develop materials having high wear and corrosion resistance so as to reduce the carbon dioxide emission from a vehicle. In the era of technology, various techniques, such as thin protective coatings and lubricants, have been successfully developed to lessen wear and friction. It is believed that by lowering friction and enhancing the tribological performance of materials, the cost of energy consumption by machines and vehicles would be dramatically reduced.

2.5 Tribological Properties of FRPs

To extend the lifespan of the materials for engineering applications, the wear resistance of the materials must be sharply enhanced. One of the greatest advantages of FRPs is that their frictions are tremendously lower than those traditional engineering metal alloys. Nonetheless, the wear mechanism of FRPs lacks understanding and is more complex than that of metal alloys. There are several established ways to enhance the tribological properties of FRPs. However, incorporating fibers in the polymer matrix is the most broadly implemented method. Fredrich [15] conducted a thorough analysis and discovered that by combining fibers with micro or nanofillers, the tribological characteristics of polymers might be enhanced. Currently, the majority of the composite market is dominated by synthetic fibers. As of today, the development of eco-friendly composites is still not properly reflected in the tribological applications in engineering sectors. Due to environmental concerns, an attempt has been made to embed the natural fibers in the polymer matrix in place of synthetic fibers. FRPs based on natural fibers have demonstrated very promising results. Yousif [16] revealed that seed oil palm-reinforced polyester composites had comparable tribological properties when the fiber content was increased to 35 vol%. Several detailed research investigations have been conducted to determine the viability of using natural fibers to replace synthetic fibers in wear and friction applications. Aside from the fiber content, the alignment of the fibers with regard to the sliding direction has a prominent impact on the tribological performance of FRPs. In particular, the fiber alignment in FRPs can be parallel, antiparallel, or normal to the sliding direction. The fiber alignment in FRPs with respect to sliding directions is illustrated in Figure 2.3. The failure mechanisms of FRPs due to wear in different sliding directions are shown in Figure 2.4. Overall, testing conditions, polymer matrix types, and fiber surface pre-treatments all influence the tribological characteristics of natural fiber-based composites. The tribological characteristics of natural fiber-reinforced composites have also been investigated in dry and wet environments to gain a better understanding.

2.5.1 Tribological Properties of Natural Fiber-Reinforced Epoxy Composites

Today, the focus has been placed on natural fibers as a contender to replace traditional fibers in FRPs. Intensive research studies have been performed to uncover the potential of natural fibers and expand their applicability in a wide range of engineering fields. Since composite materials have been widely used in a variety of technical applications, these materials might be susceptible to wear and tear

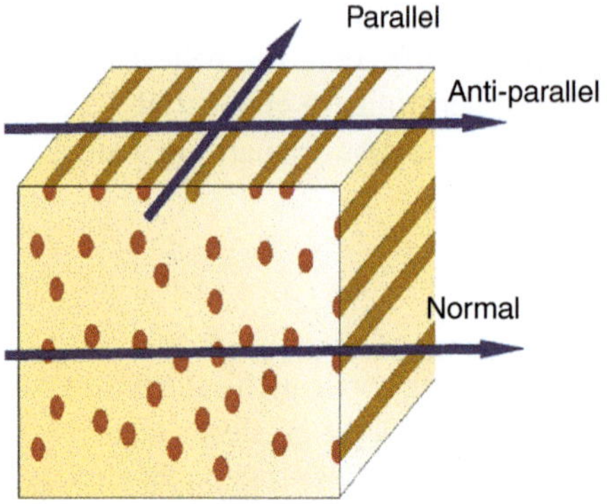

Figure 2.3 Fibre alignment with respect to sliding directions. Source: Chin and Yousif [17]/with permission of Elsevier.

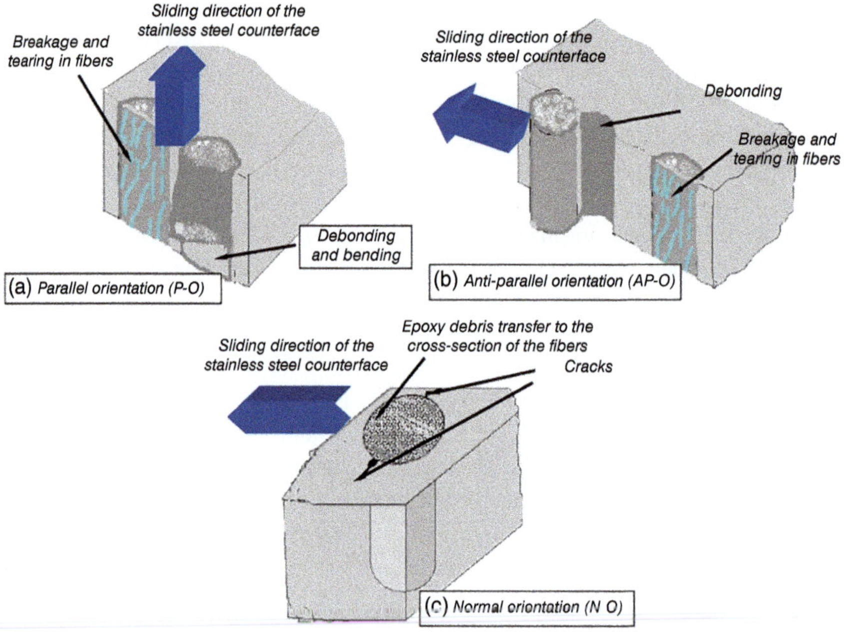

Figure 2.4 Failure mechanism of FRPs due to wear in different sliding directions. Source: Chin and Yousif [17]/with permission of Elsevier.

due to inordinate frictional force. Therefore, an in-depth understanding of the tribological properties of composites reinforced with natural fibers is absolutely essential. Chin and Yousif [17] explored the effects of fiber orientations and testing parameters on the tribological properties of kenaf fiber-reinforced epoxy composites with the aid of a block-on-disc machine. The kenaf fiber orientations were parallel, antiparallel, or normal to the sliding direction. They revealed that the addition of kenaf fiber enhanced the wear performance of neat epoxy, especially when the fiber orientation was normal to the sliding direction. Furthermore, the specific wear rate of kenaf fiber-reinforced composites was found to be less affected by the applied load and sliding velocity. The tribological properties of composites, on the other

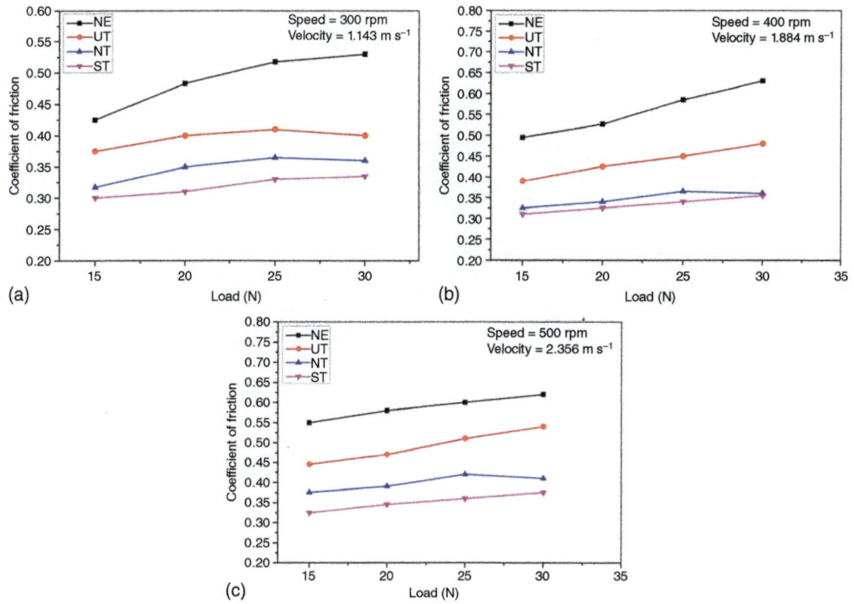

Figure 2.5 Coefficient of friction with respect to applied load under various sliding velocities (a) 1.143 m/s (b) 1.884 m/s (c) 2.356 m/s. Source: Mylsamy et al. [19]/with permission of Elsevier.

hand, were significantly influenced by the fiber orientation. The composites with fiber orientation that was normal to the sliding direction were found to have the lowest specific wear rate and highest coefficient of friction (COF). The tribological characteristics of *Syngonanthus nitens* fiber-reinforced epoxy composites with different fiber weight contents were investigated by Prakash et al. [18]. The effects of impingement angle and sliding velocity on composite wear performance were discovered. They concluded that increasing the impingement angle and sliding velocity increased the wear rate of the composites. However, it was noticed that the wear rate dropped above a certain limit of impingement angle. Overall, composites at 20 wt% possessed the highest wear resistance. The influence of chemical fiber treatment on the tribological characteristics of *Coccinia indica* fiber-reinforced epoxy composites under varying degrees of applied loads and sliding velocities was investigated by Mylsamy et al. [19]. Figures 2.5 and 2.6 demonstrate the COF and wear rate of *Indica* fiber-reinforced epoxy composites under varying degrees of applied loads and sliding velocities, respectively. From the results, it was concluded that the silane and alkali-treated composites prominently improved the wear performance of the composites. However, silane-treated composites outperformed alkali-treated composites in terms of wear resistance. The outstanding tribological properties of silane-treated composites over those of alkali-treated and nontreated composites are mainly attributed to strong fiber-matrix adhesion, which reduces the tendency of debonding. When compared to neat epoxy, untreated composites showed better wear performance, although no chemical treatment was done.

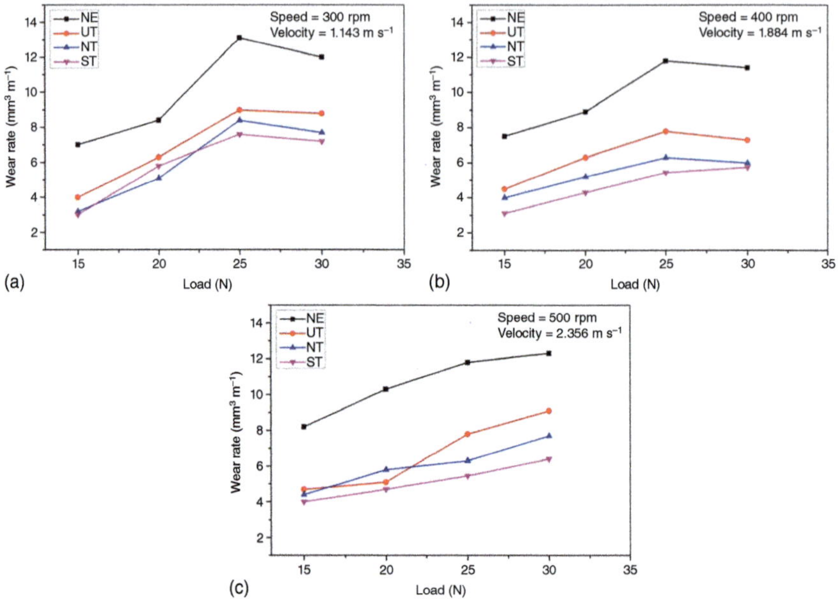

Figure 2.6 Wear rate with respect to applied load under various sliding velocities (a) 1.143 m/s (b) 1.884 m/s (c) 2.356 m/s. Source: Mylsamy et al. [19]/with permission of Elsevier.

Sahu and Gupta [20] identified the effects of surface treatment and coating on the tribological properties of sisal fiber-reinforced epoxy composites. The findings attested that the surface treatment using sodium bicarbonate and coating using poly-lactic acid enhanced the tribological properties of the composites regardless of the impingement angle and sliding velocity. In addition, the erosion rates were observed to be increased with the impingement angle and sliding velocity. Nonetheless, the erosion rate dropped once the impingement angle was increased above a certain limit. Jena et al. [21] analyzed the effect of fiber content and cenosphere filler on the wear performance of bamboo fiber-reinforced epoxy composites under varying impingement angles. Surprisingly, the data revealed that increasing the fiber content increased the wear rate of the composites. However, the incorporation of cenosphere filler augmented the wear resistance of the composites. Furthermore, the wear rate was found to increase from an impingement angle of 30° to 60°, followed by a reduction in the wear rate at 90°. Ramesh et al. [22] investigated the tribological characteristics of *Calotropis gigantea* fiber-reinforced epoxy composites after chemical treatment. Chemical treatments resulted in an improvement in the tribological properties of composites when compared to untreated composites. However, chemically treated composites using sodium hydroxide showed greater tribological properties than those using potassium permanganate. Basumatary et al. [23] evaluated the wear behaviors of *Ipomoea carnea* fiber-reinforced epoxy composites with varying fiber weight compositions. The abrasive wear test was performed at different velocities and loads. They discovered that adding up to 30% fiber to the composites increased their wear resistance. Further increase in the fiber content to 40 wt% in the composites adversely affects the wear performance due to the occurrence of fiber pull-out. Similar to other research studies, an increase in sliding velocities and loads jeopardized the wear performance of the composites. However, the wear rate gradually dropped when increasing the sliding distance during the wear tests. Pradhan et al. [24] experimentally investigated the tribological properties of *Eulaliopsis binata* fiber-reinforced epoxy composites with varying fiber contents. They discovered that increasing the fiber percentage to 20% offered the best tribological performance for the composites in terms of volumetric wear, specific wear rate, and coefficient of friction. They also came to the conclusion that increasing the applied weights and sliding velocity increased the wear rate. Nonetheless, increasing the sliding distance reduced the rate of specific wear. Chand and Dwivedi [25] determined the influence of fiber orientation, sliding distance, and load on the wear behavior of sisal fiber-reinforced epoxy composites. They came to the conclusion that adding sisal fiber to composites improves their wear resistance independent of the fiber orientations. Composites having fiber orientation that was normal to the sliding direction, in particular, tended to enhance wear characteristics the most. Furthermore, weight loss due to wear increased when the sliding distance and load increased. Mishra and Biswas [26] studied the abrasive wear properties of jute fiber-reinforced epoxy composites. The influences of fiber content, sliding velocity, and load on the specific wear rate and coefficient of friction of the composites were identified. According to the results, increasing the fiber content up to 36 wt% and the sliding velocity decreased the specific wear rate of the composites. Nonetheless, it was discovered that increasing the load increased the specific wear rate. In terms of coefficient of friction, when the sliding velocity and weight rose to a certain point, the coefficient of friction increased. The coefficient of friction decreased as the sliding velocity and load increased. Overall, the use of jute fiber in composites has lowered the coefficient of friction. A recent study conducted by Alajmi et al. [27] explored the potential of bamboo fiber as the potential reinforcement of composite materials for tribological applications under various sliding velocities, distances, and applied loads. A comparison was made between the tribological properties of bamboo fiber-based epoxy composites and neat epoxy. They revealed that the weight loss, specific wear rate, and coefficient of friction were reduced with the incorporation of bamboo fiber in the composites, irrespective of sliding velocity, distance, and applied loads. Furthermore, they concluded that the increase in the sliding velocity, distance, and applied loads increased the weight loss but decreased the specific wear rate of the composites. The reported tribological tests on natural fibre-reinforced epoxy composites are summarized in Table 2.1.

Tribological performance of natural fiber composites has been one of the essential considerations in material selection for engineering applications. Since the use of natural fiber is continuously growing, it is vital to comprehensively explore the potential of natural fibers for tribological applications. The majority of research studies have concluded that the addition of natural fiber to polymer composites could significantly enhance wear resistance. However, it was noted that the fiber content exceeding the critical limit in the composites could result in inferior properties. An excessive amount of fiber in the composites causes the

Table 2.1 Reported tribological test on natural fiber-reinforced epoxy composites.

Fiber	Matrix	Factor	Reference
Kenaf	Epoxy	Effect of fiber orientations, applied load and sliding velocity	[17]
Syngonanthus nitens	Epoxy	Effect of impingement angle, sliding velocity and fiber content	[18]
Coccinia indica	Epoxy	Effect of chemical treatment	[19]
Sisal	Epoxy	Effect of surface treatment, coating, impingement angle and sliding velocity	[20]
Bamboo	Epoxy	Effect of fiber content, cenosphere filler content and impingement angle	[21]
Calotropis gigantea	Epoxy	Effect of chemical treatment	[22]
Ipomoea carnea	Epoxy	Effect of fiber content, sliding velocities, sliding distance and applied load	[23]
Eulaliopsis binata	Epoxy	Effect of fiber content, sliding velocities, sliding distance, and applied load	[24]
Sisal	Epoxy	Effect of fiber orientation, sliding distance, and applied load	[25]
Jute	Epoxy	Effect of fiber content, sliding velocity, and applied load	[26]
Bamboo	Epoxy	Effect of fiber addition, sliding velocity, distance and applied load	[27]

agglomeration of fiber and insufficient matrix materials to provide strong structural integrity, reducing the fiber-matrix interaction and resulting in weak wear resistance. In FRPs, the prerequisite for having outstanding wear resistance is excellent fiber-matrix adhesion. Poor interfacial adhesion might lead to debonding when the composites are subjected to frictional force, leading to poor wear resistance. Apart from the fiber content, factors such as sliding velocity, sliding distance, and applied load have been identified to have a profound effect on the tribological properties of natural fiber-reinforced epoxy composites.

2.5.2 Tribological Properties of Natural Fiber-Reinforced Epoxy Hybrid Composites

Recently, researchers worldwide are showing their strong interest in hybrid composites consisting of synthetic and natural fibers embedded within a polymer matrix. Consideration of environmental friendliness and high mechanical strength in material design for industrial applications is regarded as the huge impetus for developing hybrid composites as a replacement for contemporary synthetic fiber-based composites. Research studies focusing on the tribological properties of synthetic/natural fiber-reinforced hybrid composites have continuously increased in recent years. Sneha Latha et al. [28] evaluated the tribological properties of bamboo/glass fiber-reinforced epoxy hybrid composites with varying stacking sequences. A total number of four fabrics were fixed in each composite laminate. They reported that the impingement angle significantly influences the erosion rate. Furthermore, they noticed that the non-hybrid glass and bamboo fiber-based composites showed the highest and lowest erosion rates, respectively. Meanwhile, the erosion rate of the hybrid composites was in between that of non-hybrid composites. The findings attested that the substitution of glass with bamboo fiber was beneficial to the tribological properties of hybrid composites, particularly when bamboo fiber was placed in the outermost layers of the composites. Negi et al. [29] determined the wear properties of potassium permanganate treated hemp/kevlar/carbon fiber-reinforced epoxy hybrid composites with varying hemp fiber contents. In hybrid composites, the Kevlar and carbon fiber contents were fixed at 5 wt%. The findings disclosed that the increase in the hemp fiber content in both hemp/kevlar and hemp/carbon fiber-reinforced hybrid composites reduced their specific wear rate. Overall, hemp/kevlar fiber-based hybrid composites showed the lowest specific wear rate compared to non-hybrid hemp fiber composites and hemp/carbon fiber-reinforced hybrid composites. Oladele et al. [30] analyzed the wear properties of pawpaw/glass fiber-reinforced epoxy hybrid composites. The findings disclosed that the hybrid composites possessed higher wear resistance than neat epoxy, regardless of pawpaw fiber content. However, the wear resistance of the hybrid composites decreased when the linear pawpaw fiber content increased from 3 to 15 wt%. In the case of network pawpaw/glass fiber-reinforced hybrid composites, the wear resistance dropped with the increase in the network pawpaw fiber content from 3 to 9 wt%, and it augmented with further rise of the network pawpaw to 15 wt%. Overall, linear pawpaw fiber-based hybrid composites showed greater wear resistance at the low fiber content. In comparison, those hybrid composites reinforced with network pawpaw fiber demonstrated higher wear resistance at the high fiber content. Nanda and Satapathy [31] analyzed the wear properties of hybrid composites based on human hair fiber and solid glass microspheres with different human hair fiber compositions. The solid glass microsphere content was fixed at 10 wt% in the composites. They concluded that the wear resistance of the hybrid composites could apparently be enhanced by increasing the human hair fiber content. With the addition of a solid glass microsphere to the composites, the wear resistance was improved to a greater extent. Additionally, the wear resistance of the hybrid composites was predicted using artificial neural networks and response surface methodology approaches. In comparison with response surface methodology, the findings obtained from the artificial neural network were in better agreement with the experimental data. Aslan et al. [32] made a comparison between the tribological properties of sisal/carbon and sisal/glass fiber-reinforced epoxy composites with varying sisal fiber contents. The coefficient of friction of the hybrid composites is displayed in Figure 2.7. As can be seen in Figure 2.7, the coefficient of friction showed an increasing trend with the addition of sisal fiber, irrespective of types of synthetic fiber. Figure 2.8 shows the abrasion volume of sisal/glass and sisal/carbon hybrid composites with varying sisal contents. Despite the inferior

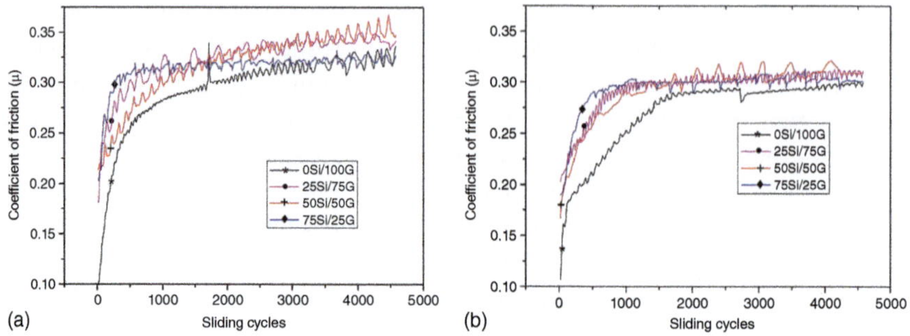

Figure 2.7 Coefficient of friction of hybrid composites (a) sisal/glass (b) sisal/carbon. Source: Aslan et al. [32]/with permission of Elsevier.

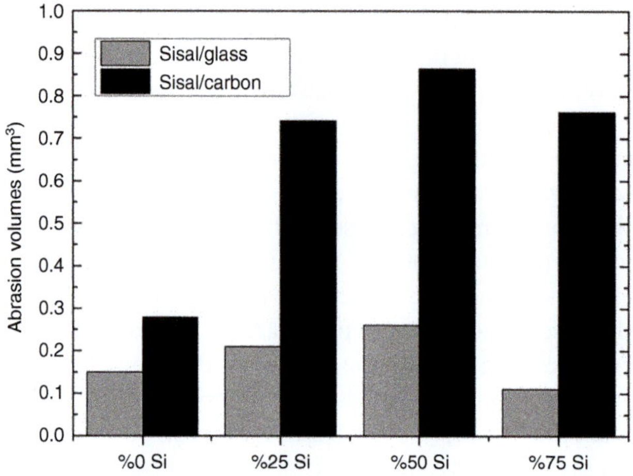

Figure 2.8 Abrasion volume of sisal/glass and sisal/carbon hybrid composites with varying sisal contents. Source: Aslan et al. [32]/with permission of Elsevier.

effect of adding sisal fiber in the hybrid composites, the incorporation of such fiber in sisal/glass hybrid composites manifested lower density and comparable abrasion volume to the non-hybrid glass fiber composites, inferring that sisal fiber possesses the high potential to supersede glass fiber. When the tribological characteristics of sisal/carbon and sisal/glass hybrid composites were investigated, sisal/carbon hybrid composites were determined to have a lower coefficient of friction than sisal/glass hybrid composites. Despite this, sisal/glass hybrid composites had a lower abrasion volume than sisal/carbon hybrid composites.

Sumesh et al. [33] identified the influence of titanium oxide nanofiller, fiber content, sliding distance, sliding velocity, and applied load on the tribological properties of pineapple/sisal fiber-reinforced epoxy hybrid composites. The relative fiber ratio of pineapple and sisal fiber was fixed at 1 : 1. According to the findings obtained, it was found that the addition of nanofiller and the increase in fiber content greatly improved the wear resistance of the hybrid composites. The presence of nanofiller occupies the vacant spaces at the fiber-matrix interface,

Table 2.2 Reported tribological test on natural fiber-reinforced epoxy hybrid composites.

Fiber	Matrix	Factor	Reference
Bamboo/glass	Epoxy	Effect of hybridization and stacking sequence	[28]
Hemp/kevlar/carbon	Epoxy	Effect of fiber type and content	[29]
Pawpaw/glass	Epoxy	Effect of fiber content and fiber structures	[30]
Human hair/solid glass microsphere	Epoxy	Effect of human hair fiber and solid glass microsphere contents	[31]
Sisal/glass/carbon	Epoxy	Effect of sisal fiber content and type of synthetic fiber	[32]
Pineapple/sisal	Epoxy	Effect of titanium oxide nanofiller, fiber content, sliding distance, sliding velocity, and applied load	[33]
Pineapple/flax	Epoxy	Effect of fiber and filler contents	[34]

thus improving the reinforcing effect. Moreover, it was also noticed that increasing the sliding distance, sliding velocity, and applied loads increased the specific wear rate of the hybrid composites. Sumesh et al. [34] optimized the wear properties of pineapple/flax fiber-reinforced epoxy hybrid composites with the addition of peanut oil cake filler using the Taguchi method. Following the findings obtained, adding the filler up to 2 wt% provided the optimum wear resistance to the hybrid composites. The filler content above the limit resulted in agglomeration, which reduced the wear resistance of the hybrid composites. Furthermore, the findings revealed a decrease in wear resistance when the fiber content increased from 20% to 40%. When looking into the testing parameters, the increase in the applied load increased the specific wear rate. However, the increase in the sliding distance undermined the specific wear rate. The reported tribological tests on natural fibre-reinforced epoxy hybrid composites are summarized in Table 2.2.

2.6 Conclusion

One of the driving factors for structural failure and equipment damage has been identified as friction and wear. In tribology, friction and wear often occur during the relative motion of two solid surfaces, and this relative motion consumes energy and promotes material surface degradation. Additionally, energy consumption, carbon dioxide emissions, and economic value of a system are all related to tribology. The energy efficiency, productivity, safety, reliability, and service life of a system can be significantly enhanced by optimizing the tribological properties. Friction and wear are two inevitable phenomena in almost all engineering applications, and hence it is vital to study the tribological properties of the materials before they are applied in industrial applications. In the future, demand for natural fiber-based composite materials is predicted to skyrocket. Since many engineering applications involve frictional forces that will cause wear and tear on the composite materials, it is essential to explore and optimize the tribological properties of natural fiber-based composites. Despite low mechanical strength of natural fibers, several research investigations have shown that they offer superior tribological performance over synthetic fibers. The combination of natural and synthetic fibers could be an alternative way to develop composite materials with promising mechanical and tribological properties.

Following the findings obtained from the previous studies, it was found that the addition of fiber content up to a certain limit is advantageous to the tribological properties of FRPs. On top of that, the fiber-matrix adhesion level must be strong enough to maximize the reinforcing effect in order to obtain remarkable tribological properties in composite materials. Therefore, it is crucial to involve fiber surface pretreatments or the inclusion of coupling agents to improve fiber-matrix compatibility. These steps trigger the morphological changes on the fiber surface and provide better mechanical anchoring between fiber and matrix. As a result, the load-carrying capacity of natural fiber-reinforced composites might be increased significantly. Furthermore, it is self-evident that increasing the sliding velocity and applied force increases the wear rate. The high applied load causes the eroded particles to exert a high force against the specimen surface, which eventually results in distortion on the composite surface. Aside from the addition of fiber, it was found that the addition of fillers could further enhance the wear resistance of the composite materials. The addition of fillers could fill up the pores and voids at the fiber-matrix interfaces, strengthening the load-transferring efficiency and thereby improving the wear resistance of the composites. Nevertheless, the amount of filler content should be well-controlled, as excessive filler content might result in agglomeration, which could adversely affect the tribological properties of the composite materials. Overall, natural fibers have been one of the emerging materials that are extensively explored in the field of composite materials, mainly due to their eco-friendly characteristics and high specific properties. To date, complete replacement of synthetic fiber with natural fiber is still impractical from the viewpoint of mechanical properties, dimensional stability, and hygrothermal properties. Therefore, the hybridization concept is speculated to provide an alternative route toward developing sustainable and advanced composite materials capable of being applied in a wide range of engineering applications.

References

1 Feng, N.L., Malingam, S.D., Subramaniam, K. et al. (2019). The influence of fibre stacking configurations on the indentation behaviour of pineapple leaf/glass fibre reinforced hybrid composites. *Def. S T Tech. Bull.* 12: 113–123.

2 Boidi, G., Krenn, S., and Eder, S.J. (2020). Identifcation of a material–lubricant pairing and operating conditions that lead to the failure of porous journal bearing systems. *Tribol. Lett.* 68: 108. https://doi.org/10.1007/s11249-020-01347-0.

3 Omrani, E., Menezes, P.L., and Rohatgi, P.K. (2016). State of the art on tribological behavior of polymer matrix composites reinforced with natural fibers in the green materials world. *Eng. Sci. Technol. Int. J.* 19: 717–736. https://doi.org/10.1016/j.jestch.2015.10.007.

4 Feng, N.L., Malingam, S.D., Subramaniam, K. et al. (2020). The investigation of the tensile and quasi-static indentation properties of pineapple leaf/kevlar fibre reinforced hybrid composites. *Def. S T Tech. Bull.* 13: 117–129.

5 Chandrasekar, M., Ishak, M.R., Sapuan, S.M. et al. (2017). A review on the characterisation of natural fibres and their composites after alkali treatment and water absorption. *Plast. Rubber Compos.* 46: 119–136. https://doi.org/10.1080/14658011.2017.1298550.

6 Vinod, A., Sanjay, M.R., and Siengchin, S. (2021). Fatigue and thermo-mechanical properties of chemically treated Morinda citrifolia fiber-reinforced bio-epoxy composite: a sustainable green material for cleaner production. *J. Clean. Prod.* 326: 129411. https://doi.org/10.1016/j.jclepro.2021.129411.

7 Sinha, E. and Panigrahi, S. (2009). Effect of plasma treatment on structure, wettability of jute fiber and flexural strength of its composite. *J. Compos. Mater.* 43: 1791–1802. https://doi.org/10.1177/0021998309338078.

8 Jagadeesh, P., Puttegowda, M., Rangappa, S.M., and Siengchin, S. (2021). A review on extraction, chemical treatment, characterisation of natural fibers and its composites for potential applications. *Polym. Compos.* 42: 6239–6264. https://doi.org/10.1002/pc.26312.

9 Feng, N.L., Malingam, S.D., and Ping, C.W. (2021). Mechanical characterisation of kenaf/PALF reinforced composite-metal laminates: effects of hybridisation and weaving architectures. *J. Reinf. Plast. Compos.* 40: 193–205. https://doi.org/10.1177/0731684420956719.

10 Feng, N.L., Malingam, S.D., Ping, C.W., and Razali, N. (2020). Mechanical properties and water absorption of kenaf/pineapple leaf fiber-reinforced polypropylene hybrid composites. *Polym. Compos.* 41: 1255–1264. https://doi.org/10.1002/pc.25451.

11 Ng, L.F., Yahya, M.Y., and Muthukumar, C. (2022). Mechanical characterisation and water absorption behaviors of pineapple leaf/glass fiber-reinforced polypropylene hybrid composites. *Polym. Compos.* 43: 203–214. https://doi.org/10.1002/pc.26367.

12 Torabinejad, V., Aliofkhazraei, M., Assareh, S. et al. (2017). Electrodeposition of Ni-Fe alloys, composites, and nano coatings—a review. *J. Alloys Compd.* 691: 841–859. https://doi.org/10.1016/j.jallcom.2016.08.329.

13 Holmberg, K. and Erdemir, A. (2017). Influence of tribology on global energy consumption, costs and emissions. *Friction* 5: 263–284. https://doi.org/10.1007/s40544-017-0183-5.

14 Holmberg, K., Andersson, P., and Erdemir, A. (2012). Global energy consumption due to friction in passenger cars. *Tribol. Int.* 47: 221–234. https://doi.org/10.1016/j.triboint.2011.11.022.

15 Friedrich, K. (2018). Polymer composites for tribological applications. *Adv. Ind. Eng. Polym. Res.* 1: 3–39. https://doi.org/10.1016/j.aiepr.2018.05.001.

16 Yousif, B.F. (2008). Replacing of glass fibres with seed oil palm fibres for tribopolymeric composites. *Tribol. Mater. Surf. Interfaces* 2: 99–103. https://doi.org/10.1179/175158308X383189.

17 Chin, C.W. and Yousif, B.F. (2009). Potential of kenaf fibres as reinforcement for tribological applications. *Wear* 267: 1550–1557. https://doi.org/10.1016/j.wear.2009.06.002.

18 Prakash, V., Pradhan, S., and Acharya, S.K. (2021). Tribological behavior of *Syngonanthus nitens* natural fiber reinforced epoxy composite. *Trans. Indian Inst. Met.* 74: 1741–1750. https://doi.org/10.1007/s12666-021-02268-9.

19 Mylsamy, B., Chinnasamy, V., Palaniappan, S.K. et al. (2020). Effect of surface treatment on the tribological properties of *Coccinia indica* cellulosic fiber reinforced polymer composites. *J. Mater. Res. Technol.* 9: 16423–16434. https://doi.org/10.1016/j.jmrt.2020.11.100.

20 Sahu, P. and Gupta, M.K. (2019). Enhancement in erosion wear resistance of sisal composites by ecofriendly treatment and coating. *Mater. Res. Exp.* 6: 085348. https://doi.org/10.1088/2053-1591/ab29b5.

21 Jena, H., Pradhan, A.K., and Pandit, M.K. (2018). Study of solid particle erosion wear behavior of bamboo fiber reinforced polymer composite with cenosphere filler. *Adv. Polym. Technol.* 37: 21718. https://doi.org/10.1002/adv.21718.

22 Ramesh, M., Deepa, C., Rajeshkumar, L. et al. (2021). Influence of fiber surface treatment on the tribological properties of *Calotropis gigantea* plant fiber reinforced polymer composites. *Polym. Compos.* 42: 4308–4317. https://doi.org/10.1002/pc.26149.

23 Basumatary, K.K., Mohanta, N., and Acharya, S.K. (2014). Effect of fiber loading on abrasive wear behaviour of *Ipomoea carnea* reinforced epoxy composite. *Int. J. Plast. Technol.* 18: 64–74. https://doi.org/10.1007/s12588-014-9065-0.

24 Pradhan, S., Acharya, S.K., and Prakash, V. (2021). Mechanical, morphological, and tribological behavior of *Eulaliopsis binata* fiber epoxy composites. *J. Appl. Polym. Sci.* 138: 50077. https://doi.org/10.1002/app.50077.

25 Chand, N. and Dwivedi, U.K. (2007). Influence of fiber orientation on high stress wear behavior of sisal fiber-reinforced epoxy composites. *Polym. Compos.* 28: 437–441. https://doi.org/10.1002/pc.20286.

26 Mishra, V. and Biswas, S. (2016). Three-body abrasive wear behavior of short jute fiber reinforced epoxy composites. *Polym. Compos.* 37: 270–278. https://doi.org/10.1002/pc.23178.

27 Alajmi, A.E., Alotaibi, J.G., Yousif, B.F., and Nirmal, U. (2021). Tribological studies of bamboo fibre reinforced epoxy composites using a BOD technique. *Polymers* 13: 2444. https://doi.org/10.3390/polym13152444.

28 Sneha Latha, P., Venkateswara Rao, M., Kiran Kumar, V.V. et al. (2016). Evaluation of mechanical and tribological properties of bamboo–glass hybrid fiber reinforced polymer composite. *J. Ind. Text.* 46: 3–18. https://doi.org/10.1177/1528083715569376.

29 Negi, A.S., Katiyar, J.K., Kumar, S. et al. (2019). Physicomechanical and abrasive wear properties of hemp/kevlar/carbon reinforced hybrid epoxy composites. *Mater. Res. Exp.* 6: 115304. https://doi.org/10.1088/2053-1591/ab438d.

30 Oladele, I.O., Ayanleye, O.T., Adediran, A.A. et al. (2020). Characterisation of wear and physical properties of pawpaw–glass fiber hybrid reinforced epoxy composites for structural application. *Fibers* 8: 44. https://doi.org/10.3390/fib8070044.

31 Nanda, B.P. and Satapathy, A. (2021). An analysis of the sliding wear characteristics of epoxy-based hybrid composites using response surface method and neural computation. *J. Nat. Fibers* 18: 2077–2091. https://doi.org/10.1080/15440478.2020.1722781.

32 Aslan, M., Tufan, M., and Küçükömeroğlu, T. (2018). Tribological and mechanical performance of sisal-filled waste carbon and glass fibre hybrid composites. *Composites,Part B Eng.* 140: 241–249. https://doi.org/10.1016/j.compositesb.2017.12.039.

33 Sumesh, K.R., Saikrishnan, G., Pandiyan, P. et al. (2021). The influence of different parameters in tribological characteristics of pineapple/sisal/TiO_2 filler incorporation. *J. Ind. Text.* https://doi.org/10.1177/15280837211022614.

34 Sumesh, K.R., Kavimani, V., Rajeshkumar, G. et al. (2020). Mechanical, water absorption and wear characteristics of novel polymeric composites: impact of hybrid natural fibers and oil cake filler addition. *J. Ind. Text.* https://doi.org/10.1177/1528083720971344.

3

Wear Properties of Flax/Epoxy-Based Composites With Different Machining Parameters

K.R. Sumesh[1], Petr Spatenka[1], and G. Rajeshkumar[2]

[1] *Czech Technical University in Prague, Faculty of Mechanical Engineering, Department of Materials Engineering, Karlovo Namesti 13, CZ-12135, Prague, Czech Republic*
[2] *PSG Institute of Technology and Applied Research, Department of Mechanical Engineering, Neelambur, Coimbatore 641 062, Tamil Nadu, India*

3.1 Introduction

Worldwide natural biodegradable products are finding huge applications in different industries due to their eco-friendly nature, availability, and lesser cost than other conventional products. In the field of polymer composites, there is a huge trend toward lignocellulosic reinforcements because these bioresources are showing better properties. Natural fibers, such as coir, flax, pineapple, banana, sisal, ramie, and kenaf, have been used in various sectors for their mechanical, wear, thermal, flammability, vibration, and damping applications. The lignocellulosic fibers from the inner bark of trees, also known as bast fibers, and from crops such as ramie, flax, and hemp have been used with different thermo- and thermosetting-based matrices for various applications. There are different factors determining the good adhesion of natural fiber-based composites. Surface treatment, fiber content, fibers used, and fabrication method are a few of them [1–4].

It has been found that water absorption is a major cause for declining the applications of natural fiber composites with different matrices. The hydrophilic content in natural fibers must be reduced with surface treatments such as NaOH, KOH, bleaching, acetylation, and isocyanate [4–6]. In the flax/epoxy-based composites, 3–5 wt% of NaOH treatment reduces the hydrophilic content in the natural fibers with good mechanical properties [4]. The flax hybrid composites with epoxy showed better adhesion at 5 wt% of NaOH. The flax fiber with 20–40 wt% showed better wear resistance. Flax fiber addition enhanced the adhesion between fiber/matrix phase, which led to improved wear resistance. The natural fiber content above 50 wt% observed declined wear resistance due to surface deformations in the polymer combinations [5, 6].

Applied load, sliding distance (SD), and sliding speed are the different machining parameters in pin-on-disc wear apparatus determining the properties. Increase

Tribological Properties, Performance, and Applications of Biocomposites, First Edition.
Edited by Chandrasekar Muthukumar, Senthilkumar Krishnasamy, Senthil Muthu Kumar Thiagamani, and Ganesan Chinnachamy.

in load from 5 to 15 N created abnormalities in the combinations by creating a high-pressure zone, thus leading to declining wear resistance in natural fiber epoxy combinations [7, 8]. The SD (500–1500 m) does not create any changes to the surface of natural fiber combinations, while the sliding speed (0.5–1.5 m s^{-1}) has a negative impact on the wear resistance of the epoxy-based combination [9–12]. It has been found that fiber content has a major influence in determining the wear properties of epoxy-based natural fiber composites than other machining parameters. In this research work, flax fiber content and pin-on-disc machining parameters such as SD, sliding speed, and applied load will be varied to find the optimum condition of combination.

3.2 Materials and Methods

Natural fiber flax was selected as the reinforcement and was purchased from Tamil Nadu, India. Epoxy resin (LY556) with suitable hardener (HY951) was used as the matrix material in a 10:1 ratio. Initially, alkali treatment (5% NaOH) was provided for the fibers to reduce the hydrophilic nature of natural fibers and, thus, to improve the adhesion with epoxy-based composites during the fabrication. After the surface treatment, it was heated in the oven for three hours, and finally cropped at 1 cm in length.

3.2.1 Method

The compression molding method was used in fabricating flax/epoxy-based composites. The compression mold had a shape with a $300 \times 300 \times 3\ \text{mm}^3$ size and was used with 20–40 wt% of flax fibers. A temperature of 120 °C and pressure of 14 MPa were applied in the combination.

3.2.2 Wear Testing

The pin-on-disc equipment was used for testing wear properties of natural fiber composites. The ASTM G 99 was used as the standard, with EN 31 steel counter disc and surface for sliding with the composite sample. The testing was carried out by varying flax fiber content (wt%), sliding speed (Vs), applied load (L), and SD in this research. The following equation was used in determining the specific wear rate (SWR) of the composites:

$$\text{SWR} = \frac{\text{WL}}{\rho \times L \times \text{SD}}\ \text{mm}^3\ \text{Nm}^{-1} \tag{3.1}$$

where WL is weight loss of combination after wear testing in grams, ρ is the density of flax/epoxy sample in g cm^{-3}, L is load applied during testing in Newtons, and SD is sliding distance in meter.

3.3 Results and Discussion

3.3.1 Wear Results

The SWR was taken from the wear loss of the specimen after the experiment on the pin-on-the-disc machine. The first machining parameter, applied load, observed high wear rate from 0.42 to 0.545×10^{-5} mm^3 Nm^{-1} (Figure 3.1) with the incorporation of flax fiber from 20 to 40 wt%. In Figure 3.1, applied load is varying from 5 to 15 N, and other factors such as SD and sliding speed are fixed constant as 500 m and 0.5 m s^{-1}, respectively. It was already observed that a load from 5 to 15 N creates a high-pressure zone in the composites that will turn into surface deformations in the combination [13–15]. The SD does not cause any increase in the SWR (Figure 3.2) due to lesser impact of the factor with epoxy-based composites [5, 6]. In this data, SD was varied from 500 to 1500 m, and applied load and sliding speed were fixed constant as 5 N and 0.5 m s^{-1}, respectively. SD does not cause any further damage to the composite combination; however, it results in slight improvement in wear resistance in all the combinations. The wear rate decreased from 0.52 to 0.4×10^{-5} mm^3 Nm^{-1} with the increase in SD up to 1500 m.

The final machining parameter sliding speed (0.5–1.5 m s^{-1}) created an increase in the wear rate for the flax/epoxy combination (Figure 3.3). In this data, sliding speed is varied from 0.5 to 1.5 m s^{-1} and applied load and SD are fixed constant as 5 N and 500 m, respectively. High sliding speed created more contact with the counter surface and, thus, the speed in the rotation of the disc created damage to the specimen, which led to increase in wear rate [16–18]. The SWR increases from 0.41 to 0.53×10^{-5} mm^3 Nm^{-1} with the increase in sliding speed up to 1.5 m s^{-1}.

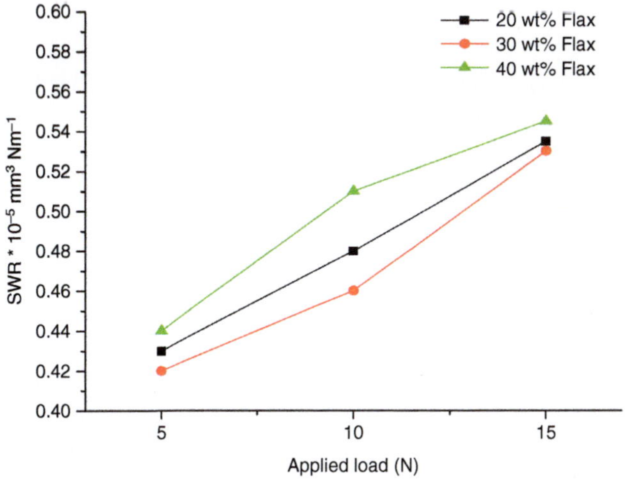

Figure 3.1 SWR vs applied load.

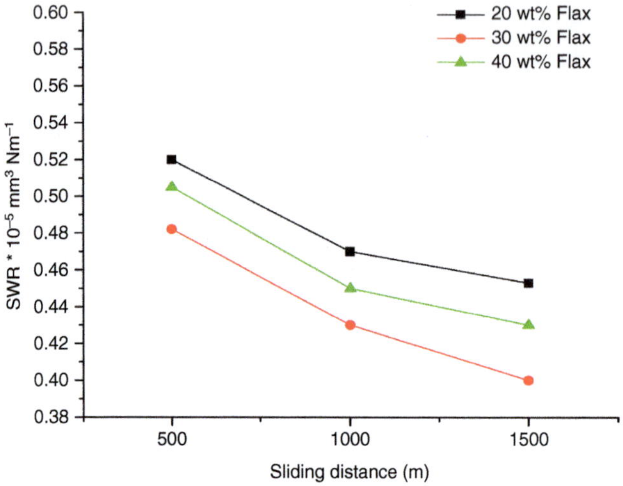

Figure 3.2 SWR vs sliding distance.

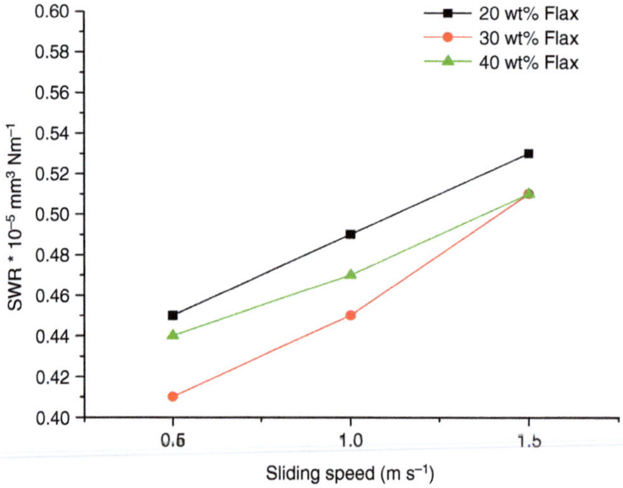

Figure 3.3 SWR vs sliding speed.

In all machining parameters, applied load has a huge impact in increasing the SWR of the flax/epoxy-based combination, followed by sliding speed and SD.

The natural fiber addition of flax wt% at 30 observed lesser wear rate due to good adhesion of flax fiber with epoxy-based composites. The high increase in variations of applied load, SD, and sliding speed up to 15 N, 1500 m, and 1.5 m s^{-1}, respectively, did not reduce the wear resistance of these 30 wt% flax/epoxy combinations. The high incorporation at 40 wt% increases the wear rate of the pull-out mechanism due to the damage in mechanical bonding between epoxy resin and flax fiber reinforcement. The reinforcement addition below the optimum range created nonhomogeneous fiber distribution and made a weak point in the matrix, creating a high-stress concentration by applying load [19–21]. Fiber breakages, fiber pull-outs, matrix breakages, voids, and fiber debonding occurred with higher incorporation of natural fibers with epoxy-based composites [21].

It has been previously found that 20–30 wt% flax fiber addition created high increase in tensile and flexural properties of epoxy matrix composites [2]. Good bonding of the combination increases the hardness and density of the combination, which also results in good wear resistance of the combination. The alkali treatment of 5% NaOH also created a high crystalline structure with high crystallinity index (CI); the increase in cellulosic content over lignin and hemicellulose content is the main reason for the increased crystalline nature [5, 6]. This result also states that wear rate is directly proportional to applied load (N) and sliding speed ($m\,s^{-1}$), but inversely proportional to SD (m). More the contact between counter specimen and composite samples, higher will be the wear rate.

3.4 Conclusions

The combination of flax/epoxy-based composites showed lower wear rate of $0.4 \times 10^{-5} mm^3\,Nm^{-1}$ with 30 wt% flax, machining parameters of SD 1500 m, applied load of 5 N, and sliding speed 0.5 m/s^{-1}. High wear rate of $0.545 \times 10^{-5}\ mm^3\ Nm^{-1}$ was observed at 40 wt% flax, SD 500 m, and load 5 N. Alkali treatment with 5% NaOH also has a high influence on the compatibility of matrix/fiber phases that adds to the wear resistance of the combination. In all the machining parameters, applied load has a huge impact on increasing the SWR of the flax/epoxy-based combination, followed by sliding speed and SD. It is observed that loads from 5 to 15 N create a high-pressure zone in the composites, which will create surface deformations in the combination. The SD does not cause any increase in the SWR due to its lesser impact on the factor with epoxy-based composites. High sliding speed created more contact with the counter surface, and the speed in the rotation of the disc created damage to the specimen, which led to increase in wear rate. This result also states that wear rate is directly proportional to applied load (N) and sliding speed ($m\,s^{-1}$), but inversely proportional to SD (m). More the contact between counter specimen and composite samples, higher will be the wear rate.

References

1 Nirmal, U., Yousif, B.F., Rilling, D., and Brevern, P.V. (2010). Effect of betelnut fibres treatment and contact conditions on adhesive wear and frictional performance of polyester composites. *Wear* 268: 1354–1370. https://doi.org/10.1016/j.wear.2010.02.004.

2 Mittal, V., Saini, R., and Sinha, S. (2016). Natural fiber-mediated epoxy composites—a review. *Composites, Part B* 99: 425–435. https://doi.org/10.1016/j.compositesb.2016.06.051.

3 Chand, N. and Dwivedi, U.K. (2006). Effect of coupling agent on abrasive wear behaviour of chopped jute fibre-reinforced polypropylene composites. *Wear* 261: 1057–1063. https://doi.org/10.1016/j.wear.2006.01.039.

4 Oushabi, A. (2019). The pull-out behavior of chemically treated lignocellulosic fibers/polymeric matrix interface (LF /PM): a review. *Composites, Part B* 174: https://doi.org/10.1016/j.compositesb.2019.107059.

5 Sumesh, K.R., Kavimani, V., Rajeshkumar, G. et al. (2020). Mechanical, water absorption and wear characteristics of novel polymeric composites: impact of hybrid natural fibers and oil cake filler addition. *J. Ind. Text* 1–28: https://doi.org/10.1177/1528083720971344.

6 Sumesh, K.R., Kavimani, V., Rajeshkumar, G. et al. (2020). An investigation into the mechanical and wear characteristics of hybrid composites: influence of different types and content of biodegradable reinforcements. *J. Nat. Fibers* 1–13: https://doi.org/10.1080/15440478.2020.1821297.

7 Nirmal, U., Hashim, J., and Low, K.O. (2012). Adhesive wear and frictional performance of bamboo fibres reinforced epoxy composite. *Tribol. Int.* 47: 122–133. https://doi.org/10.1016/j.triboint.2011.10.012.

8 Shuhimi, F.F., Bin, A.M.F., Kalam, M.A. et al. (2016). Tribological characteristics comparison for oil palm fibre/epoxy and kenaf fibre/epoxy composites under dry sliding conditions. *Tribol. Int.* 101: 247–254. https://doi.org/10.1016/j.triboint.2016.04.020.

9 Sumesh, K.R., Saikrishnan, G., Pandiyan, P. et al. (2021). The influence of different parameters in tribological characteristics of pineapple/sisal/TiO_2 filler incorporation. *J. Ind. Text.* https://doi.org/10.1177/15280837211022614.

10 Correa, C.E., Betancourt, S., Vázquez, A., and Gañan, P. (2017). Wear performance of vinyl ester reinforced with Musaceae fiber bundles sliding against different metallic surfaces. *Tribol. Int.* 109: 447–459. https://doi.org/10.1016/j.triboint.2017.01.009.

11 Kumar, S., Patel, V.K., Mer, K.K.S. et al. (2019). Himalayan natural fiber-reinforced epoxy composites: effect of *Grewia optiva/Bauhinia vahlii* fibers on physico-mechanical and dry sliding wear behavior. *J. Nat. Fibers* https://doi.org/10.1080/15440478.2019.1612814.

12 Yousif, B.F., Lau, S.T.W., and McWilliam, S. (2010). Polyester composite based on betelnut fibre for tribological applications. *Tribol. Int.* 43: 503–511. https://doi.org/10.1016/j.triboint.2009.08.006.

13 Akpan, E.I., Wetzel, B., and Friedrich, K. (2018). A fully biobased tribology material based on acrylic resin and short wood fibres. *Tribol. Int.* 120: 381–390. https://doi.org/10.1016/j.triboint.2018.01.010.

14 Joseph, J., Munda, P.R., Kumar, M. et al. (2020). Sustainable conducting polymer composites: study of mechanical and tribological properties of natural fiber reinforced PVA composites with carbon nanofillers. *Polym. Technol. Mater.* https://doi.org/10.1080/25740881.2020.1719144.

15 Nanda, B.P. and Satapathy, A. (2020). An analysis of the sliding wear characteristics of epoxy-based hybrid composites using response surface method and neural computation. *J. Nat. Fibers* 0478: https://doi.org/10.1080/15440478.2020.1722781.

16 Fei, J., Zhang, C., Luo, D. et al. (2018). Vertically aligned TiO_2 nanorods-woven carbon fiber for reinforcement of both mechanical and anti-wear properties in resin composite. *Appl. Surf. Sci.* 435: 156–162. https://doi.org/10.1016/j.apsusc.2017.10.182.

17 Kumar, M.N.S., Yaakob, Z., Mohan, N., and Babu, S.P.K. (2010). Mechanical and abrasive wear studies on biobased *Jatropha* oil cake incorporated glass—epoxy composites. *J. Am. Oil Chem. Soc.* 87: 929–936. https://doi.org/10.1007/s11746-010-1575-0.

18 Aslan, M., Tufan, M., and Küçükömeroğlu, T. (2018). Tribological and mechanical performance of sisal-filled waste carbon and glass fibre hybrid composites. *Composites, Part B* 140: 241–249. https://doi.org/10.1016/j.compositesb.2017.12.039.

19 Maurya, H.O., Jha, K., and Tyagi, Y.K. (2017). Tribological behavior of short sisal fiber reinforced epoxy composite. *Polym. Polym. Compos.* 25: 215–220. https://doi.org/10.1177/096739111702500306.

20 Rajeshkumar, G. (2020). A new study on tribological performance of *Phoenix* sp. fiber-reinforced epoxy composites. *J. Nat. Fibers* https://doi.org/10.1080/15440478.2020.1724235.

21 Nirmal, U., Hashim, J., and Megat Ahmad, M.M.H. (2015). A review on tribological performance of natural fibre polymeric composites. *Tribol. Int.* 83: 77–104. https://doi.org/10.1016/j.triboint.2014.11.003.

4

Polyester-Based Biocomposites for Tribological Applications

Anand Gobiraman[1], Santhosh Nagaraja[2], and Vishvanathperumal Sathiyamoorthi[3]

[1] *Achariya College of Engineering Technology, Department of Mechanical Engineering, Achariyapuram, Villianur, Uruvaiyar, Pondicherry 605110, India*
[2] *MVJ College of Engineering, Department of Mechanical Engineering, Near ITPL Main Road, 3G Homes Crimson Layout, Kadugodi, Bangalore, Karnataka 560067, India*
[3] *S.A. Engineering College, Department of Mechanical Engineering, Poonamallee, Avadi Road Veeraraghavapuram, Thiruverkadu, Chennai, Tamil Nadu 600077, India*

4.1 Introduction: Background and Driving Forces

Since the beginning of time, natural fibers have been employed for tribological applications, and composites reinforced with natural fibers in particular have grown in significance for real-time applications. Natural fiber-reinforced composites are often light, robust, non-abrasive, and have good mechanical qualities, as well as being environmentally beneficial. These composites have excellent qualities, which are mostly determined by their composition. Many scientists are looking into natural fibers, including sisal, jute, and kenaf. Excellent mechanical qualities are offered by natural fiber composites, which are improved by the inclusion of polyester polymers. Furthermore, they have great biodegradability and are environmentally friendly. Due to their better qualities, such as high strength, low density with great thermal insulation, environmental sustainability, recyclability, and higher mechanical capabilities, natural fiber composites can be utilized in place of synthetic fibers. Natural fibers have been used in the manufacture of biodegradable composites by a number of researchers, who have discovered that they have superior qualities. Some researchers have also explored natural fiber-reinforced composites chemically, thermally, and physically. High-performance products that need to be light but robust enough to sustain dynamic loading are increasingly using natural fiber-reinforced polyester composites. Furthermore, traditional composites are not eco-friendly. The existence of two different components at the same time is one of the major disadvantages of adopting this vast family of composites. This makes reuse and recycling difficult at the conclusion of the composite's life cycle. As a result, it is often preferable to dispose of them directly in a landfill or incinerate these composites. Also, the fabrication of plastic calls for terrific intake of fossil fuel-based resources,

Tribological Properties, Performance, and Applications of Biocomposites, First Edition.
Edited by Chandrasekar Muthukumar, Senthilkumar Krishnasamy,
Senthil Muthu Kumar Thiagamani, and Ganesan Chinnachamy.

which are nonrenewable. Polymer composites constitute an incredible possibility for a variety of applications; however, their effect on the surroundings limits their degree of diffusion. However, as the usage of composites for tribological applications grows, so does research into natural fibers in polyester composites. In fact, the automotive parts subjected to wear are being made using natural fiber-reinforced composites, and numerous pilot initiatives have been taken forward to evolve a new processing methodology. Furthermore, a completely new methodology for both product and manufacturing must be proposed, integrating modern environmental evaluation methods such as risk assessment and life cycle assessment.

The use of natural fiber-reinforced polyester composites in high-wear applications, including automotive, marine, submersible, and aviation structural components, is very promising. Therefore, natural fibers are generally considered to have higher durability and require little or no maintenance over long periods of time. Additionally, lightweight and wear resistance are a must for ships and aircraft to meet design requirements and operate at impeccable speed and reliability. The composites, with special reference to natural fiber-reinforced polymers, have found extensive use in the defense industry after their first use during World War II, especially to overcome problems of wear in steel, aluminum, and their alloys. Weight reduction is an important consideration, especially with the higher weight found on commercial aircraft and marine structures. Over the years, natural fiber-reinforced composites have evolved to meet the growing and variable demand of the aviation sector while maintaining a relevant use for the manufacture of consolidated elements of ships and vessels that include grates, conduits, culverts, pipes, and hulls. Composite materials are currently utilized in every aspect of engineering and for a variety of components and structures in marine field, especially the propeller blades, ship hulls, upper frames, radomes, railings, and decks of fishing boats and other underwater structures. In recent years, polyester composites have been extensively used owing to their high performance and safety considerations.

However, it's far vital to advocate a totally new technique for each product and manufacturing, including cutting-edge technologies for environmental evaluation, such as characterization of the mechanical properties and lifecycle evaluation [1].

Also, polyester composites are generally used in recent times in bearings, brake pads, and so on. Further procedures for designing polymeric composites with a view to low friction and low wear are being researched. Polyester composites with graphite reinforcement and polymer fibers impregnated with vegetable oils as self-lubricating agents are proposed, lowering the friction coefficient and wear rates dramatically.

Due to principal technological modifications in the last decade, numerous additives are crafted from engineering polymers and composites for their use as antifriction agents. These composites with polyester resins are known to replace conventional materials, especially for self-lubricating bearings, linear guides, mechanical seals, bushings, belts, gears, and pulleys. These additives are even used for contact mechanics to enhance durability, fatigue strength, and wear resistance.

Polyester composites with graphite reinforcement and polymer fibers impregnated with vegetable oils as self-lubricating agents are proposed, lowering the friction coefficient and wear rates dramatically.

4.2 Materials and Methods

For the synthesis of polyester composites, a variety of material combinations are used, and various researchers from across the world have worked on polyester-natural fiber composites and hybrid polyester composites. In this regard, the manufacturing methods for the synthesis of polyester-natural fiber composites, the tribological test methods utilized by numerous other researchers, and an extensive review of the natural fibers used in the synthesis of the composites are described in this part.

4.2.1 Natural Fibers

Natural fibers are increasingly being used to make composites due to their environmental sustainability. The natural fibers of hemicellulose, lignin, and cellulose chemical combinations can be obtained from seeds, leaves, basts, and fruits. Natural fibers have a number of environmental benefits, including the fact that they are biodegradable and carbon positive [2]. They are also nonabrasive, which means less wear on tools and industrial equipment. Enhanced strength, ductility, and stiffness are a few more major advantages that may be found when looking at the material's unique qualities. Further, by using natural fibers instead of synthetics, the weight can be reduced by 40%, and the flexural strength can also be improved [3]. The qualities of thermoplastic natural fiber composites can also be improved by physical and chemical treatments [4]. Synthetic fibers generally outperform natural fibers in terms of strength properties, but not always. According to Lopresto et al. [5], composites containing some natural fibers can approach the tensile and tensile strength modulus of synthetic fibers. There are two main categories of natural fibers utilized in the creation of composites: non-wood and wood fibers. In addition, whereas wood fibers are classified according to how hard or soft they are, non-wood fibers are divided based on the origin of the plant from which they are collected. The various natural fiber types used in the current investigation are summarized in Figure 4.1.

4.2.2 Polyester–Natural Fiber Composites

Pineapple leaf fiber-reinforced polyester composites were created by Senthilkumar et al. [6], who also carried out mechanical and free vibration characterization experiments on the composites. Their research found that increasing the pineapple leaf fiber content in composites up to 55% increased bending, compressive strength, and tensile properties, but that beyond that point, the qualities deteriorated due to void development and debonding. Uma Devi et al. [7] investigated the optimum volume and lengths of pineapple leaf fibers in polyester matrix composites, finding that the typical length of the pineapple fibers for improved bending rigidity of the composite materials is 35 mm, and the optimum lengths of the composites are about 2.2 times that of the polyester resin. The impact strength of treated and untreated sisal-reinforced polyester composites was evaluated by Prasad et al. [8]. The researchers found that the bulk stiffness and strength of the composite laminates

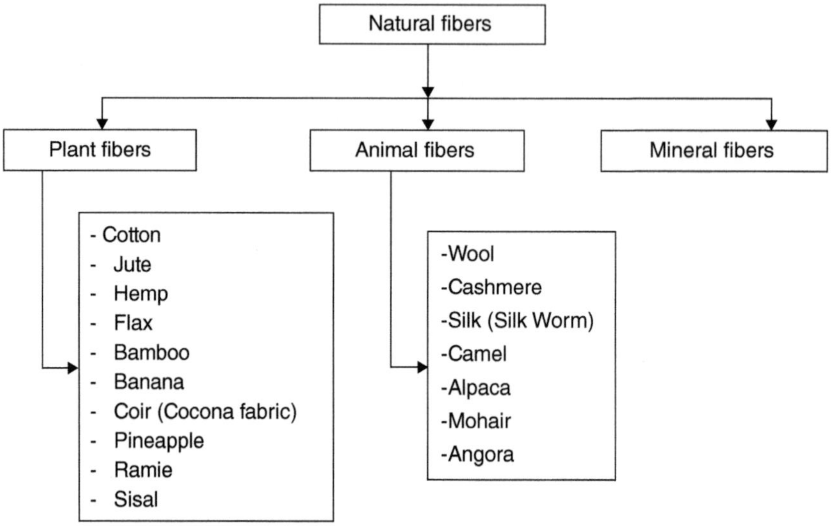

Figure 4.1 Types of natural fibers used for the synthesis of biocomposites.

were enhanced by the polyester matrix composite with embedded sisal fibers treated with 4% sodium hydroxide (NaOH). The adhesion between the polyester matrix and jute fibers was also surface changed using fluorocarbon bases, microemulsion silicon, and alkaline agents, according to Sever et al. [9], who evaluated the treated jute/polyester composites. The findings showed that treating the fibers with NaOH increased the composites' flexural, tensile, and interlaminar shear strengths. Kenaf fiber, NaOH, and polyester resin were combined using a vacuum infusion technique by Mohd Yuhazri et al. [10]. The alkalization process enhanced the mechanical properties of composites as compared to composites constructed with kenaf fiber that had not been treated. Alkalized kenaf fiber composites' measured mechanical efficiency also improved as NaOH levels rose.

4.2.3 Hybrid Polyester–Composites

By fusing two or more different types of fibers together in a matrix, hybrid polyester composites are created. Different scholars have offered various definitions of hybrid composites. According to Kumar and Bhanuprakash [11], hybrid composites are reinforced materials incorporated into a variety of various matrices. Furthermore, according to Fu et al. [12], these composites are a combination of two reinforcements that have been incorporated into the matrix. In order to generate hybrid composites, Nurazzi et al. [13] did study on polyester matrix by reinforcing it with sugar palm yarn and glass fabric. These composites' increased tensile strength, modulus, flexural characteristics, impact properties, and compressive strength are shown by the results of their characterization. The results showed that the treated sugar palm fibers also had better mechanical properties than the raw composites. The mechanical characteristics of hybrid woven jute-roselle-reinforced polyester composites were studied by Mohamad Hamdan et al. [14]. The study found that composite hybridization significantly affects the mechanical characteristics of the composites. According to Nimanpure et al. [15], positive hybridization improves the tensile strength and increases the flexural strength of sisal fiber/kenaf hybrid polyester composites by 24% and 36%, respectively, when compared to neat composites. Jamal et al. [16] explored the hybrid woven kenaf/glass composite. The findings show that, in comparison to plain and kenaf composites, hybrid composites have better flexural strength. Samyn [17] created a composite material using a polyester matrix, polyester fabric with a plain weave, and pineapple fiber. Coefficient of friction and wear for thermosetting polyester composite material Composites reinforced with plain-weave polyester fabric fared better than hybrid composites. In comparison to neat composites, the hybrid composite with a natural and synthetic fiber interface showed a lower wear rate.

4.2.4 Methods of Production of Biocomposites

There are various ways to make green composites with natural fibers, and some of them are explained in this chapter.

4.2.4.1 Stratification

Layers (also known as lamina) are used to produce composites. Fibers arranged in one or two orientations make up single-ply composites. Depending on the grades of fiber and resin, the necessary qualities, and the application, the layer thickness varies substantially. In advanced composite research, such as those carried out by Ghelli and Minak [18], natural fibers of 0.1–0.5-mm thickness are often used to obtain the composite lamina. Boria et al. [19] and Poodts et al. [20] all focused on the thickness of natural fiber composites for low-cost applications. Natural fiber alignment produces composites with exceptional qualities. When natural fibers are unidirectional, the composite exhibits strong strength and stiffness qualities in the direction of fiber orientation. Because most constructions are not loaded in a single direction, fiber orientation in multiple directions is crucial. This is because each layer is made of unidirectional fabric and the orientation of the fibers can be changed from one layer to another, or by using a multidirectional fabric, and multiple layers are stacked together in the stratification technique to achieve higher strength characteristics. Laminating describes such a stack, and layering describes the process. The quantity of laminates is controlled by the characteristics of the layers in order to ensure enough mechanical strength and weight. These composites typically have four to forty layers, but there are known cases of naval vessels with more layers [21]. Laminated composites can be made with either long or short staple fibers, depending on their fiber characteristics. Composites with short, discontinuous fibers are lengthy and continuous fibers, even though they are beautiful and useful in some respects (cost, process automation, material reuse potential, etc.). In actual life, however, short strands and discontinuous fibers are not always oriented randomly. In some situations, the mechanical properties of the composite may be enhanced by the proper layer orientation, even for short and discontinuous fibers.

4.2.4.2 Hand Lay-Up Method

The Handy lay-up method, one of the oldest composite production techniques, dates from the 1800s. Molding is part of the composite manufacturing process. It uses a straightforward process of impregnating the fibers with hands before placing them on the mold. Hand lay-ups are still commonly employed today, particularly for the synthesis of natural fiber biocomposites. The process is labor intensive and low-volume, which makes it particularly suitable for large components such as the hull of boats. A natural fiber reinforcing mat, cloth, or roving is placed in a manually opened mold, and resin is poured over it before being brushed or sprayed. To finish the laminate structure, the trapped air is manually evacuated with a doctor's blade or roller. The most commonly utilized matrix resins are room temperature-cured polyester and polyester. Without the use of external heat, the catalyst in the resin system commences curing to cure the fiber-reinforced resin composite. A colored gel coat is initially applied to the surfaces of high-quality parts. Using this method, components of any size may be produced from composites quickly and affordably. It provides a practical answer to technological limitations. A "solid laminate," a fiber laminated to the correct shape, or a "sandwich," consisting of an outer composite layer and an inner core made of light material, are both suitable methods for achieving the artifact. The inner and outer skin serve extra functions in terms of aesthetics and core protection in addition to their primary role of bearing loads.

For the purpose of creating natural fiber-reinforced polyester composites, notably for tribological applications, Fragassa et al. [22] have examined the hand layup procedures. The hand layup method is the greatest methodology out of all the others for creating a high-performance composite, but it has some limitations.

4.2.4.3 Vacuum Bagging Technique

During the curing process, vacuum bag formation is employed to impart mechanical pressure to the laminate. Several purposes are served by pressurizing composite laminates. The air trapped between the layers is first removed throughout the operation. Second, it densifies the fiber layer, allowing for effective force transfer between fiber bundles and preventing the fibers' orientation from moving during the curing process. It also reduces atmospheric humidity. Last but not least, vacuum bag technology improves the fiber-to-resin ratio of composite products. The physical qualities of advanced composites are enhanced by these benefits. Other manufacturing procedures, like vacuum injection, are favored in the boat industry over vacuum packaging. Bank [23] provides a high-level summary of the methodologies and procedures.

4.2.4.4 Tribological Tests on Natural Fiber-Reinforced Polyester-Based Biocomposites

The adhesive wear test and the abrasive wear test are the two main types of wear tests used on polyester-based biocomposites. The methodology adopted for the wear tests is defined by ASTM based on the specimen type, specimen geometry, configuration of the sliding disc/rubbing wheel, adhesive disc/abrasive particles used, and the applications for which the tests have to be simulated. Figure 4.2 provides an overview

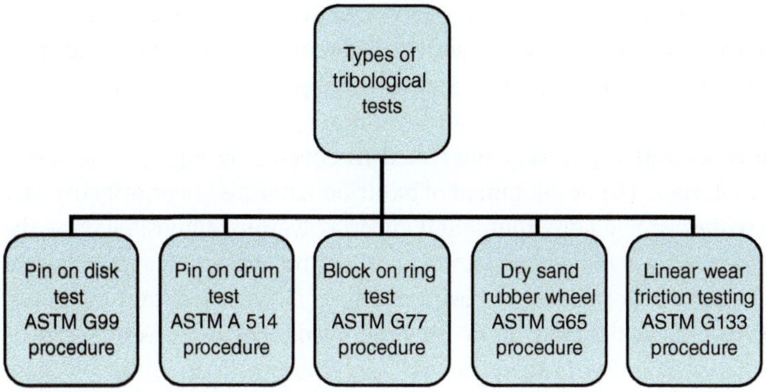

Figure 4.2 Types of tribological tests for natural fiber-reinforced biocomposites.

of the tribological tests used by the researchers to evaluate natural fiber-reinforced polyester-based biocomposites.

A pin specimen of 6–8 mm in diameter and 25–30 mm in length is pressed vertically or horizontally against a disc made of case-hardened steel during the POD wear test (ASTM G99 method). The operational parameters are applied load, sliding distance and velocity, and contact circumstances. In this tribological test, the specimen's contact area is kept constant with respect to the sliding time.

When performing an ASTM A514 pin-on-drum wear test, the specimen moves horizontally against a revolving drum. Testing can be either sticky or abrasive, as in the case of the pin-on-drum wear test. The pin-on-drum wear test replicates wear, particularly on conveyor belts, in some applications.

The block-on-ring wear test is conducted according to ASTM G77 guidelines. The specimen is clamped at 90° to the axis of rotation against a spinning ring. The sliding distance, temperature, applied weight, wet/dry sliding conditions, and sliding velocity are examples of typical operating parameters. For this kind, a sample size of 102,050 mm^3 is advised.

The dry sand rubber wheel test is accomplished in accordance with the ASTM G65 procedure. The test coupon is exposed to abrasive wear against the sand grains introduced between the rubber wheel and the specimen. This type of wear test is especially used to simulate the wear on rollers and tyres.

To mimic wear on sliding surfaces, window panels, drawers, etc., linear wear friction testing is done. The specimen is inserted in a container containing abrasive particles and moved linearly against the counterface during linear wear friction testing. The ASTM G133 protocol is used for performing the wear testing.

4.3 Tribological Characteristics of Polyester-Based Biocomposites

The advantages of natural fibers and those of a synthetic matrix phase are combined in natural fiber-reinforced polyester composites to provide characteristics that set them apart from their constituent parts. This is one of the factors contributing to the superior tribological characteristics and wear resistance of natural fiber-reinforced polyester composites. There are also further changes in the biocomposites' characteristics.

The modern tribo industry is very interested in developing high-performance, wear-resistant materials. The development of bio-tribomaterials is a priority in order to save energy while reducing environmental effect. Polyester composites show the possibility of doing this, as they demonstrate excellent mechanical and tribological performance. Therefore, polyester composites are used as wear-resistant materials. Evaluating the tribological behavior of polyesters and polyester composites is known to be challenging due to their complex polymorphism structure. Because of this, friction and wear in polyester composites are not always proportionate to one another. Before offering these materials for actual usage, it is critical to assess their tribological behavior and surface contact [24, 25]. Adhesion, tillage, and bump deformation are considered to be the three most important friction mechanisms of polyesters and their related composites. These three mechanisms contribute differently depending on the context, material qualities, operating conditions, and contact surfaces. Furthermore, in polyesters and composites, adhesive wear is the major wear mechanism. When the mating surface is smooth, adhesive wear occurs, but when the polyester composites have a rough surface, polishing wear occurs. When the surface of tougher polyesters scrapes against smooth surfaces and is repeatedly stressed during rolling and sliding motions, fatigue wear occurs. Ploughing wear is also regarded as the primary mechanism of elastic wear in polyesters and composites. Furthermore, viscoelasticity and temperature-related features can be used to describe the tribological behavior of polyesters and their associated composites. Adhesive wear resistance of these materials is critical in this regard. As a result, the quantity of heat generated by sliding contact between two materials, which results in heat dissipation and material fracture, is a key source of worry. The high-temperature integrity that develops has a significant impact on the material's tribological performance The material's tribological performance is determined by a variety of factors divided into categories such as material composition and surface geometry. Surface contact, hardness, roughness, and friction forces may all influence the material's tribological properties. The fundamental reason for this is the ridge's deformation. One of the key factors influencing the tribological characteristics of polyester composites is the ensuing surface roughness. The effect of surface hardness, on the other hand, can be explained by one of the following: The existence of wear asperities on the mating surfaces causes elastic wear. Material-related issues such as polymer blends, filler additives, interfacial adhesions, chemical treatments, and manufacturing processes can all affect tribological performance.

There have been numerous investigations on the tribological interactions of polyester-based biocomposites, particularly the dry sliding wear behavior of the composites with natural fiber reinforcement. Studies show that adding natural fibers to polyester reduces the wear rate of biocomposites. In this context, a thorough analysis of bio- tribomaterials that require little to no external lubricating is carried out. Due to the great volatility of several "natural" element properties, this has occurred (e.g. the percentage of water absorbed by flax). To eliminate the substantial variability that is employed to differentiate the green composite, preliminary experimental material testing is necessary. The hand-molded laminates were the subject of substantial studies by Pavlovic et al. [26]. The polyester matrix mixed with natural fibers results in natural fiber-reinforced polyester composites, which have better mechanical properties. They looked at the effects of natural reinforcements on two composite material combinations: one that combined natural and synthetic fibers (flax and carbon), and the other that included two different natural fibers (Gauze and flax). These eco-friendly materials were contrasted with polyester resin matrix composites bonded with synthetic fibers.

Additional details on the investigations and results may be found in the most recent papers, Fragassa [27], Hyseni et al. [28], Pavlovic et al. [26], or Pavlovic et al. [26]. The mechanical and tribological characteristics of flax and basalt fiber are the main topic of the research by Fragassa et al. [22].

In-depth research on composites was conducted by Zivkovic et al. [29] using vinylester composites and hybrids to assess the impact of fiber length and orientation on tribological properties. The findings demonstrate that choosing the proper fibers and matrix for the project in order to produce a green composite is influenced by the tribomechanical properties.

In accordance with the findings of numerous researchers, green composites (such as basalt fiber and a vinylester green matrix) can offer a workable solution in terms of mechanical properties, comparable to other widely used synthetic materials, while also incorporating the crucial property of eco-sustainability. They also confirm that basalt and hybrid laminates perform better in terms of mechanical strength than traditional fibers. As expected, when their Young modulus is included, the maximum strain increases. Combining basalt with flax fibers reduces the stiffness and brittleness of the material, as seen by plastic behavior after yielding.

In a fretting wear test on polyester composites, Veazie et al. [30] discovered a considerable decrease in interfacial strength as a result of fractures brought on by tribological degradations. The interfacial fracture strength was lowered in specimens that already had cracks prior to wear (by about 50%). Specimens subjected to extreme wear and tear usually show signs of degradation, particularly at the face sheet/middle interface. Under the impact of nanosilica, the fracture strength of nanocomposites was measured with different wear durations. The accurate load-crack displacement curves were plotted using CT scans performed at room temperature. Each neat polyester and composite specimen is susceptible to brittle fracture, and for nanocomposites, natural fibers typically have an impact on how well the materials perform under tribological loading conditions and display superior properties to their neat counterparts. When comparing the four-factor load circumstances to the three-factor tribological loading, the cross-ply laminate fretting wear characteristics are more prominent. The specimens with higher percentage of fiber exhibit better strength and wear-resistant characteristics, and they turned out to withstand the tribological load even after 10^6 cycles.

In order to conduct statistical validations on the tribological properties of bidirectional jute/carbon fiber polyester composites, Ravikumar et al. [31] used the RSM

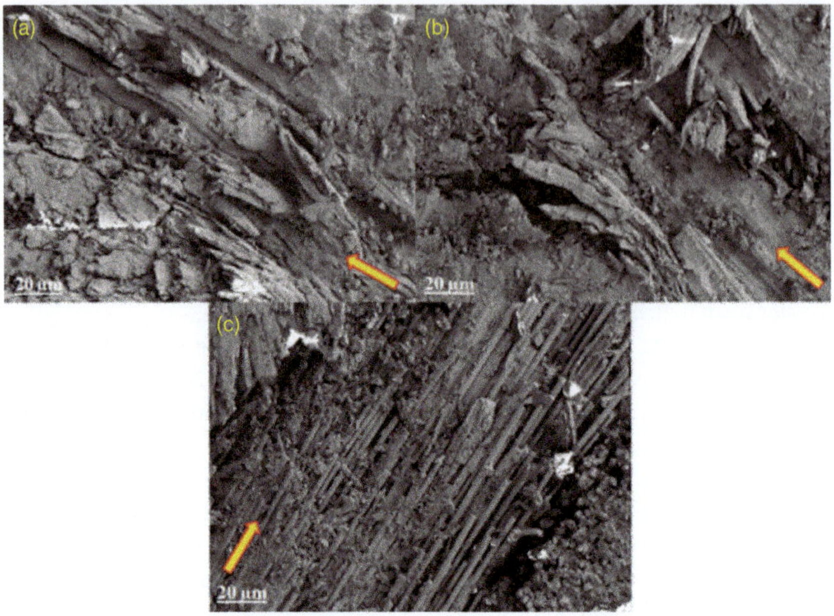

Figure 4.3 Micrograph of worn surface of the polyester hybrid composites. Source: Ravikumar et al. [31]/with permission of Taylor and Francis Group.

approach. The effects of load, fiber weight fraction, and sliding velocity on wear and friction values were examined. A dry sliding test on a pin-on-disc (POD) machine was used to evaluate the tribological performance of these composites. To assess the interactions between process factors and composite wear, the RSM technique was used. Due to less contact between the composite and rotating disc, wear loss increased with increasing sliding velocity and load and decreased with increasing volume percentages of the reinforcements. The shape of worn-out surfaces was also investigated. It was revealed from the micrographs that the fiber delamination occurred due to asperity contacts between the composite and disc. The furrow and ridge type of wear was observed in the micrographs, and the fiber debonding occurred due to the matrix delamination and ploughing due to the plastic deformation. Figure 4.3 shows micrographs of the composite surface's worn-out state.

The tribological properties of the biocomposites reinforced with oil palm fibers were examined by Khalil et al. [32]. According to reports, adding natural fibers to biocomposites improves their tribological properties. On the test coupons, they have run POD wear tests at varied levels of load, sliding distance, and sliding velocity. The research has demonstrated the usefulness of biocomposites in the biomedical and automotive industries. Oil palm fiber/polyester composite's tribological characteristics were researched by El-Tayeb and Yousif [33]. Results compared to neat polyester revealed a notable improvement. The POD tribo-tester machine was used to carry out the tribological characterization, which involved a

Table 4.1 Brief survey of composites with different fibers with its major findings.

Reference	Fiber type	Major findings
	Oil palm	Abrasive wear test was conducted against SiC abrasive paper. The oil palm-reinforced polymer composite exhibited wear resistance in comparison with the neat polyester for different loads and rotational speeds. The weight loss for neat polyester was found to be 28 mg for 50 rpm for 3 min under a load of 20 N as against 22 mg for the same set of test parameters for oil palm- reinforced polyester composite.
Chandrasekar et al. [35]	Sisal fiber and coconut sheath fiber	The impact of sisal fiber on the polyester composite reinforced with coconut sheath fiber was investigated. With increasing sliding distance, the wear-related mass loss decreased. This was attributed to the sisal fiber's role in facilitating a strong connection between the matrix and the coconut sheath fiber.
	Betelnut fiber	In an effort to do research, Gill tried using betelnut fibers in a polyester matrix. The testing involved sliding at a speed of 2.8 m/s, applying stresses between 5 and 30 N, and sliding over distances between 0 and 6.72 km. Additionally, the effect of the fiber mat orientation on the wear behavior of the composites was investigated. In comparison to neat polyester and composites with various orientations, it was claimed that betelnut mat reinforced in the parallel orientation (p-o) displayed better wear resistance.
Ibrahem [36]	Date palm seeds	Researchers studied the impact of date palm seed powder at various levels of normal load, sliding speed, and sliding distance on the wear conductivity of composite materials. Compared to the clean polyester composite, the coefficient of friction for composites reinforced with date palm seed powder dropped by 0.16 for 15% weight composition.
El-Tayeb [37]	Sugarcane fibers	El-Tayeb looked into the possibility of utilizing sugarcane fibers in a polyester matrix. According to the study's findings, the neat polyester composite was less resistant to wear than the composite reinforced with sugar cane fiber. Additionally, the wear rate of the composite fell by 20% with an increase in fiber length to 5 mm, indicating that the fiber length has an effect on the composite's wear resistance.
Dilli Babu et al. [38]	*Calotropis gigentea* fruit fiber	The purpose of the wear experiments was to investigate how the weight loss and coefficient of friction of the polyester-based biocomposites were affected by the volume fraction of the *C. gigentea* fruit fiber. It was noted that at a load of 3 kg and a volume percent of 0.29, the specimen's wear loss in terms of weight and coefficient of friction was less. Compared to neat polyester composites, the *C. gigentea* fruit fiber-reinforced polyester composite demonstrated improved wear behavior.

variety of parameters and levels, including sliding distances between 0 and 4.5 km, sliding velocities between 1.6 and 3.8 m $second^{-1}$, and loads between 10 and 30 N. The results showed that the composites performed better in the tribological tests than neat polyester, particularly in terms of reduced wear loss and coefficient of friction due to the tight link between the matrix and the reinforcement. Basic research on the tribological properties of palm fiber-reinforced biocomposites was done by Yousif and Nirmal [34]. In comparison to the pure polymer, the results showed a significant improvement in the wear characteristics of the composites.

The wear behavior of polyester composites reinforced with natural fibers has been the subject of numerous studies. Table 4.1 summarizes the type of natural fiber utilized as well as the key findings of the research carried out.

The wear mechanisms in polyester-based biocomposites are primarily caused by abrasive or adhesive wear, according to Chandramohan et al. [39] assessment of the tribological performance of natural fiber-reinforced polymer composites. When there is adhesive wear, one surface slides over another, wearing down the composites. When two sliding surfaces are forced against one another, the asperity contacts—which have a relatively small surface area—carry the contact loads, increasing the contact pressure over the asperity and causing adhesion between the two surfaces. The contact layer may break as a result of moving between the two surfaces, which will then cause the adhesive to wear. The specimen rubs up against abrasive particles, which causes abrasive wear. Due to sliding contact between the grain's rough surface and the specimen's relatively softer surface, abrasive wear causes material loss through the elimination of the contact surface. The mechanisms of the abrasive and adhesive wear losses in the natural fiber-reinforced polyester biocomposites are schematically shown in Figures 4.4 and 4.5.

Significant research articles on the tribological performance of polyester composites reinforced with natural fibers have been evaluated by Nirmal [40]. From their review, it is reported here that the wear of polyester composites is primarily caused by abrasion, adhesion, corrosion, fretting, melting, oxidation, or cavitation effect. The wear loss in the polyester composites is majorly due to fiber debonding, bending, fragmentation, oxidation, and erosion of the matrix bonding the fibers. However, the use of natural fibers provides substantial resistance against wear by the above mechanisms due to the stronger bonding and microcoring.

Figure 4.6 gives the wear classification for polymers reported by Briscoe and Sinha [41] in his article on the tribology of biopolymer composites.

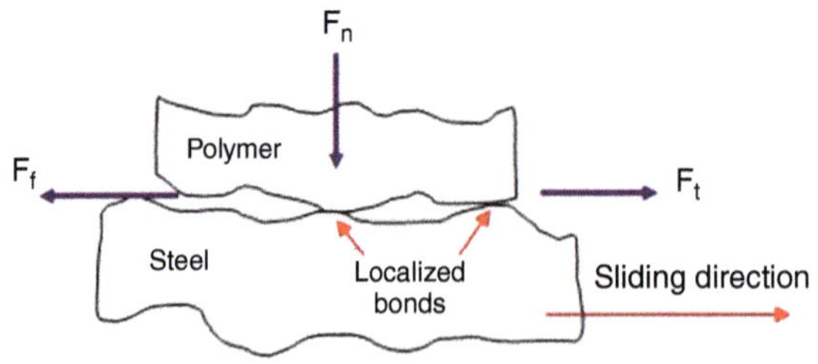

Figure 4.4 Mechanism of adhesive wear.

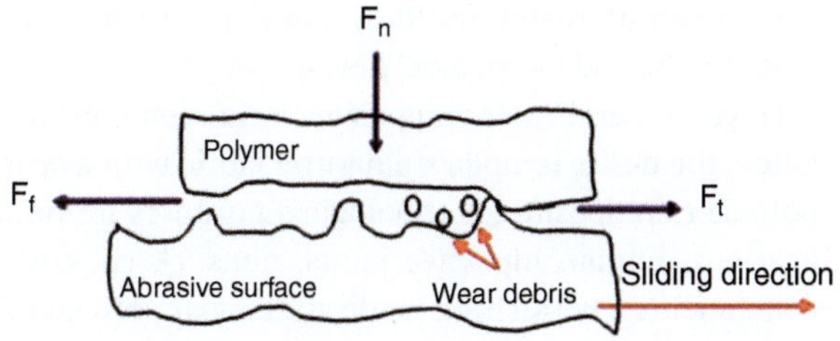

Figure 4.5 Mechanism of abrasive wear.

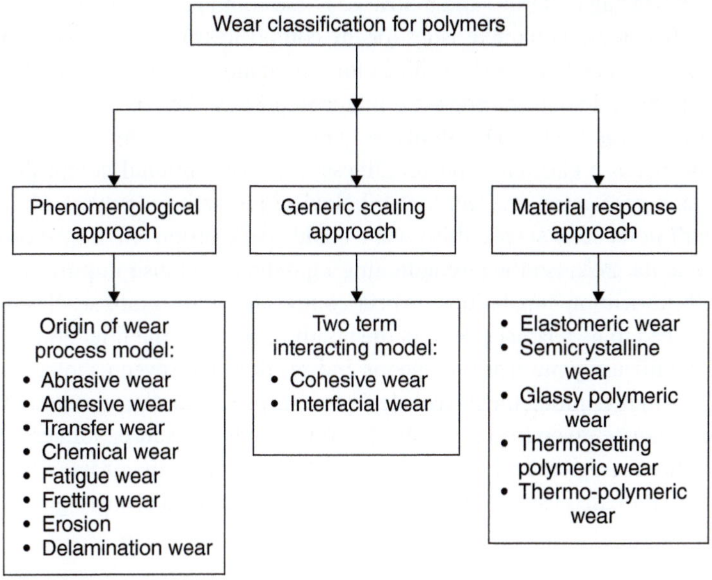

Figure 4.6 Schematic for wear classification of polymers. Source: Adapted from Briscoe and Sinha [41].

The wear of polyesters and polyester-based composites is a complex process, and the wear of these composites can be classified based on the interacting models, material response approach, or wear model. It is evident from the reviews of Briscoe and Sinha [41] that the wear process model may be abrasive, adhesive, fretting type, fatigue type, erosive, or delamination type based on the phenomenology between the mating surfaces; cohesive or interfacial wear based on the interacting model; elastomeric, semicrystalline, thermosetting polymeric, or thermopolymeric wear based on the material response approach.

4.4 Polyester-Based Biocomposites for Tribological Applications

When compared to composites made of other polymers, polyester-based biomaterials exhibit exceptional tribological characteristics. Since they are good wear-resistant materials, they have a variety of uses in the aviation, marine, automotive, and biomedical fields.

Polyesters and their composites have seen considerable surge in technical uses. Often, the desire to update a natural fabric with a synthetic fabric to create a hybrid polyester composite for tribological purposes influences such applications. Gears, bearings, human hip/knee joints, nonstick cooking pans, and other tribological components have all been made with linear polyesters' self-lubricating components. In composite brake systems for automobiles, polyesters and phenolic compounds are employed as load-bearing materials and matrices. These combinations are required to sustain better tribological characteristics for the composites. Materials that may withstand excessive stress and may be utilized for withstanding fretting wear also. The recently emerging domain of polyester nanocomposites may be a promising factor to consider in this regard. The ability of nanocomposites to simultaneously and isotopically increase elasticity and toughness over conventional composites is their key advantage. Despite the fact that tribological research on polyester biocomposites is still in its early stages, this review could spark various investigations in the next few years. Because the strengthening ingredient will also improve the composites' self-lubrication capabilities, hybrid composites with nanoparticles or nanofibers may need to be reconsidered. As the bonding between the filler and the matrix may be significantly more advantageous and result in improved tribological characteristics, various reinforcements (covalent grafting of a natural precursor with the nanoparticle) can be taken into consideration for nanodebris and nanofibers of various compositions. Agglomeration of nanoparticles is another major issue that can be effectively addressed using chemical inoculants or mechanical stirring.

4.5 Conclusions

To attain higher mechanical and tribological qualities, there is an increasing demand for superior composite structure performance in the modern world. Utilizing natural fibers to enhance the surface qualities of polyester-based composites is one potential growth direction. The previously mentioned characteristics of composites can be enhanced even more by the use of natural fibers. The natural fibers are taking center stage among the numerous reinforcements used in polyester composites. This chapter summarizes earlier research on the tribological characteristics of polyester composites reinforced with natural fibers. The tribological characteristics of natural fiber-reinforced polyester composites are influenced by modifications in the synthesis methods and chemical composition. The use of natural fiber-reinforced polyester composites for multifunctional applications, notably tribological applications, has also been covered in numerous research studies. The importance of natural fibers in the synthesis of composites, particularly for automotive parts, aerospace components, and marine constructions needing improved tribological performance, has been highlighted by the thorough investigations. With hybridization and the use of natural fibers, the usage of polyester-based biocomposites, particularly for bearing elements, fretting wear applications, and biomedical components, is on the rise.

References

1 Barberio, G., Buttol, P., Righi, S., and Andretta, M. (2010). Combined approach of Risk Assessment and Life Cycle Assessment for the environmental evaluations: an overview. In: *Proceeding of Ecomondo Congress*, 1262–1268, 3–6. November Rimini: Maggioli.

2 Bogoeva-Gaceva, G., Avella, M., and Errico, M. (2007). Natural fiber eco-composites. *Polym. Compos.* 28: 98–107.

3 De Paola, S., Minak, G., Fragassa, C., and Pavlovic, A. (2013). Green composites: a review of state of art. In: *Proceedings of 30th Danubia Adria Symposium on Advanced Mechanics* (ed. Croatian Society of Mechanics), 77–78. Primosten, Croatia (25–28 September 2013).

4 Zivkovic, I., Pavlovic, A., and Fragassa, C. (2016). Improvements in wood thermoplastic composite materials properties by physical and chemical treatments. *Int. J. Q. Res.* 10 (1): 205–218.

5 Lopresto, V., Leone, C., and De Iorio, I. (2011). Mechanical characterisation of basalt fibre reinforced plastic. *Composites, Part B* 42: 717–723.

6 Senthilkumar, K., Saba, N., Chandrasekar, M. et al. (2019). Evaluation of mechanical and free vibration properties of the pineapple leaf fibre reinforced polyester composites. *Constr. Build. Mater.* 195: 423–431. https://doi.org/10.1016/j.conbuildmat.2018.11.081.

7 Uma Devi, L., Bhagawan, S.S., and Thomas, S. (1997). Mechanical properties of pineapple leaf fiber-reinforced polyester composites. *J. Appl. Polym. Sci.* 64 (9): 1739–1748.

8 Prasad, G.L.E., Gowda, B.S.K., and Velmurugan, R. (2017). Comparative study of impact strength characteristics of treated and untreated sisal polyester composites. *Procedia Eng.* 173: 778–785. https://doi.org/10.1016/j.proeng.2016.12.096.

9 Sever, K., Sarikanat, M., Seki, Y. et al. (2012). Surface treatments of jute fabric: the influence of surface characteristics on jute fabrics and mechanical properties of jute/polyester composites. *Ind. Crops Prod.* 35 (1): 22–30. https://doi.org/10.1016/j.indcrop.2011.05.020.

10 Mohd Yuhazri, K.R., Phongsakorn, P.T., Sihombing, H. et al. (2011). Mechanical properties of kenaf/polyester composites. *Int. J. Eng. Technol. IJET-IJENS* 11 (1): 127–131.

11 Kumar, K. and Bhanuprakash, N. (2017). A review on different types of hybrid composite materials with different matrix proportions. *Int. Res. J. Eng. Technol.* 4 (5): 1089–1093.

12 Fu, S.Y., Xu, G., and Mai, Y.W. (2002). On the elastic modulus of hybrid particle/short-fiber/polymer composites. *Composites, Part B* 33 (4): 291–299. https://doi.org/10.1016/S1359-8368(02)00013-6.

13 Nurazzi, N.M., Khalina, A., Sapuan, S.M., and Ilyas, R.A. (2019). Mechanical properties of sugar palm yarn/woven glass fiber reinforced unsaturated polyester composites: effect of fiber loadings and alkaline treatment. *Polimery* 64 (10): 12–22. https://doi.org/10.14314/polimery.2019.10.3.

14 Mohamad Hamdan, M.H., Siregar, J.P., Thomas, S. et al. (2019). Mechanical performance of hybrid woven jute–roselle-reinforced polyester composites. *Polym. Polym. Compos.* 27 (7): 407–418. https://doi.org/10.1177/0967391119847552.

15 Nimanpure, S., Hashmi, S.A.R., Kumar, R. et al. (2019). Mechanical, electrical, and thermal analysis of sisal fibril/kenaf fiber hybrid polyester composites. *Polym. Compos.* 40 (2): 664–676. https://doi.org/10.1002/pc.24706.

16 Jamal, S.K., Hassan, S.A., Wong, K.J., and Bahru, J. (2011). Mechanical properties of hybrid woven kenaf/recycled glass fibre reinforced polyester composites. *J. Built Environ. Technol. Eng.* 1: 1–10.

17 Samyn, P. (2017). Reciprocating sliding of polyester textile fabric composites along different fabric orientations. *J. Compos. Mater.* 51 (2): 221–240. https://doi.org/10.1177/0021998316642830.

18 Ghelli, D. and Minak, G. (2011). Low velocity impact and compression after impact tests on thin carbon/polyester laminates. *Composites, Part B Eng.* 42 (7): 2067–2079.

19 Boria, S., Pavlovic, A., Fragassa, C., and Santulli, C. (2016). Modeling of falling weight impact behavior of hybrid basalt/flax vinylester composites. *Procedia Eng.* 167: 223–230. https://doi.org/10.1016/j.proeng.2016.11.691.

20 Poodts, E., Minak, G., Mazzocchetti, L., and Giorgini, L. (2014). Fabrication, process simulation and testing of a thick CFRP component using the RTM process. *Composites, Part B* 56: 673–680.

21 Mouritz, A.P., Gellert, E., Burchill, P., and Challis, K. (2001). Review of advanced composite structures for naval ships and submarines. *Compos. Struct.* 53 (1): 21–42.

22 Fragassa, C., Pavlovic, A., and Santulli, C. (2017). Mechanical and impact characterisation of flax and basalt fibre vinylester composites and their hybrids. *Composites, Part B* https://doi.org/10.1016/j.compositesb.2017.01.004.

23 Bank, L.C. (2006). *Composites for Construction: Structural Design with FRP Materials*. London: Wiley.

24 Alotaibi, J.G., Alajmi, A.E., Mehoub, G.A., and Yousif, B.F. (2021). Epoxy and polyester composites' characteristics under tribological loading conditions. *Polymers* 13 (14): 2230. https://doi.org/10.3390/polym13142230.

25 Artemenko, S.E. (2003). Polymer composite materials made from carbon, basalt, and glass fibres structure and properties. *Chem.* 35 (3): 226–229.

26 Pavlovic, A., De Paola, S., Fragassa, C., and Minak, G. (2013). Mechanical characterization of green composites: earliest experimental evidences by flexural tests. In: *Proceedings of 21th Composite/Nano Engineering (ICEE)*, 273–276. Tenerife, Spain.

27 Fragassa, C. (2016). Effect of natural fibers and bio-resins on mechanical properties in hybrid and non-hybrid composites. In: *Proceedings of the 8th Conference on Times of Polymers & Composites: From Aerospace to Nanotechnology*. Ischia (Italy): American Institute of Physics (AIP) *AIP Conference Proceedings* (19–23rd June 2016). vol. 1736, No. 4949693. https://doi.org/10.1063/1.4949693.

28 Hyseni, A., De Paola, S., Minak, G., and Fragassa, C. (2013). Mechanical characterization of ecocomposites. In: *Proceedings of 30th Danubia Adria Symposium* *on Advanced Mechanics*. Primosten, Croatia (25–28 September 2013), 175–176. Croatian Society of Mechanics.

29 Zivkovic, I., Pavlovic, A., Fragassa, C., and Brugo, T. (2017). Influence of moisture absorption on the impact properties of flax, basalt and hybrid flax/basalt fiber reinforced green composites. *Composites, Part B* 111: 148–164. https://doi.org/10.1016/j.compositesb.2016.12.018.

30 Veazie, D.R., Robinson, K.R., and Shivakumar, K. (2004). Effects of the marine environment on the interfacial fracture toughness of PVC core sandwich composites. *Composites, Part B* 35 (461): 466.

31 Ravikumar, P., Suresh, A.R., and Rajeshkumar, G. (2020). An investigation into the tribological properties of bidirectional jute/carbon fiber reinforced polyester hybrid composites. *J. Nat. Fibers* https://doi.org/10.1080/15440478.2020.1764444.

32 Khalil, H.A., Bhat, A.H., and Yusra, A.I. (2012). Green composites from sustainable cellulose nanofibrils: a review. *Carbohydr. Polym.* 87 (2): 963–979.

33 El-Tayeb, N.S.M. and Yousif, B.F. (2007). The effect of oil palm fibres as reinforcement on tribological performance of polyester compposite. *Surf. Rev. Lett.* 14 (6): 1095–1102.

34 Yousif, B.F. and Nirmal, U. (2011). Wear and frictional performance of polymeric composites aged in various solutions. *Wear* 272 (1): 97–104.

35 Chandrasekar, M., Senthilkumar, K., Kumar, T.S.M. et al. (2021, 2021). Chapter 6—Effect of adding sisal fiber on the sliding wear behavior of the coconut sheath fiber-reinforced composite. In: *Elsevier Series on Tribology and Surface Engineering, Tribology of Polymer Composites* (ed. S.M. Rangappa, S. Siengchin, J. Parameswaranpillai, and K. Friedrich), 115–125. Netherlands: Elsevier.

36 Ibrahem, R.A. (2015). Effect of date palm seeds on the tribological behaviour of polyester composites under different testing conditions. *J. Mater. Sci. Eng.* 2015: 1–5.

37 El-Tayeb, N.S.M. (2008). A study on the potential of sugarcane fibers/polyester composite for tribological applications. *Wear* 265 (1–2): 223–235.

38 Dilli Babu, G., Sivaji, B.K., and Nanda, K.P. (2014). Tensile and wear behavior of Calotropis gigentea fruit fiber reinforced polyester composites. *Procedia Eng.* 97: 531–535.

39 Chandramohan, D., Ravikumar, L., Sivakandhan, C. et al. (2018). Review on tribological performance of natural fibre-reinforced polymer composites. *J. Bio Tribo Corros.* 4: 55. https://doi.org/10.1007/s40735-018-0172-x.

40 Nirmal, U., Jamil Hashim, M.M.H., and Ahmad, M. (2015). A review on tribological performance of natural fibre polymeric composites. *Tribol. Int.* 83: 77–104.

41 Briscoe, B.J. and Sinha, S.K. (2002). Wear of polymers. *Proc. Inst. Mech. Eng. J. Eng. Tribol.* 216: 401–413.

5

Tribological Properties of the Natural Fiber-Reinforced Vinyl Ester Composites

Krushna Gouda[1], Muthukumar Chandrasekar[1], Vellaichamy Parthasarathy[2], Senthilkumar Krishnasamy[3], and Senthil Muthu Kumar Thiagamani[4]

[1]*Hindustan Institute of Technology and Science, Simcrash Centre, Department of Aeronautical Engineering, Rajiv Gandhi Salai, OMR, Padur, Chennai, Tamil Nadu 603103, India*
[2]*Hindustan Institute of Technology and Science, School of Humanities and Science, Kelambakkam, Chennai, Tamil Nadu 603103, India*
[3]*PSG Institute of Technology and Applied Research, Department of Mechanical Engineering, Salem-Coimbatore Highway, Avinashi Road, Neelambur, Coimbatore, Tamil Nadu, 641062, India*
[4]*Kalasalingam Academy of Research and Education, Department of Mechanical Engineering, Krishnankoil, Tamil Nadu 626126, India*

5.1 Introduction

Advanced composite materials possess good wear resistance and mechanical properties. The different kinds of thermosetting polymers are used as a matrix to develop composites. The various types of synthetic and natural fibers (NFs) are used as reinforcing agents to develop polymer composites. However, the usage of NFs has increased for the development of eco-friendly composites. The fibers derived from plant and animal resources are termed as NFs. Moreover, NFs have outstanding properties in comparison with synthetic fibers. Hence, NFs are the right choice to develop advanced composites.

The utilization of NFs has increased in various fields such as automobile industries, building construction, and commercial appliances, as they are considered a potential alternative for artificial fibers like Kevlar, carbon, glass, polyamide, and aramid fiber. Various NFs obtained from plant sources are show in Figure 5.1. These natural fibers have been used as reinforcing agents to develop composite materials with enhanced properties. NFs possess low density, corrosion resistance, and high specific strength, and they are also abundantly available in nature. The bio-based polymer composites also exhibit better mechanical and thermal resistance owing to the addition of NFs to the polymer matrix.

Polymer-based composites are considered promising tribo-engineering materials in different mechanical applications. Wear performance is the most important factor for tribo-material selection. The wear performance of the polymers considerably

Tribological Properties, Performance, and Applications of Biocomposites, First Edition.
Edited by Chandrasekar Muthukumar, Senthilkumar Krishnasamy,
Senthil Muthu Kumar Thiagamani, and Ganesan Chinnachamy.

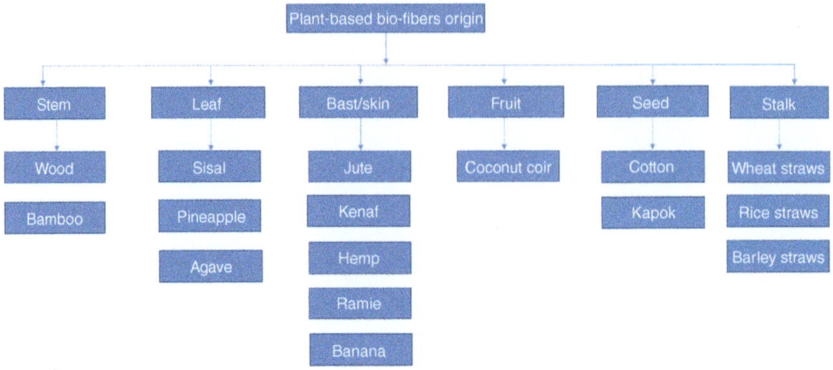

Figure 5.1 Plant-based bioresources.

increases after reinforcing them with fibers [1, 2]. The polymer composites loaded with either natural or synthetic fibers are used to make tribo-components for aircraft structures since they can withstand even harsh environmental conditions like high stress, external loading, high-temperature variation, and speed [3–5]. Among various fibers, carbon fiber shows better strength and wear prevention with improved thermal conductivity and stability, which are essential for better tribological properties. The quick dissipation of generated heat due to friction needs to be expelled as soon as possible to reduce the maximum longevity of polymer matrix composite. The fiber delamination rate also increases with increasing friction temperature, and the composite materials will retain their properties for a longer period even at rapid heat dissipation [6]. The carbon fiber-based composites exhibit better lubrication behavior. The advanced polymer composites have been prepared by reinforcing them with natural fillers and fibers. The matrix material plays a key role in the development of advanced composite materials. The selection of the matrix, filler, and NF is based on the requirement for specific applications. It was reported that the use of short fiber led to a decrease in the wear performance of the polymer composite [7]. Researchers revealed that there was an enhancement in the performance of the polymer composites with fiber and filler reinforcement. Thermoset polymers such as vinyl ester (VE), polyester, epoxy, polyurethane, and phenolic have been used as a matrix to develop polymer composites due to their high strength, lightweight, and reliable cleaning process. Polymer composites are used to make various mechanical components such as bearing cages, bearing bushes, rollers, seals, cams, wheels, gears, transmission belts, and clutches [8–11]. The non-lubricated condition is an ideal way to test and understand the wear behavior of the materials.

The VE resin is employed as a matrix for developing composites since it creates a strong bond with fiber/filler. The VE resin shows intermediate bonding strength among the epoxy resin and unsaturated polyester resin. Researchers suggested that there was an improvement in the mechanical properties of the composites when using VE as a matrix rather than by using polyester resin as a matrix. VE was also suggested to be used over polyester resin as it has better toughness, wear resistance, and corrosion prevention [12, 13]. The experimental investigation suggested that the addition of secondary filler and fiber helps enhance the mechanical properties of the

VE composite. The VE finds applications in automobile and aeronautical industries owing to its high stiffness and strength [14].

The VE resin is bonded with comonomers such as butyl acrylate, methyl acrylate, and styrene to extend its use in the development of composite materials with improved properties. Several research groups investigated to identify suitable comonomers, which could improve the wear resistance and mechanical properties of VE [15]. It was reported that the VE resin with styrene comonomers exhibited better mechanical properties. Abrasive wear is possible only when the material surface is rubbed against another body under fixed or different loading conditions. The major abrasive wear arises in chute liners, mining fields, power plants, heavy equipment, and earthmovers [16]. The three body abrasives are often considered of practical importance. Heavy machinery equipment such as power plants, industrial slurry handling gear pumps, and agricultural machine parts have gained less attention than two body abrasives [17, 18]. The literature on the wear performance of natural filler or fiber-reinforced polymer composites is scarce. In the last few decades, researchers have shown interest in developing thermosetting polymer-based wear resistance composites.

Among various thermosetting polymers, VE resin exhibits strong chemical resistance and excellent mechanical properties. The VE resin is more economical than epoxy resins and stronger than polyester resins. The VE resin is more hydrophobic as compared to polyester resins, and it also bonds well with fiberglass. Both NFs and fillers are used as reinforcing agents to develop biocomposites. The VE resin reinforced with natural fillers exhibited an improvement in wear resistance. The natural filler-reinforced VE resin composites are more economical and eco-friendly. The introduction of fillers into VE resin enhances the tribo-mechanical behavior of the developed materials. The report on polymer composites reinforced with natural and woven fibers had been investigated by many research groups owing to their ease of preparation and availability.

Composite materials are considered promising alternative to conventional structural materials such as steel, iron, ceramics, wood, and timber owing to their enhanced material strength and minimal specific weight. Among the wide variety of composite materials, the glass fiber-based composite possesses superior thermo-mechanical properties. However, the disposal of some restrictions on reusability of conventional composites creates major environmental issues. Therefore, bio-based composites are a suitable alternative for various conventional structural materials due to their biodegradability. Hence, the development of bio-based composites has received much attention from material scientists to overcome the environmental issues associated with conventional structural materials. The graphite fillers and silicon carbide fillers were incorporated into glass-vinyl ester (GVE) composites to study the influence of fillers on three-body abrasive wear behavior [19]. The GVE composites loaded with the fillers were subjected to both rubber wheel/dry sand abrasion tests at a speed of 200 rpm under two different loading conditions (22 and 32 N) by varying the rubbing distance from 270 to 100 m. The worn surfaces of the particulate-loaded GVE composites were examined using SEM.

Suresha and Kumar [20] examined the mechanical properties and two-body abrasive wear characteristics of VE composite reinforced with glass and carbon fabric. The wear loss increased with the increasing abrasive particle size and rubbing distance. The specific wear rate was estimated as 10.89×10^{11} m^3 Nm^{-1} for VE/glass fabric composite, which was found to be higher as compared to that of VE/carbon fabric composite (4.02×10^{-11} m^3 Nm^{-1}). The SEM micrographs of the worn surfaces revealed broken glass fibers in the VE/glass fiber composite. However, the carbon fibers were seen to retain their structure in the SEM micrographs of worn surfaces for carbon/vinyl ester composite. It proved a strong interfacial adhesion between the carbon fiber and VE resin in the carbon/vinyl ester composite. In another study, the dry sliding wear and friction characteristics of glass/carbon-vinyl ester composites were assessed by using a block-on-ring wear tester [21]. The wear rate and coefficient of friction (COF) were observed to increase while increasing the sliding velocity and applied load. The types of fabrics incorporated into the matrix and temperature at the interface affected the friction coefficient and wear rate. The specific wear rate of the VE/carbon composite was increased by 184% as compared to the VE/glass fiber composite. Also, the sliding wear and friction response of the VE/carbon fiber are much better than those of the VE/glass fiber composite and pristine VE resin.

The VE/*Polyalthia longifolia* seed (PLS) composites were prepared by incorporating the various weight percentages of PLS fillers using the compression molding technique [22]. The wear characteristics of VE/PLS composites were examined using a pin-on-disc wear tester. The VE/PLS composite loaded with 25 wt% of PLS exhibited a minimum COF at the applied load of 5 N and 700 rpm and minimum wear loss at the applied load of 10 N and 300 rpm. The SEM micrograph was analyzed to comprehend the uniform dispersion of PLS in the VE matrix. The VE/PLS composites are suitable for developing small load-bearing components like a brake pad, seat pad, and clutch plate. The effect of micro- and nano-SiC particles on the mechanical and tribological properties of VE/basalt bidirectional mat composites was studied by Thooyavan and his research team [23]. The enhancement in the wear resistance of VE/basalt bidirectional mat composite was observed only at 4 wt% SiC filler loading in the composite matrix. The VE composite reinforced with chemically treated chopped Roselle fiber was examined to assess the mechanical properties and wear performance by Manickam et al. [24]. The VE composite reinforced with 10 wt% Roselle fiber exhibited the highest wear rate under the load of 5 and 10 N.

The VE was reinforced with 10 wt% of Kevlar fiber and also filled with different wt% (4–16 wt%) of cement by-pass dust (CBPD) to study the sliding wear performance, mechanical, and thermo-mechanical properties of the VE composites [25]. The sliding wear test was performed on pin-on-disc rig apparatus for the developed VE/Kevlar fiber/CBPD composites under different loading conditions (10–25 N), sliding velocities (1.5–4.5 m s^{-1}), abrasive grit size (18–48 μm), and sliding distance (1–4 km). The statistics-based Taguchi experimental design was employed to analyze the wear characteristics, and the significant factors associated with wear characteristics were also analyzed statistically. The hardness of VE/Kevlar fiber/CBPD composites increased with the increasing CBPD filler concentration. The VE composite reinforced with 12 wt% CBPD exhibited good wear resistance with the minimum COF even at higher sliding velocities.

Natrayan and Merneedi [26] utilized various bio-wastes like hibiscus, human hair, epoxy glass fusion fiber, and coir pith particulate to reinforce the VE resin. The wear characteristics of epoxy glass fusion fiber/hibiscus/human hair reinforced VE and coir pith particulate reinforced VE composites were evaluated using a pin-on-disc apparatus to compare their wear performance under different loading conditions (5–20 N in steps of 5 N) at constant sliding velocity (1.5 $m\,s^{-1}$) and distance (1500 m). The specific wear rate and COF were compared to be lower for epoxy glass fusion fiber/hibiscus/human hair reinforced VE composite than for coir pith particulate reinforced VE composite. The random orientation of epoxy glass fusion fiber created defects in the composite, which provided a higher load-bearing capacity to the composite and also ensured less friction between the surface of the sample and abrasive disc during the wear test. The COF and wear rate were increased by 44% and 59%, respectively, for glass fusion fiber-reinforced composites in comparison with coir pith-reinforced composites. The SEM micrographs of the worn-out surface of epoxy glass fusion fiber/hibiscus/human hair-reinforced VE composite revealed pullout of fibers, micro-cracks, and bonding loss, which were reported as responsible for the reduction of wear behavior. The hybrid synthetic Kevlar and natural bagasse fibers were used as reinforcing agents to prepare graded and homogeneous VE composites to evaluate their wear performance using pin-on-disc apparatus [27]. The different wt.% of bagasse fiber (5, 10, and 15 wt%) was incorporated into VE resin by maintaining the wt% of Kevlar fiber as constant (5 wt%) for preparing the VE composites. The VE composites reinforced with 15 wt% bagasse fiber recorded a lower specific wear rate at higher velocity and load.

The wear resistance and friction behavior of *Musaceae* fiber (MF) bundle-reinforced VE composites were analyzed using a pin-on-disc machine [28]. The COF value was estimated to be in the range of 0.25–0.5 for VE composites incorporated with different wt% of fiber (0–60%). The incorporation of MF bundles into VE matrix improved the wear performance by restricting the propagation of cracks under loading conditions. The MF bundle-reinforced composites exhibited a specific wear rate in the range of 0.4×10^{-4} to 1×10^{-4} $mg\,Nm^{-1}$. The reinforcement of jute/epoxy composites with two different ceramic fillers, such as Al_2O_3 and SiC, was carried out by hand lay-up techniques by Ahmed et al. [29]. The dry sliding wear test was carried out to estimate the frictional coefficient and wear loss of the Al_2O_3 and SiC-reinforced jute/epoxy composites using pin-on-disc machine by varying normal load (30–50 N), sliding velocities (3 $m\,s^{-1}$) at a constant sliding distance (1800 m). The jute/epoxy composite reinforced with Al_2O_3 showed better wear resistance in comparison with SiC-filled composites. The wear mechanism of the Al_2O_3 and SiC-filled jute/epoxy composite was understood by analyzing the SEM micrographs of the worn-out surfaces. The jute/epoxy composites incorporated with 15 wt% of Al_2O_3 ceramic fillers exhibited lower wear loss and COF than SiC-filled jute/epoxy composites.

The surface of the areca sheath (AS) fiber was modified by treating it with benzoyl chloride to reinforce the polyvinyl alcohol (PVA) [30]. The PVA composites were reinforced with different wt% of surface-modified AS fiber (10%, 20%, 30%, and 40%) by injection molding to study their erosion properties. The silica sand particle (200 μm) was employed as an erodent to study wear behavior by varying impact angles (15 N to 90° in steps of 15°) and impact velocities (48, 70, 82, and 109 $m\,s^{-1}$). The influence of impact angle and impact velocity on the erosion rate of PVA/AS composites was also studied. The PVA/AS composites exhibited a peak rate of erosion at an impact angle of 45° due to semi-ductile behavior. However, the maximum erosion rate was observed at a 30° impact angle for pristine PVA. The removal of the material in the composite was due to plastic distortion, micro-cutting, and micro-ploughing. The erosion rate of the AS fiber-reinforced composites was reported to be increased while increasing the fiber content. The reinforced PVA/AS composites with 10 wt% AS fiber showed minimum erosion as compared to other composites. The eroded surface texture was investigated by SEM analysis to comprehend erosion mechanism. This study also suggested that 27 wt% of AF fiber-reinforced PVA composite can be used in the erosion environment due to its excellent mechanical properties.

In another study, Correa and his same research group studied the effect of fiber size (152, 287, and 638 μm) on the wear behavior and friction coefficient of MF bundle-reinforced VE composites [31]. The sliding wear test was performed against the steel counter body for all the developed composites under fixed test parameters using pin-on-disc equipment. The VE composites reinforced with an MF size of 287 μm showed the highest wear resistance. The morphological analysis of the worn surfaces of the composite revealed that the wear mechanisms associated with all the samples were crazing, surface fatigue, and adhesion. The glass fiber with Grewia optiva and jute fiber was used to reinforce three different polymer resins, such as polyester, VE, and epoxy, by land lay-up method to investigate their erosive wear characteristics [32]. The silica particles of different sizes (100–250 μm) were forced to strike the hybrid fiber-reinforced composites for 15 min under different experimental conditions through a tungsten carbide nozzle to carry out erosive wear analysis. The erosive wear of the hybrid fiber-reinforced composites was influenced by the impingement angle, fiber content, and velocity. The synthetic/NF-reinforced epoxy composite showed the lowest wear rate, and the highest wear rate was recorded for the synthetic/NF-reinforced polyester composite among all the fabricated composites. The inclusion of synthetic/NF into the VE composites reduced erosive wear even at higher angles of impingement (75° and 90°).

5.2 Natural Fiber-Based VE Composite

The fiber or fibrils extracted from the various natural resources are called NFs, and they are converted into filament, strips, thread, and mat types based on the requirements of specific applications. NFs are broadly classified into three different types, such as animal, plant, and mineral-based fibers. The VE composite incorporated with NF has high modulus and specific strength in comparison to metal and metal composites. The bio-based composites are used in the aeronautical and aerospace industries due to their properties. NFs such as jute, kenaf, hemp, coir, sisal, flax, ramie, and abaca have been extensively investigated to extent their use in various technological applications. The major reasons for selecting NFs over synthetic fibers are that they are renewable, less in cost, noncorrosive, nonabrasive, lightweight, good mechanical properties, and have good crash absorbance. NFs have few limitations, such as low thermal resistance, low processing temperature, crack generation, and material brittleness while exposed to excess sunlight, besides their wide usability.

5.3 Problems Associated with Natural Fiber-Based Composite

The biocomposites are environmentally friendly materials since their constituents are derived from natural resources. Hence, they are biodegradable after disposal without creating any environmental pollution. Biocomposites find numerous applications in various fields and enable the production of sustainable materials by converting agricultural wastes. However, the uniform dispersion of NFs into the polymer matrix is quite challenging. The NFs are reinforced with synthetic or biopolymers to improve the compressible, tensile, flexural, and tribological properties of the biocomposites. The fiber impurity and poor dispersion of fibers lead to delamination and quick crack generation in the developed composites.

5.4 Conclusion

The combination of synthetic and NFs-reinforced VE composites is termed "hybrid composites," which are suitable for load-bearing tribological applications due to their excellent mechanical properties. The VE resin is suggested as promising matrix to fabricate composites with better mechanical properties. The wear performance of the polymers considerably increases with the NF reinforcements. The wear mechanism of the VE composites is understood by investigating the SEM micrographs of the worn surface. An improvement in the wear resistance of VE/Kevlar fiber at higher sliding velocity is achieved by incorporating 12 wt% of CBPD. The chemically treated Roselle fiber is recommended as reinforcing agent to improve the wear performance of VE composites because the treated Rosella fiber enables a strong interface with VE matrix. The combination of Kevlar/bagasse fibers as a hybrid reinforcement offers a low-specific wear rate at higher loads and velocity. The incorporation of MF into VE restricts crack propagation under loading conditions, resulting in better wear resistance. However, the selective approach for reinforcing the NFs with the thermosetting matrix needs further investigation to extend their applications.

References

1 Shivakumar, K.N., Swaminathan, G., and Sharpe, M. (2006). Carbon/vinyl ester composites for enhanced performance in marine applications. *J. Reinf. Plast. Compos.* 25 (10): 1101–1116.

2 Suresha, B., Chandramohan, G., Samapthkumaran, P. et al. (2006). Friction and wear characteristics of carbon-epoxy and glass-epoxy woven roving fiber composites. *J. Reinf. Plast. Compos.* 25 (7): 771–782.
3 Soutis, C. (2005). Carbon fiber reinforced plastics in aircraft construction. *Mater. Sci. Eng., A* 412 (1–2): 171–176.
4 Bijwe, J., Rattan, R., Fahim, M., and Tiwari, S. (2008). Erosive wear of carbon fabric reinforced polyetherimide composites: role of amount of fabric and processing technique. *Polym. Compos.* 29 (3): 337–344.
5 Rattan, R. and Bijwe, J. (2006). Carbon fabric reinforced polyetherimide composites: influence of weave of fabric and processing parameters on performance properties and erosive wear. *Mater. Sci. Eng., A* 420 (1-2): 342–350.
6 Chand, N. and Gautam, K.K.S. (1994). Influence of load on abrasion of fly ash-glass fibre-reinforced composites. *J. Mater. Sci. Lett.* 13 (4): 230–233.
7 Rajesh, J.J., Bijwe, J., and Tewari, U.S. (2002). Friction and wear studies of short glass fiber reinforced polyetherimide composites. *Wear* 252: 769–776.
8 Friedrich, K.B.A.U.K., Lu, Z.B.A.U.K., and Hager, A.M. (1995). Recent advances in polymer composites' tribology. *Wear* 190 (2): 139–144.
9 Lancaster, J.K. (1968). The effect of carbon fibre reinforcement on the friction and wear of polymers. *J. Phys. D: Appl. Phys.* 1 (5): 549.
10 Lu, Z., Friedrich, K., Pannhorst, W., and Heinz, J. (1993). Wear and friction of a unidirectional carbon fiber-glass matrix composite against various counterparts. *Wear* 162: 1103–1113.
11 Vishwanath, B., Verma, A.P., and Rao, C.K. (1993). Effect of reinforcement on friction and wear of fabric reinforced polymer composites. *Wear* 167 (2): 93–99.
12 Nabinejad, O., Debnath, S., and Mohsen Taheri, M. (2017). Oil palm fiber vinylester composite; effect of bleaching treatment. *Mater. Sci. Forum* vol. 882: 43–50. Trans Tech Publications Ltd.
13 Umachitra, C., Palaniswamy, N.K., Shanmugasundaram, O.L., and Sampath, P.S. (2017). Effect of mechanical properties on various surface treatment processes of banana/cotton woven fabric vinyl ester composite. *Appl. Mech. Mater.* vol. 867: 41–47. Trans Tech Publications Ltd.
14 Torres, J.P., Vandi, L.J., Veidt, M., and Heitzmann, M.T. (2017). The mechanical properties of natural fibre composite laminates: a statistical study. *Composites Part A* 98: 99–104.
15 Johnson, R., Arumugaprabu, V., and Ko, T.J. (2019). Mechanical property, wear characteristics, machining and moisture absorption studies on vinyl ester composites–a review. *Silicon* 11 (5): 2455–2470.
16 Harsha, A.P. and Tewari, U.S. (2002). Tribo performance of polyaryletherketone composites. *Polym. Test.* 21 (6): 697–709.
17 Budinski, K.G. (1997). Resistance to particle abrasion of selected plastics. *Wear* 203: 302–309.
18 Chand, N., Naik, A., and Neogi, S. (2000). Three-body abrasive wear of short glass fibre polyester composite. *Wear* 242 (1–2): 38–46.
19 Suresha, B. and Chandramohan, G. (2008). Three-body abrasive wear behaviour of particulate-filled glass–vinyl ester composites. *J. Mater. Process. Technol.* 200 (1-3): 306–311.

20 Suresha, B. and Kumar, K.N.S. (2009). Investigations on mechanical and two-body abrasive wear behaviour of glass/carbon fabric reinforced vinyl ester composites. *Mater. Des.* 30 (6): 2056–2060.

21 Suresha, B., Kumar, K.S., Seetharamu, S., and Kumaran, P.S. (2010). Friction and dry sliding wear behavior of carbon and glass fabric reinforced vinyl ester composites. *Tribol. Int.* 43 (3): 602–609.

22 Nagaprasad, N., Stalin, B., Vignesh, V. et al. (2021). Applicability of cellulosic-based Polyalthia lon gigolia seed filler reinforced vinyl ester biocomposites on tribological performance. *Polym. Compos.* 42 (2): 791–804.

23 Thooyavan, Y., Kumaraswamidhas, L.A., Raj, R.E., and Binoj, J.S. (2022). Influence of SiC micro and nano particles on tribological, water absorption and mechanical properties of basalt bidirectional mat/vinyl ester composites. *Compos. Sci. Technol.* 219: 109210.

24 Manickam, C., Kumar, J., Athijayamani, A., and Diwahar, N. (2015). Mechanical and wear behaviors of untreated and alkali treated roselle fiber-reinforced vinyl ester composite. *J. Eng. Res.* 3 (3): 1–13.

25 Gangil, B., Patnaik, A., Kumar, A., and Biswas, S. (2013). Thermo-mechanical and sliding wear behaviour of vinyl ester–cement by-pass dust particulate-filled homogenous and their functionally graded composites. *Proc. Inst. Mech. Eng., Part J: J. Eng. Tribol.* 227 (3): 246–258.

26 Natrayan, L. and Merneedi, A. (2021). Experimental investigation on wear behaviour of bio-waste reinforced fusion fiber composite laminate under various conditions. *Mater. Today Proce* 37: 1486–1490.

27 Singh, T., Gangil, B., Singh, B. et al. (2019). Natural-synthetic fiber reinforced homogeneous and functionally graded vinylester composites: effect of bagasse-Kevlar hybridization on wear behavior. *J. Mater. Res. Technol.* 8 (6): 5961–5971.

28 Correa, C.E., Betancourt, S., Vázquez, A., and Gañan, P. (2017). Wear performance of vinyl ester reinforced with *Musaceae* fiber bundles sliding against different metallic surfaces. *Tribol. Int.* 109: 447–459.

29 Ahmed, K.S., Khalid, S.S., Mallinatha, V., and Kumar, S.A. (2012). Dry sliding wear behavior of SiC/Al_2O_3 filled jute/epoxy composites. *Mater. Des.* 1980–2015 (36): 306–315.

30 Nayak, S. and Mohanty, J. (2020). Erosion wear behavior of benzoyl chloride modified areca sheath fiber reinforced polymer composites. *Compos. Commun.* 18: 19–25.

31 Correa, C.E., Betancourt, S., Vázquez, A., and Gañan, P. (2015). Wear resistance and friction behavior of thermoset matrix reinforced with *Musaceae* fiber bundles. *Tribol. Int.* 87: 57–64.

32 Khare, J.M., Dahiya, S., Gangil, B. et al. (2021). Comparative analysis of erosive wear behaviour of epoxy, polyester and vinyl esters based thermosetting polymer composites for human prosthetic applications using Taguchi design. *Polymers* 2021 (13): 3607.

6

Friction and Sliding Wear Properties of the Natural Fiber-Reinforced Polypropylene Composites

Emel Kuram

Gebze Technical University, Department of Mechanical Engineering, 41400 Gebze, Kocaeli, Turkey

6.1 Introduction

In the 1980s, asbestos was banned as a component of friction composites because of its negative impact on health. Since it has been banned, its constituents have been evaluated in the composition of friction composites. A sample is aramid fibers for non-asbestos, less metallic (or nonmetallic) friction composite [1, 2]. Though there are no reports showing aramid fibers as hazardous, they are synthetic in origin and therefore must be developed via chemical synthesis [3], which can hardly be considered environmentally friendly procedure. Natural materials are viable options, while natural biodegradable and renewable materials are the best options for use as constituents in different types of composites [4].

Polymer and its composites find increasing use in several applications in rolling/sliding parts such as cams, bearings, clutches, gears, grinding mills, piston rings, rollers, seals, transmission belts, and wheels in which self-lubricating characteristics are exploited to eliminate requirement for grease or oil lubrication, along with contamination issues [5]. However, wear and friction problems arise when there is contact between the sliding pairs [6]. Researchers [7–9] declared that by choosing right material combinations, friction coefficients could be decreased and wear resistance enhanced.

Friction composites developed for brake linings must meet a variety of demands, as they must exhibit acceptable coefficient of friction values, stability at high temperatures, and better resistance to wear. They must be quiet and vibration-free during braking and should include environmentally friendly constituents [4].

Natural fibers have been better option of reinforcement for polymeric matrix than man-made fibers owing to their benefits such as abundant, great modulus, high specific strength, less price and less density [10–15]. Additionally, these fibers also show biodegradability, environmentally friendly and recyclability behavior [16, 17]. However, these fibers also have some drawbacks such as larger variability in properties, limiting the matrix selection options, lower durability, lower processing temperatures and lower strength in comparison to synthetic fibers. Also, main

Tribological Properties, Performance, and Applications of Biocomposites, First Edition.
Edited by Chandrasekar Muthukumar, Senthilkumar Krishnasamy,
Senthil Muthu Kumar Thiagamani, and Ganesan Chinnachamy.

drawbacks of fibers are greater water sorption and weak interfacial adhesion with polymers [18–23]. In friction composites, temperature at friction surface during intense braking could locally reach several hundred °C and there is question whether plant fibers could be employed for application at high temperatures [4].

Several researchers reported that wear rate and friction force depend on lubrication, material type, normal force, relative humidity, relative motion, roughness of rubbing surfaces, stick slip, temperature, and vibration [24–26]. Parameters that determine tribological characteristics of polymers and their composites also include processing, molecular structure, characteristics, surface texture, and viscoelastic behavior [27–30]. A number of works have also been conducted investigating influence of contact geometry, environment, and test conditions on wear and friction behavior of polymers and composites [6].

Composites are extensively employed for structural parts exposed to friction and wear under normal loads. Factors affecting tribological performance of composites include chemical treatment and operating conditions (fiber orientation, applied load, sliding distance, and sliding velocity) [31].

Owing to relative movement of plastic components, there is always wear and friction, which strongly affects service life [32]. Wear and friction characteristics of polymers are affected by many factors: additives amount and type [33–35], crystallinity [36], material selection (polymer-on-polymer or non-polymer-on-polymer) [37], sliding speed, service temperature [38], specific pressure [35, 39], surface treatment of the part, surface roughness [38], use of regrinding [40, 41].

This chapter provides a comprehensive coverage of the various research works on natural fiber based polypropylene (PP) composites related to tribological studies. Various parameters affecting friction and wear, failure causes of parts and effect of material parameters are basically discussed in chapter. Primary target of chapter is to give information about tribological characteristic of natural fiber reinforced composites and demonstrate usability in several industries where tribology plays dominant role.

6.2 Polypropylene

PP is a thermoplastic widely utilized since it is cheap, with relatively good strength and ease of processing. It is one of the best options as matrix for natural fiber reinforced composite due to less melting temperature below thermal degradation temperature of natural fibers. Hashmi et al. [42] declared lining material from PP is employed in chemical sector because of great wear and chemical resistance.

6.3 Natural Fibers

If fiber is derived from various sections of plant or living species, is called natural fiber. There is great variety of natural fibers with various chemical compositions, mechanical properties, and physical properties, such as bamboo, abaca, banana, coconut, jute, hemp, sisal, kenaf, etc. [43].

Natural fibers have less price, density, and weight, are environmentally friendly nature with low health hazards and pollution. It is believed that usage of natural fibers in composite design is not only valuable in terms of economic and environmental aspects, but also as renewable raw materials for novel green products from a sustainability perspective [44]. Natural fibers have great potential to not only reduce CO_2 emissions, also protect non-renewable sources with replacing glass fiber in polymer composite [45].

Natural fiber could be achieved from sources such as animals, minerals or plants [31]. There are three main classes of them namely animal, mineral and plant based (vegetable) fibers. Categories of natural fibers are depicted in Figure 6.1. Since the most widely used natural fibers are plant origin fibers, natural fibres of plant origin will be briefly mentioned as natural fibers in this chapter. Plant fibers are significant kinds of natural fibers and they usually consist chiefly of lignin, hemicellulose, cellulose and pectin. Plant fiber is classified as stem fiber/bast fiber, leaf fiber, fruit fiber, seed fibers and stem fiber [31]. Vegetable fibers possess high fiber strength therefore are preferred by automotive sector. Vegetable fibers are renewable, non-abrasive and have so much calorific value that they could be burned at end of life for energy recovery. They are much safer to handle and less likely to affect lungs during processing and usage [45].

Natural fiber possesses multi-walled structures and longitudinal form is nearly cylindrical. Fiber cell consists of four constituents: primary, secondary, and tertiary wall and lumen. Representation of natural fiber cell is shown in Figure 6.2. Cell wall consists of many layers of fibrillar structures with fibrillae [43].

In general, natural fibers are ligno-cellulose and include lignin, hemicellulose, cellulose, pectin and wax as primary constituents [43]. Chemical composition of natural fibers varies in accordance with age, cultivation method, location, rainfall during growth, soil in which it is cultivated and species [43]. Chemical formulations of natural fibers influence characteristics of their composites. The chemical structure of the fiber varies according to the origin—hemicellulose, cellulose, and lignin are more common, other additives such as ash, wax, and pectin are also present [44]. Chemical and mechanical characteristics of natural fibers declared by various researchers are summarized in Table 6.1.

Cellulose: Cellulose is the most available renewable source and is available in greater proportion than any component in natural fiber because of the feature natural fiber, also named cellulose fiber. Cellulose is β-1,4-linked linear polymer of glucose unit and insoluble in water [111]. But existence of cellulose causes fiber to be hydrophilic results in great affinity to water resulting in weak compatibility to the hydrophobic polymer. It causes high mechanical strength of the fibers [112].

Hemicellulose: While hemicellulose includes many various sugar units, cellulose does not; it possesses chain branching, while cellulose possesses a linear chain and possesses a smaller polymerization degree than cellulose [112]. It behaves as compatibilizer between lignin and cellulose [113].

Lignin: Lignin provides compressive strength and rigidity to fibers. Thanks to lignin, plants reach great heights. Lignin is like chemical adhesive within and among

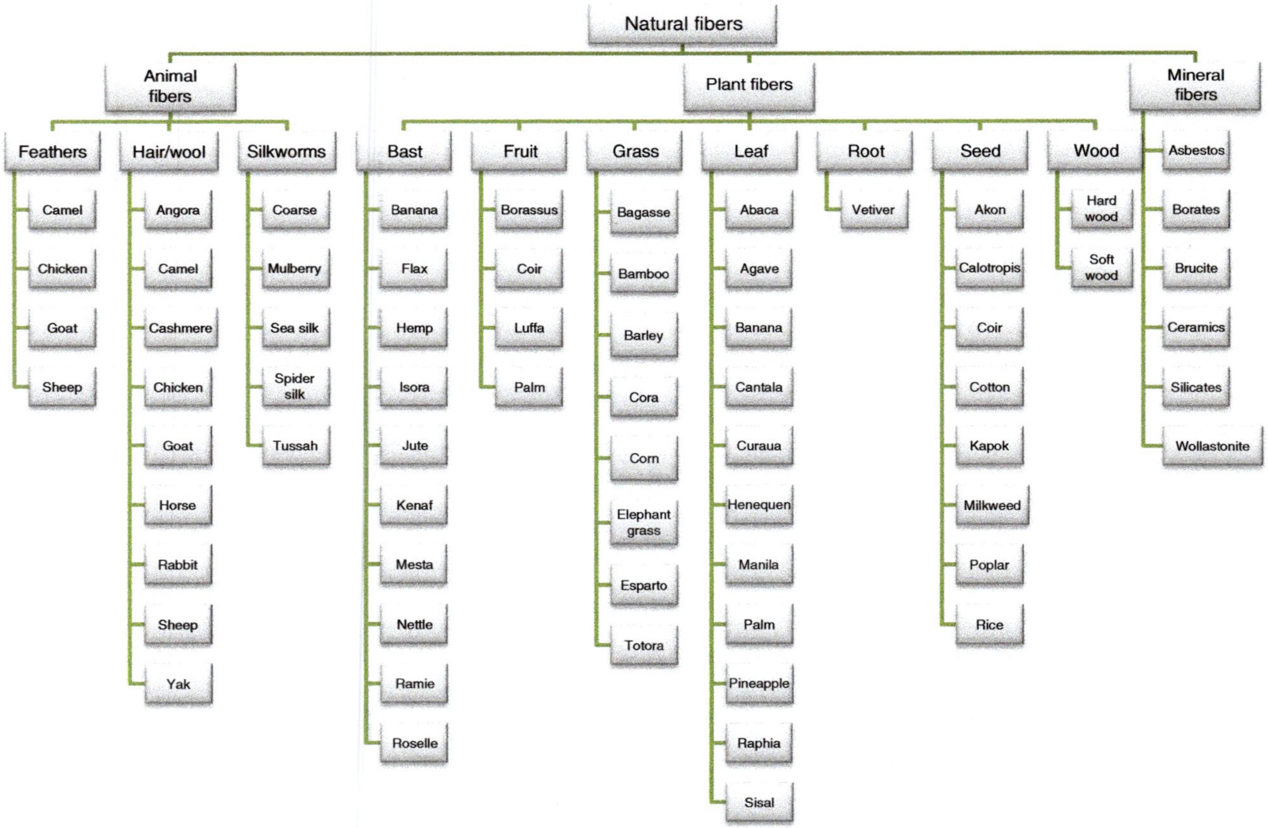

Figure 6.1 Categories of natural fibers.

Table 6.1 Mechanical and chemical characteristics of natural fibers.

	Chemical formulation			Mechanical properties			
Fiber	**Cellulose (%)**	**Hemicellulose (%)**	**Lignin (%)**	**Tensile strength (MPa)**	**Tensile modulus (GPa)**	**Elongation (%)**	**References**
Abaca	56–70	15–25	6.0–12.4	400–980	12.0–31.1	2.9–10.0	[11, 46–52]
Bamboo	26.00–73.83	12.49–30.00	10.15–32.20	140–1930	11.0–46.3	1.30–6.48	[50, 52–60]
Coir	32–50	0.15–25.00	30–45	15–593	1.269–6.000	14.21–75.00	[11, 50–53, 56, 61–82]
Date palm	32.0–35.8	24.1–28.1	26.5–28.9	97–309	2.5–12.0	2.0–10.0	[50, 56, 82–84]
Flax	64–81	11.62–20.60	2–5	300–1500	24–100	1.2–10.0	[11, 47, 50–52, 57, 62, 63, 65, 66, 71–73, 79, 85–91]
Hazelnut shell	25.9	29.9	42.5	—	—	—	[92]
Hemp	57–90	14.0–22.4	3.7–13.0	19.9–1726.1	0.9938–9970.0000	1.3–8.1	[11, 19, 50–52, 62, 63, 66, 71, 72, 82, 89, 93]
Jute	33.4–73.2	13.0–22.7	5–28	29–800	2.2–55.0	0.97–19.00	[11, 47, 50–53, 55, 57, 62, 63, 65, 66, 69, 71–73, 75, 79, 85, 87, 90, 94–97]
Kenaf	31–72	20.3–23.0	8–19	208–1191	2.86–53.00	1.26–7.50	[11, 50, 52, 71, 72, 82, 90, 93, 96, 98–101]
Sisal	38.2–80.0	5.63–26.00	3.92–26.00	31–840	1.46–920.00	2–25	[11, 47, 49–53, 56, 62, 63, 65–69, 71–73, 79, 82, 85, 87, 90, 94, 102–109]
Wood	45–50	23	27	90–180	10–70	—	[59, 73, 110]

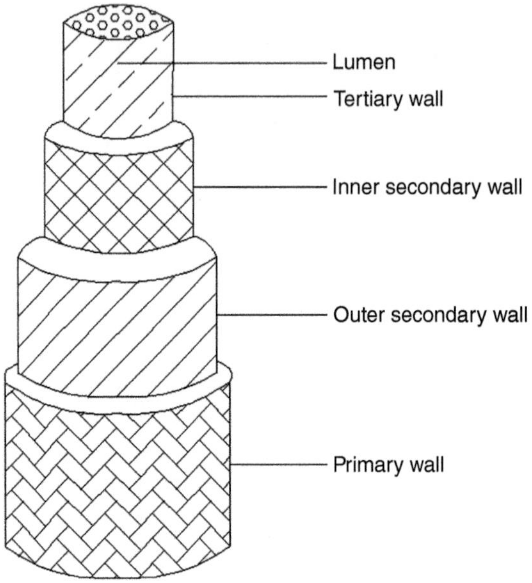

Figure 6.2 Natural fiber structure.

fibers and helps maintain water inside fibers. Lignin acts as protection against biological attack and gives strength to the stem to resist gravity and wind [113].

Pectin: Pectin is chemical compound known by the common name heteropolysaccharide that gives flexibility to plants [111]. Pectin is usually present in fruit and bast fibers. If pectin deteriorates, there is decrement in fiber strength. It dissolves in water after neutralization with ammonium hydroxide or alkali [43].

Wax: Wax is combination of substituted long chains of aliphatic hydrocarbons. Wax possesses alcohols, alkaline, aldehydes, fatty acid, ketone, and other components [114].

Several researchers have reported that tribobehavior of composites is enhanced by addition of natural fiber to polymeric matrix, but this behavior is dependent on fiber orientation, length, fiber size, volume fraction, and conditions (speed, load, and temperature) [115]. Natural fibers of animal and vegetable origin possess good effect on dynamic and static friction coefficients [32]. Various techniques such as additives, coatings, and surface treatment have been used to obtain good tribological characteristics of natural fibers [86, 116–118]. Generally, incorporation of natural fibers to matrix enhances wear performance of natural fiber reinforced composite as compared with wear property of neat polymer [45, 119].

6.4 Natural Fiber-Reinforced PP Composites

Environmental awareness and increasingly stringent political regulations have forced both academia and industry to consider novel materials that will meet all the required specifications [119]. As petroleum-based polymers become scarce and environmental concerns rise, it is important to optimize both issues. Environmentally friendly approach is to mix the eco-friendly agricultural waste filler with nonbiodegradable polymer, which could diminish price of production, decrease the environmental effect, and enhance characteristics of the developed composite [120]. Natural fiber composites (green composites) have high interest owing to their abundant availability, biodegradability, and recyclability. Relatively less density and adaptability to have stacking sequences give great stiffness and strength in direction of great loading, making materials attractive in comparison to traditional metals [121].

For composites, primary issue is heterogeneity of material. Therefore, surface finishing operations are often needed to achieve desired smoothness of composites. In addition, improper bonding between fibers and matrix often causes high wear in sliding contact [45].

While natural fiber-reinforced composites retain great stiffness-to-weight ratio and specific strengths found in synthetic fiber-reinforced composites, they also have beneficial properties such as biodegradability, CO_2 neutrality, low material cost, nonabrasiveness and recyclability [122–125]. However, using natural fibers as reinforcement have some disadvantages. Aggregate formation during processing, weak resistance to moisture, and incompatibility with the hydrophobic polymeric matrix are problems associated with negative effects of tribo-mechanical characteristics of composites [86].

Natural fibers are reinforced in matrix in different forms, such as different orientation, long fiber, short fiber, and mat form [44]. The choice of manufacturing method is mostly based on kind of matrix, the shape and size of composite, and application. Table 6.2 presents natural fiber reinforced composites fabricated by various methods. Natural fiber composites could be categorized according to kind of reinforcement configuration (Figure 6.3).

Fiber: Fibers may be short or long and random or unidirectional.

Mat fiber: Fibers are woven mats form and arranged in layers.

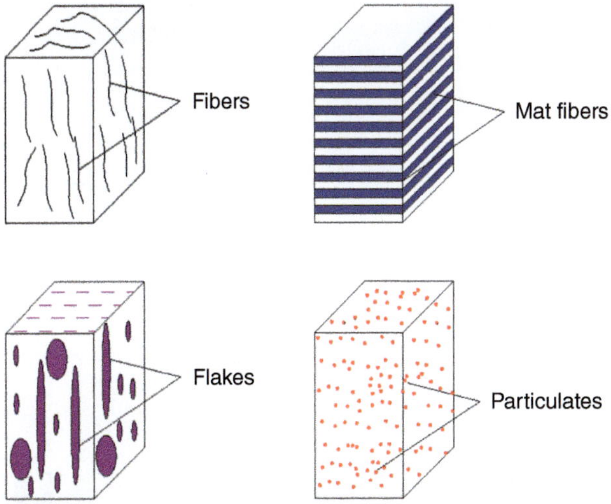

Figure 6.3 Various kinds of reinforcement configurations.

Table 6.2 Natural fiber-reinforced PP composites fabricated by various methods.

Fiber	Fiber amount (wt%)	Production method	Reference
Borassus fruit	5	Corotating twin-screw extruder	[126]
Cellulose Coconut Flax Fleece	10–30	Extruder	[32]
Coir	2.5, 5, 7.5	Melt-blending	[115]
Coir fiber Coir shell particle Coir fiber/coir shell particle (hybrid)	20 15/5, 10/10, 5/15 (for hybrid)	Twin-screw extruder	[127]
Hay Lemon Palm Reed	10, 20, 30	Mixing and dual-directional screw extruder	[128]
Hemp fiber Hemp particle Hemp fiber/hemp particle (hybrid)	20, 30 Fiber/powder ratio: 80/20, 60/40, 33/67 (for hybrid)	Twin-screw extruder	[129]
Hemp Hemp/glass fiber (hybrid)	40 35/5, 30/10, 25/15 (for hybrid)	Mixing	[130]
Hemp Hemp/glass fiber (hybrid)	40 35/5, 30/10, 25/15 (for hybrid)	Mixing	[131]
Jute	23	Mixing	[132]
Jute	5, 7.5, 10, 12.5, 15	Injection molding	[133]
Jute Kenaf	10, 20, 30, 40, 50, 70	Hot pressing	[96]
Jute	40	Compression molding	[134]
Sisal	23	Mixing	[135]
Sisal/glass fiber (hybrid), sisal/carbon fiber (hybrid)	10, 19, 26, 30, 31 (sisal content, vol%)	Corotating single-screw extruder	[136]
Turmeric spent	10, 20, 30, 40	Twin-screw corotating extruder	[120]
Wood flour	5, 15, 25, 35, 45, 55	Injection molding	[137]

Flake: Reinforcement is flakes form.

Particulate: Reinforcement is small particles form.

Natural fiber-reinforced polymer composites are replacing traditional materials in several industries, particularly automobile manufacture, where less wear and a lower friction coefficient are significant [31, 134]. Finally, given its less price, great deal of development in jute/PP composite can be seen in a range of industries, such as constituents for artificial human joints, automobile brake pads, surface flooring, and shoe soles [134].

6.5 Tribological Properties of Natural Fiber-Reinforced PP Composites

Tribology is study of interacting surfaces in relative motion [138]. Tribology includes work and implementation of fundamentals of friction, wear, and lubrication since functioning of several biological, electromechanical, and mechanical systems depends on suitable wear and friction values [45]. Tribology involves rolling and sliding surfaces. Tribologically, the material must be designed so that its surface is both hard and smooth. So, wear and friction are decreased. Between two degrading phenomena, friction is crucial to determining the efficiency of mechanical assemblies involving sliding surface contact. It is also reason for wear, which is often mechanism that limits the service life of device. Therefore, tribological loads must be considered for an optimized part design [45]. In the industry, most energy is consumed by friction. That's why tribologists want to reduce friction by designing materials or adopting surface finishing technique [45].

Among different properties, the tribological properties of composites aid in determining wear and friction behavior [139]. Tribological characteristics are determined by means of two parameters, namely friction coefficient and wear rate.

$$k = \frac{\Delta m}{\rho L F_n}$$

$$k = \frac{2a\Delta h}{nIA\omega^2}$$

$$\mu = \frac{F}{R}$$

where k is specific wear rate ($m^3\ J^{-1}$), Δm is loss in mass (g), ρ is sample density ($g\,cm^{-3}$), L is travel length, sliding distance (m), F_n is applied normal load (N), a is apparent area of contact (m^2), Δh is decrement in thickness because of wear (m), n is number of engagement and disengagement, I is moment of inertia of apparatus considering configuration inertia and spindle ($kg\,m^2$), ω is angular velocity of apparatus ($rad\,s^{-1}$), μ is friction coefficient, F is frictional force against motion (N), and R is reaction offered against load (N).

To determine wear and friction characteristics of materials, there are tribotesting devices called tribometers or tribotesters, which enable predetermined loads and lubrication and achieve wear and friction behavior from specimens [45]. Tribology testing techniques for fiber-reinforced plastics are summarized as follows:

Dry sand rubber wheel test: When abrasive wear on bearings, bushes, tire treads, etc., needs to be measured, dry sand rubber wheel can be viable choice. The setup consists of rubber wheel that rotates against loaded sample via appropriate brackets. Hopper is employed to introduce sand particles into the rubbing zone. Flow rate of sand particles can be controlled manually. Sample dimension in accordance with ASTM G65 standard is 70 mm × 20 mm × 70 mm [140].

Pin-on-drum test: In pin-on-drum, drum is rotated with aid of shaft and loaded sample is moved linearly on drum by a screw mechanism. The speeds of sample and drum are controlled by the appropriate control mechanism in motor. When drum is coated by abrasive, test is abrasive otherwise test is adhesive. Test standard for the technique is ASTM A514 and drum is composed of aluminum, cast iron, and steel. In this setup, the sliding of rollers and conveyor belts are evaluated [141].

Block-on-ring test: Block-on-ring apparatus is utilized for testing rolling or sliding wear of components (bearings, pulleys, camshafts, and tire treads) [142]. Technique employs rotating ring against, which sample is pressed by aid of dead weight. Contact area of sample could be changed and load cell is utilized to measure friction force induced during testing. Counterweight is attached to end of arm to balance applied load. Test is performed according to ASTM G135-95 standard and sample dimension is 10 mm × 20 mm × 50 mm [139].

Pin-on-disc test: Pin-on-disc tribometer is employed to analyze sliding wear [143]. It utilizes a pin-shaped sample with dimensions of 10 mm × 10 mm × 20 mm according to ASTM G99 standard. Sample is pressed against rotating disc, and components like dead weight and load cell are similar to block on ring tester. If shape of sample is block, test is named as block on disc setup. Temperature of friction region is determined by an infrared thermometer. Erosion study is performed employing an air jet erosion setup in which hard particles are allowed to strike the sample surface at an angle. Because of the continuous particle flow, surface is eroded, causing weight loss and wear. Loss in weight is determined by weighing sample [139].

According to sliding direction, three fiber orientations can be observed: (a) parallel, (b) antiparallel, and (c) normal orientation (Figure 6.4).

Tribological characteristics for materials are based on several parameters that fall under category of surfaces and materials [144]. Friction and wear occur for several

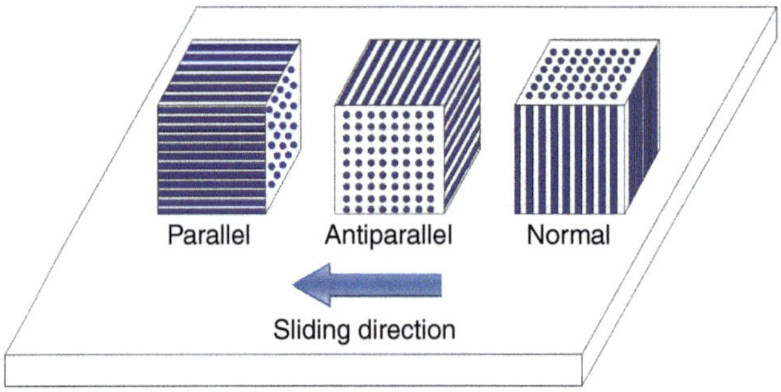

Figure 6.4 Fiber orientations: Parallel, antiparallel, and normal according to sliding direction.

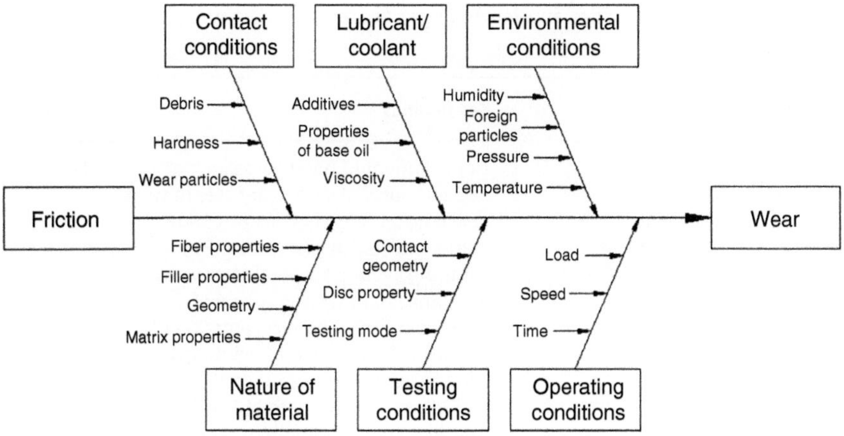

Figure 6.5 Causes of wear and friction.

reasons. Environmental conditions, lubricants, nature of material, and operating parameters are significant factors to be considered with respect to wear. Different reasons of wear and friction are illustrated in fishbone diagram (Figure 6.5). Several engineering parts such as bearings, shafts and brake shoes are exposed to wear and friction because of different loads during movement. Therefore, determining tribological characteristic of material is crucial for manufacturers of tribology-based parts [139]. Wear and friction characteristics are affected by properties of fillers and reinforcements incorporated to composite matrix [139]. Tribological characteristics of natural fiber-reinforced composites are influenced with many conditions such as chemical treatments, fiber content, fiber orientation, interfacial adhesion, manufacturing techniques, matrix type, and operating conditions [119, 144]. Natural fibers-reinforced composites have enhancement in tribological characteristics and their properties are comparable to traditional fibers. Also, fiber orientation and fiber treatment are two significant parameters that may influence tribological performances where treated fibers and normally oriented fibers show better wear and friction behavior [145].

Wear and friction are significant tribological factors during the relative motion of surfaces, and they generally tend to dissipate energy and degrade material [146, 147]. Generally, there are several approaches to enhancing the tribological behavior of neat polymers. Popular technique is addition of fibers to polymers to develop composites [148, 149].

Materials are subject to friction and wear in relative motion with other materials. It causes tribological damage to materials. Surface of the material very often deteriorates owing to tribological damage. Wear damage to material system influences reliability and durability of material, which can cause failure and is also costly. Material properties (hardness, chemical composition, and toughness) and surface properties have important effects on wear [44]. Matějka et al. [4] stated that choosing an appropriate reinforcement combination in polymeric matrix can improve tribology (friction-wear) performance. Friction material with optimal weight percent of fiber

Table 6.3 Tribological studies and key findings of natural fiber- reinforced polypropylene (PP).

Type of wear	Fiber	Key findings	Reference
Two body abrasive wear	Jute	Development of bond between matrix and fiber played significant role in wear resistance. Using maleic anhydride grafted polypropylene (MA-g-PP) coupling agent provided better wear resistance than without using of a coupling agent. The incorporation of MA-g-PP during melt blending gave better wear resistance in comparison to jute PP composites with MA-g-PP solution-treated jute fibers, which was due to the greatest thermal stability of jute/PP composite with melt-blended MA-g-PP.	[132]
Two body abrasive wear	Sisal	The incorporation of MA-g-PP greatly influenced wear resistance of sisal PP composites.	[135]
Three-body abrasive wear	Turmeric-spent	Turmeric filler reduced the abrasion resistance of PP/turmeric-spent composites.	[120]

reinforcement presents better wear resistance [115]. The incorporation of wood flour into PP is beneficial according to its tribological properties [137]. Various tribological studies and key findings for natural fiber PP composites are presented in Tables 6.3 and 6.4.

6.5.1 Friction Coefficient of Natural Fiber-Reinforced PP Composites

The friction coefficient diminished when both load and sliding speed increased [134]. At load of 30 N and speed of 3 m s^{-1}, sample produced with pure PP and PP/jute showed higher friction force values [134]. Friction coefficient of jute fiber-reinforced PP reduced with increasing normal loads. Incorporation of jute filler up to 12.5% increased the friction coefficient. Adding more filler, i.e. 15%, reduced the friction coefficient. Friction coefficient diminished as load increased and increased with addition of jute filler at 75 and 100 cm $second^{-1}$ sliding speeds [133]. Nasir and Ghazali [151] investigated the friction properties of glass-pultruded kenaf and paddy straw-reinforced PP employing pin-on-disc tester at various loads and speeds. Outcomes indicated that friction coefficient diminished with an increase in load and speed, but normal load was more influential. However, Nijjar et al. [126] declared that friction coefficient increased proportionally with applied load.

Friction coefficients of untreated fibers were higher than those of treated fiber-reinforced polymers. Treated fiber-reinforced polymers presented a lower friction

Table 6.4 Specific wear rate and friction coefficient for natural fiber PP composites.

		Operating parameters						
Fiber	Tester	Sliding speed	Applied load	Sliding distance	Friction coefficient	Specific wear rate	Weight loss	Reference
Cellulose Coconut Flax Fleece	—	100 mm min^{-1}	106.6 N	—	*Static friction coefficient* 10 wt% cellulose = 0.09 20 wt% cellulose = 0.05 30 wt% cellulose = 0.05 10 wt% coconut = 0.10 20 wt% coconut = 0.06 30 wt% coconut = 0.06 10 wt% flax = 0.06 20 wt% flax = 0.05 30 wt% flax = 0.06 10 wt% fleece = 0.11 20 wt% fleece = 0.08 30 wt% fleece = 0.08 *Dynamic friction coefficient* 10 wt% cellulose = 0.04 20 wt% cellulose = 0.04 30 wt% cellulose = 0.04 10 wt% coconut = 0.05 20 wt% coconut = 0.05 30 wt% coconut = 0.04 10 wt% flax = 0.04 20 wt% flax = 0.04 30 wt% flax = 0.04 10 wt% fleece = 0.08 20 wt% fleece = 0.04 30 wt% fleece = 0.06	—	—	[32]

(Continued)

Table 6.4 (Continued)

		Operating parameters						
Fiber	**Tester**	**Sliding speed**	**Applied load**	**Sliding distance**	**Friction coefficient**	**Specific wear rate**	**Weight loss**	**Reference**
Flax	Atomic force microscopy	2, 4, 6, 8, 10 μm s^{-1}	10, 20, 30 μN	—	<0.015	—	—	[150]
Hay Lemon Palm Reed	Pin-on-disc	—	—	—	10 wt% hay = 0.77 20 wt% hay = 0.76 30 wt% hay = 0.82 10 wt% lemon = 0.67 20 wt% lemon = 0.76 30 wt% lemon = 0.76 10 wt% palm = 0.59 20 wt% palm = 0.68 30 wt% palm = 0.68 10 wt% reed = 0.59 20 wt% reed = 0.64 30 wt% reed = 0.65	—	—	[128]
Jute	Pin-on-disc wear testing machine	50, 75, 100 cm s^{-1}	4, 6, 8 kg	1500 m	0.33200–0.47245	—	0.001–0.020 g	[133]
Paddy straws Unidirectional glass-pultruded-kenaf	Pin-on-disc machine	1.178–2.749 m s^{-1}	9.82–19.64 N	50 cm	Paddy straws reinforced PP = 0.5–4 Unidirectional glass-pultruded-kenaf reinforced PP = 2.76–4.54	Paddy straws reinforced PP = 0.5–4 × 10^{-5} mm^3 Nm^{-1} Unidirectional glass-pultruded-kenaf reinforced PP = 0.8–1.79 × 10^{-5} mm^3 Nm^{-1}	*Volume loss*: Unidirectional glass-pultruded-kenaf reinforced PP = 0.00404 cm^3 *Volume loss*: Unidirectional glass-pultruded-kenaf reinforced PP = 0.00201 cm^3	[151]

coefficient than the others, causing less wear. This is due to strong bond between treated fibers and matrix. Interface temperature also increased because of variations in the friction coefficient [126].

Frictional heat occurred by the sliding of surfaces causes the temperature of thermoplastic composites to increase. Hence, PP allows greater surface contact area between composite surface and countersurface for cycle of softening and solidification stages. Therefore, increment in surface area causes increment in friction coefficient [136].

Friction coefficient was concluded to be variable with fiber content as well as fiber kind. Maximum value of the friction coefficient was obtained with composites containing 30 wt% of hay because of brittleness of hay fiber comparable to other fiber types and the brittleness of the hay fiber facilitates breakage during friction and therefore allows more material to be removed in such a composite. The lowest friction coefficient was achieved with composites containing 10 wt% of palm and reed. The friction coefficient for each composite was greater than pure PP polymer, as reinforcing fibers increase surface roughness during wear and therefore more material removal will occur in the composite [128]. Hay fiber composites tend to increase the friction coefficient with fiber loading more than other fiber types. But composites with other fiber kinds did not exhibit trend with increasing fiber amounts. Friction coefficients were nearly constant for both palm and lemon fibers, increasing only slightly for reed composites, where fiber amounts changed from 20 to 30 wt%. It is owing to brittleness of hay fiber, which increases the removal of mass during the application of frictional force on the composite [128]. The addition of borassus fruit fibers into PP significantly decreased friction coefficient of neat PP [126]. Increasing sisal fiber loading in composites increased friction coefficient [136]. Increasing proportion of sisal fibers in glass and carbon hybrid composites resulted in great voids because of large sisal bundles. This reduces the actual area of contact surface owing to discontinuous phase of the big sisal fiber bundles and thus the greater sisal amount indicated the greater coefficient of friction as demonstrated in the first phase of the friction cycles [136]. The increased sisal fiber amount in hybrid composites increased friction coefficient of composites at beginning of their unstable state, but the sisal amount of 50% gave the greatest friction coefficient for hybrid composites. As sisal fiber amounts increased in hybrid composites, number of sliding cycles reaching steady state diminished [136]. The incorporation of sisal fiber to carbon fiber PP composites increased friction coefficient, while incorporation of sisal to glass PP composites gave better tribological characteristics [136]. With sisal/glass and sisal/carbon composites, hybrid ratio of 25/75 wt% indicated the best friction performance [136]. Depended on comparison of friction characteristic, composite containing 5.6% by volume of jute fiber and 8.4% by volume of powdered hazelnut shell was chosen as the best combination. Assuming that maximal temperature in braking will be up to 290 °C, composite containing 23.6% by volume of jute fiber and 6.3% by volume of powdered hazelnut shell gives acceptable stability of friction coefficient [4]. Chegdani et al. [150] investigated multiscale frictional properties of natural fibers. Micro-friction of each composite was determined with atomic force microscopy instrumentation for the scratch test technique. For dynamic friction coefficient, tribological results showed that flax fibers caused greater friction than glass fibers, while PP created the greatest friction. The outcome is due to the mechanical behavior of each material at same scale. Viscoelastic behavior of the flax fibers and plastic behavior of PP add complementary components to dynamic friction coefficient. Dynamic friction coefficient of flax fibers increased with load increment.

Chegdani and El Mansori [152] fabricated flax fiber PP composites and determined tribological properties by double-crossed helix and concluded that resulting composite by improved surface quality showed decreased friction.

6.5.2 Wear Behavior of Natural Fiber-Reinforced PP Composites

Wear is inevitable and is responsible for huge economic loss. Wear should be minimized so that life is not endangered and replacement prices are diminished [45]. It is undesirable material removal from components surfaces. Some mechanisms for wear are adhesive, abrasive, corrosive, cavitation, erosive, and fretting [153]. Adhesive wear occurs when particles of one material adhere to the other when load is implemented between two sliding materials. Abrasive wear is caused by sliding of hard particles between surfaces under pressure. Abrasive wear occurs between two mating gears in motion and can be alleviated by eliminating contaminants between friction surfaces [154]. Corrosive wear occurs because of chemical reactions between the surface and surrounding medium [139]. Erosive wear occurs owing to impact of hard particles on solid surface; cavitation wear occurs owing to impact of the liquid because of the collapse of bubbles on surface [155]. Fretting wear occurs due to sliding and vibration. In fatigue wear, cycles are needed to create debris. Fatigue of parts induces surface and subsurface cracks. After repeated cycles, parts suffer damage from the removal of big fragments from the surface, a process named fatigue wear [139].

Wear mechanism has the following stages [135]:

1. plastic deformation of the matrix,
2. microcutting,
3. microploughing of fiber and matrix,
4. debonding of fiber from the matrix.

After debonding, the abrasive particle can separate fiber debris from specimen. Hard abrasive particles plowed out debris from specimen and then deepened groove. Interaction of matrix with the fiber is important factor in the wear of fiber-reinforced polymer composites [135].

Combined influences of fiber/matrix adhesion properties, critical fiber length, and fiber mechanical characteristics are parameters decreasing wear rate and weight loss [126].

Weight loss rose with the sliding distance for loads owing to the progressive material removal with the sliding distance. During abrasion, contact zone temperature increased with an increment in sliding distance, which loosened bonding and thus increased material removal [135]. Wear volume increased with rising sliding distance, which is owing to gradual material removal with distance [132]. Because material removal increases with sliding distance, weight loss increases with rising sliding distance for applied loads. During abrasion, the temperature in contact zone increased with increment in the sliding distance, resulting in further loosening of bonding and thus greater material removal. PP/turmeric-spent composites with less turmeric-spent amount showed minimal weight loss at loads. Specific wear rates show sharp decrease in smaller abrading distances and then remain relatively stable.

As the load increases and the speeds change, penetration and thus weight loss increase proportionally. Weight loss was lower for treated fiber-reinforced plastic for subsequent loads and speeds than for raw fiber-reinforced plastic [126]. The wear behavior changed with speed. It has been found that as the speed increases, the corresponding wear rate and weight loss increase [126]. For PP and jute/PP composites, specific wear rate increased with an increment in load. Wear rate of PP and composite increased with load independent of the sliding velocity. An increment in specific wear rate by an increment in load means a greater sample weight loss [134]. The wear loss of jute fiber PP decreased with increasing normal loads. The weight loss diminished with rising sliding speed, and weight loss increased with incorporation of filler [133]. At various applied loads, treated specimens showed greater wear resistance as compared to untreated composites. Wear rate increased with increasing load because deeper grooves on specimen surface caused more material removal from specimen. When polymer comes into contact with abrasive particle, it penetrates into PP, creating tracks. Tracks induced by abrasive particles become deeper at load of 7 N. Microcutting and microploughing processes are seen in jute PP. Interaction between abrasive and composite surfaces causes stresses at contact point. The extent of material deformation is mainly based on stress values [132]. Dwivedi and Chand [135] declared that increment in load decreased wear rate of sisal fiber PP composite. For all composites, weight loss increased with increment of load, which manifested itself in opening deeper grooving on specimen surface and removing greater material from specimen with a rising load. The abrasive particles create a series of tracks by penetrating deeper into PP. Wear tracks created by abrasive particles became deeper as load increased from 1 to 7 N. Interaction between composite surfaces and abrasives causes stresses at contact point. The extent of material deformation mainly depends on stress values [135]. Syed and Syed [120] studied physio-mechanical and wear characteristics of the turmeric-spent filler-reinforced (10, 20, 30, and 40%w/w) PP composite for tribological and load-bearing applications. For all composites, increased wear volume with increasing load is manifested by deeper grooves on specimen surface and greater material being removed from specimen by rising load. Abrasive particles penetrate deeper into PP, creating deeper furrow track. The track created with abrasive particles gets deeper as the load rises from 23.54 to 33.54 N. Nasir and Ghazali [151] investigated the wear properties of glass-pultruded kenaf and paddy straw-reinforced PP employing pin-on-disc setup at different loads and speeds. Outcomes showed that specific wear rate diminished with increments in load and speed, but load was more influential. Failure mode seen was micro-buckling followed by splitting while fiber/matrix interface failure formed. Incorporation of bio-fiber reduced wear rate of PP.

Tribological characteristics of Borassus fruit fine fiber- reinforced PP were explained according to fiber matrix surface modifications. Outcomes demonstrated that reinforcement of Borassus fruit fine fiber to PP matrix and modifications enhanced wear properties of pure PP [126]. Alkali-treated fiber reinforcement showed good wear characteristics in composites than untreated reinforcement. Due to alkali treatment, the removal of surface impurities from fiber promoted better adhesion of fiber/matrix and hence its wear characteristics [126]. The greatest wear loss was obtained with untreated jute fiber/PP because of weak interfacial bond between jute fiber and PP. Maleic anhydride grafted polypropylene (MA-g-PP) solution-treated composites showed less volume loss in comparison with untreated composites. It is because of enhanced fiber/matrix adhesion that MA-g-PP is incorporated into fiber surface, which reduces wear. Incorporation of coupling agent into composite during blending provided higher wear resistance with respect to MA-g-PP solution-treated jute fiber PP, due to modification of matrix during blending of MA-g-PP at 170 °C. MA-g-PP melt-treated jute fiber PP can possess more entanglement, which increases wear resistance for composites. Another is enhancement between the fiber and matrix interface with melt mixing [132]. Grooving occurs in PP as abrasive particles slide on surface. In microploughing, the material is forced to sides. For the worn surface of untreated jute fiber-reinforced PP, debonding, microcutting, and microploughing were primary wear processes. After debonding among fiber and matrix, abrasive particles remove fragmented fibers that appear as debris from surface. Generally, polymer composite wear proceeds in four stages: fiber wear, matrix wear, fiber fracture, and fiber/matrix interfacial debonding. After modification by MA-g-PP, interfacial debonding among jute fiber and PP improved, and the fourth mechanism dominated other ones. After successive passes of specimen, debris is formed and eliminated from surface. Abraded surfaces demonstrate various microfracture and micro-deformation mechanisms that operate as fiber is incorporated [132]. The improved wear properties with incorporation of MA-g-PP might be due to less fiber/matrix debonding and a reduction in fiber pullout. The unmodified jute fiber in PP does not possess sufficient bonding between the fiber and the matrix [132]. Untreated sisal fiber-reinforced PP caused the greatest weight loss because of poor interfacial bonding between matrix and fiber. Unmodified sisal fiber PP showed large deformation on surface. Composites with MA-g-PP exhibited a drop in weight loss in comparison to unmodified sisal fiber PP owing to enhanced fiber/matrix adhesion caused by the incorporation of coupling agents, which decreased material wear [135].

An increase in friction temperature causes more wear and plastic deformation [126].

The researchers reported that adding natural fibers in various forms such as fibers, particles, or fabrics, to thermoplastics improved wear resistance [123, 134, 156]. Incorporation of Borassus fruit fibers into PP significantly decreased adhesive wear and weight loss compared to pure PP. Borassus fruit fibers increased the wear resistance of neat PP [126]. With increment of coir content, wear rate of composite decreased from 9.6×10^{-7} to 5.2×10^{-7} g r^{-1}, because of increased rigidity of composite [115]. 5 wt% coir composites have streak flaws on the surface, which are common patterns for abrasive wear and are in good agreement with relationship between hardness and wear rate. Additionally, several microscale cracks and cavities could be observed in 5-wt% coir composites. Abrasive resistance is poor mainly because of low modulus, as proved by great wear rate of 6.5×10^{-7} g r^{-1} [115]. Jute fiber PP composites were examined by Yallew et al. [134] for wear and sliding performances. Jute reinforcement decreased friction coefficient and specific wear rate by 3.5–45% and 65%, respectively. After wear, micro-cracks, ploughing, and fiber and matrix damages were seen on surface of composites. Syed and Syed [120] studied physio-mechanical and wear characteristics of the turmeric-spent filler-reinforced (10, 20, 30, and 40%w/w) PP for tribological and load- bearing applications. Wear volume increased with an increment in filler loading. The result could be due to better dispersion of turmeric spent in PP at lower filler content and thus better wetting with PP matrix, thus resisting the debonding of the particle from the matrix. At higher filler content, the turmeric-spent particles agglomerate because of the hydrogen bonding between them (strong filler–filler interaction), while interaction between the turmeric-spent particles and the PP decreases as a result, and the turmeric- spent particles could be easily debonded from matrix. Plastic deformation of PP decreased with increasing filler amounts. This is because the soft PP is constrained between the solid boundaries of the rigid filler particles. Aurrekoetxea et al. [157] examined wear resistance of wood-reinforced PP and reported that wear rate for the pine wood/PP was the smallest; while the pure PP had the greatest since wear mechanism of neat PP was combination of melting and plastic flow. For pine wood, delamination and micro-fissuring caused by deformation were primary factors in the wear mechanism. Load carrying capability of PP was enhanced by incorporation of wood flour under the identical sliding distance, which was because of good wood flour crack initiation and propagation resistance of PP. The smallest weight loss was obtained with 25 wt% by 60% drop in comparison to only PP. But, in the case of composites with greater wood flour amounts (>25 wt%), wood flour particles gave maximum steel counterface roughening in the sliding wear mode, and thus composites provided better weight loss values. This may be attributed to absence of binder resins followed by rapid bonding peeling off among wood flour and PP under shear loading. While optimal wear resistance was obtained with 25 wt% of wood flour composite, smaller values were achieved with greater filler content. Composites containing 25 wt% wood flour gave minimum wear rate, whereas maximum was found at 55 wt%. It may be due to increased wear debris at higher wood flour loadings. Accordingly, abrasive wear mode is associated with wood flour particles acting as wear debris [137]. Increased fiber amounts in carbon and glass hybrid composites created deeper and wider worn patterns. For both composites, ratio of 75/25 indicated homogenous distribution and the greatest wear valley area along wear pattern circle [136]. It was observed that glass hybrid composite showed smaller abrasion volume than carbon hybrid composite. While abrasion volume of the sisal/glass hybrid composite increased slightly by incorporation of sisal fiber, abrasion volume of the carbon hybrid composite increased considerably, being approximately three times greater than the sisal/glass composites [136]. Adhesion, adhesive, and abrasive wear mechanisms are typical in hybrid fiber-reinforced PP. Increased sisal fiber amounts caused more fibrillation and greater transfer films on worn surfaces [136].

6.6 Conclusions

Main target of current chapter is to give information about tribological behavior of natural fiber-reinforced PP and demonstrate various parameters affecting friction and wear. Following results were provided from literature works.

- Several researchers have reported that tribobehavior of composites is enhanced by addition of natural fiber to polymer matrix; but this behavior is dependent on fiber orientation, fiber size, volume fraction, length, and test conditions. Natural fibers-reinforced composites have enhancement in tribological characteristics and properties are comparable to traditional fibers.
- Natural fiber-reinforced composites are replacing traditional materials in several industries, particularly in automobile manufacture in which low sliding wear and friction coefficient are significant. Jute/PP composite can be seen in tribological parts (automobile brake pads, artificial human joints, surface flooring, shoe soles, etc.).
- Friction coefficients of untreated fibers were higher than those of treated fiber-reinforced polymers.
- Friction coefficient was concluded to be variable with fiber amount as well as fiber kind.
- Weight loss increased with an increment in sliding distance for loads owing to progressive material removal with the distance.
- An increase in friction temperature causes more wear and plastic deformation.
- Adding natural fibers in various forms (fibers, particles, or fabrics) into PP improved wear resistance.
- Wear mechanism has stages: matrix plastic deformation, microcutting, microploughing of the fibers and matrix, and debonding fiber from the matrix.

Acknowledgments

A word of "thank you" is simply not enough. However, there are no words to describe how grateful I am to my interventional neurologist, Prof. Hasan Huseyin Karadeli, for providing me with a second chance in life after my brain operation. I am sure that without his operation and treatment, I would be unable to do everything I am able to do now. So, I will stick to this simple word. I would like to thank my interventional neurologist Prof. Hasan Huseyin Karadeli, who saved my life in August 2019. I want to dedicate this book chapter to my family and Prof. Hasan Huseyin Karadeli.

References

1 Gopal, P., Dharani, L.R., and Blum, F.D. (1996). Hybrid phenolic friction composites containing Kevlar® pulp Part 1. Enhancement of friction and wear performance. *Wear* 193: 199–206.

2 Satapathy, B.K. and Bijwe, J. (2006). Composite friction materials based on organic fibres: sensitivity of friction and wear to operating variables. *Composites Part A* 37: 1557–1567.

3 Yang, H.H. (1988). Aramid fibers. In: *Fibre Reinforcements for Composite Materials* (ed. A.R. Bunsell), 249–327. Amsterdam, The Netherlands: Elsevier.

4 Matějka, V., Fu, Z., Kukutschová, J. et al. (2013). Jute fibers and powderized hazelnut shells as natural fillers in non-asbestos organic non-metallic friction composites. *Mater. Des.* 51: 847–853.

5 Zhang, S.W. (1998). State-of-the-art of polymer tribology. *Tribol. Int.* 31: 49–60.

6 Nuruzzaman, D.M. and Chowdhury, M.A. (2012). Friction and wear of polymer and composites. In: *Composites and Their Properties* (ed. N. Hu), 299–330. Croatia: IntechOpen.

7 Lawrence, C.C. and Stolarski, T.A. (1989). Rolling contact wear of polymers: a preliminary study. *Wear* 132: 183–191.

8 Yamaguchi, Y. (1990). *Tribology of Plastic Materials: Their Characteristics and Applications to Sliding Components*. Netherlands: Elsevier.

9 Hooke, C.J., Kukureka, S.N., Liao, P. et al. (1996). The friction and wear of polymers in non-conformal contacts. *Wear* 200: 83–94.

10 Abdelmouleh, M., Boufi, S., Belgacem, M.N., and Dufresne, A. (2007). Short natural-fibre reinforced polyethylene and natural rubber composites: effect of silane coupling agents and fibres loading. *Compos. Sci. Technol.* 67: 1627–1639.

11 Li, X., Tabil, L.G., and Panigrahi, S. (2007). Chemical treatments of natural fiber for use in natural fiber-reinforced composites: a review. *J. Polym. Environ.* 15: 25–33.

12 Gupta, M.K. and Srivastava, R.K. (2014). Tensile and flexural properties of sisal fibre reinforced epoxy composite: a comparison between unidirectional and mat form of fibres. *Procedia Mater. Sci.* 5: 2434–2439.

13 Maurya, H.O., Gupta, M.K., Srivastava, R.K., and Singh, H. (2015). Study on the mechanical properties of epoxy composite using short sisal fibre. *Mater. Today Proc.* 2: 1347–1355.

14 Gupta, M.K. and Srivastava, R.K. (2015). Effect of sisal fibre loading on wear and friction properties of jute fibre reinforced epoxy composite. *Am. J. Polym. Sci. Eng.* 3: 198–207.

15 Gupta, M.K. and Srivastava, R.K. (2017). Mechanical, thermal and dynamic mechanical analysis of jute fibre reinforced epoxy composite. *Indian J. Fibre Text. Res.* 42: 64–71.

16 Tserki, V., Zafeiropoulos, N.E., Simon, F., and Panayiotou, C. (2005). A study of the effect of acetylation and propionylation surface treatments on natural fibres. *Composites Part A* 36: 1110–1118.

17 Gupta, M.K. and Gond, R.K. (2017). Influence of concentrations of alkali treatment on mechanical and dynamic mechanical properties of hemp/polyester composite. *Am. J. Polym. Sci. Eng.* 5: 24–33.

18 Kalia, S., Kaith, B.S., and Kaur, I. (2009). Pretreatments of natural fibers and their application as reinforcing material in polymer composites—a review. *Polym. Eng. Sci.* 49: 1253–1272.

19 Shahzad, A. (2012). Hemp fiber and its composites—a review. *J. Compos. Mater.* 46: 973–986.

20 Kabir, M.M., Wang, H., Lau, K.T., and Cardona, F. (2012). Chemical treatments on plant-based natural fibre reinforced polymer composites: an overview. *Composites Part B* 43: 2883–2892.

21 Gupta, M.K., Srivastava, R.K., and Bisaria, H. (2015). Potential of jute fibre reinforced polymer composites: a review. *Int. J. Fiber Text. Res.* 5: 30–38.

22 Bajpai, P.K., Singh, I., and Madaan, J. (2013). Tribological behavior of natural fiber reinforced PLA composites. *Wear* 297: 829–840.

23 Gupta, M.K. and Srivastava, R.K. (2016). A review on characterization of hybrid fibre reinforced polymer composite. *Am. J. Polym. Sci. Eng.* 4: 1–7.

24 Archard, J.F. (1980). Wear theory and mechanisms. In: *Wear Control Handbook* (ed. M.B. Peterson and W.O. Winer), 35–80. United States: ASME.

25 Tabor, D. (1987). Friction and wear—developments over the last 50 years, keynote address. In: *Proceedings, International Conference of Tribology—Friction, Lubrication and Wear, 50 Years on*, 157–172. London: Institute of Mechanical Engineering.

26 Chowdhury, M.A. and Helali, M.M. (2006). The effect of frequency of vibration and humidity on the coefficient of friction. *Tribol. Int.* 39: 958–962.

27 Bahadur, S. and Zheng, Y. (1990). Mechanical and tribological behavior of polyester reinforced with short glass fibers. *Wear* 137: 251–266.

28 El-Tayeb, N.S.M. and Mostafa, I.M. (1996). The effect of laminate orientations on friction and wear mechanisms of glass reinforced polyester composite. *Wear* 195: 186–191.

29 El-Tayeb, N.S. and Gadelrab, R.M. (1996). Friction and wear properties of E-glass fiber reinforced epoxy composites under different sliding contact conditions. *Wear* 192: 112–117.

30 Bahadur, S. and Polineni, V.K. (1996). Tribological studies of glass fabric-reinforced polyamide composites filled with CuO and PTFE. *Wear* 200: 95–104.

31 Shuhimi, F.F., Bin Abdollah, M.F., Kalam, M.A. et al. (2017). Effect of operating parameters and chemical treatment on the tribological performance of natural fiber composites: a review. *Part. Sci. Technol.* 35: 512–524.

32 Běhálek, L., Lenfeld, P., Seidl, M. et al. (2012). Friction properties of composites with natural fibres, synthetic and biodegradable polymer matrix. In: *2012 Nanocon*. Brno, Czech Republic (23–25 October, 2012).

33 Friedrich, K. and Reinicke, P. (1998). Friction and wear of polymer-based composites. *Mech. Compos. Mater.* 34: 503–514.

34 Suchánek, J., Kuklík, V., and Zdravecká, E. (2007). *Abrazivní opotřebení materiálů (Abrasive Wear of Materials)*. Praha: ČVUT v Praze.

35 Capandis, D. and Klebanska, M. (2011). Badania tribologiczne kompozytów POM z włóknem aramidowym i z proszkiem PTFE współpracujących ze stalą. *Tribologia* 42.

36 Bely, V.A., Savkin, V.G., and Sviridyonok, A.I. (1971). Effect of structure on polymer friction. *Wear* 18: 11–18.

37 Rymuza, Z. (2007). Tribology of polymers. *Arch. Civ. Mech. Eng.* 7: 177–184.

38 Pouzada, A.S., Ferreira, E.C., and Pontes, A.J. (2006). Friction properties of moulding thermoplastics. *Polym. Test.* 25: 1017–1023.

39 Myshkin, N.K. and Kovalev, A.V. (2009). Adhesion and friction of polymers. In: *Polymer Tribology* (ed. S.K. Sinha and B.J. Briscoe), 3–37. Singapore: Imperial College Press.

40 Capandis, D., Wieleba, W., Wožniak, J., and Wierciak, M. (2007). Właściwości tribologiczne wybranych materiałów polimerowych wytworzonych z recyklatu. *Tribologia* 2: 117–126.

41 Dulebová, L. and Greškovič, F. (2011). Influence of regrind on properties of plastics produced by injection moulding. *Mater. Eng.* 18: 44–48.

42 Hashmi, S.A.R., Dwivedi, U.K., Naik, A., and Chand, N. (2002). Effect of chemical treatments on abrasive wear behaviour of PP based lining materials. *Metals Mater. Processes* 14: 235–240.

43 Sahu, P. and Gupta, M.K. (2017). Sisal (*Agave sisalana*) fibre and its polymer-based composites: a review on current developments. *J. Reinf. Plast. Compos.* 36: 1759–1780.

44 Vigneshwaran, S., Sundarakannan, R., John, K.M. et al. (2020). Recent advancement in the natural fiber polymer composites: a comprehensive review. *J. Cleaner Prod.* 277: 124109.

45 Sahoo, P., Das, S.K., and Acharya, S.K. (2017). Tribological aspects of natural fiber composites. In: *Green Composites: Materials, Manufacturing and Engineering* (ed. J.P. Davim), 22–57. Germany: Walter de Gruyter GmbH.

46 Sun, R., Fang, J.M., Goodwin, A. et al. (1998). Fractionation and characterization of polysaccharides from abaca fibre. *Carbohydr. Polym.* 37: 351–359.

47 Shibata, M., Takachiyo, K.-I., Ozawa, K. et al. (2002). Biodegradable polyester composites reinforced with short abaca fiber. *J. Appl. Polym. Sci.* 85: 129–138.

48 Le Digabel, F. and Avérous, L. (2006). Effect of lignin content on the properties of lignocellulose-based biocomposites. *Carbohydr. Polym.* 66: 537–545.

49 Munawar, S.S., Umemura, K., and Kawai, S. (2007). Characterization of the morphological, physical, and mechanical properties of seven nonwood plant fiber bundles. *J. Wood Sci.* 53: 108–113.

50 John, M.J. and Anandjiwala, R.D. (2008). Recent developments in chemical modification and characterization of natural fiber-reinforced composites. *Polym. Compos.* 29: 187–207.

51 Jawaid, M. and Abdul Khalil, H.P.S. (2011). Cellulosic/synthetic fibre reinforced polymer hybrid composites: a review. *Carbohydr. Polym.* 86: 1–18.
52 Faruk, O., Bledzki, A.K., Fink, H.-P., and Sain, M. (2012). Biocomposites reinforced with natural fibers: 2000–2010. *Prog. Polym. Sci.* 37: 1552–1596.
53 Jain, S., Kumar, R., and Jindal, U.C. (1992). Mechanical behaviour of bamboo and bamboo composite. *J. Mater. Sci.* 27: 4598–4604.
54 Ismail, H., Edyham, M.R., and Wirjosentono, B. (2002). Bamboo fibre filled natural rubber composites: the effects of filler loading and bonding agent. *Polym. Test.* 21: 139–144.
55 Okubo, K., Fujii, T., and Yamamoto, Y. (2004). Development of bamboo-based polymer composites and their mechanical properties. *Composites Part A* 35: 377–383.
56 Murali Mohan Rao, K. and Mohana Rao, K. (2007). Extraction and tensile properties of natural fibers: vakka, date and bamboo. *Compos. Struct.* 77: 288–295.
57 Wang, Y., Wang, G., Cheng, H., and Tian, G. (2010). Structures of bamboo fiber for textiles. *Text. Res. J.* 80: 334–343.
58 Abdul Khalil, H.P.S., Bhat, I.U.H., Jawaid, M. et al. (2012). Bamboo fibre reinforced biocomposites: a review. *Mater. Des.* 42: 353–368.
59 Stokke, D.D., Wu, Q., and Han, G. (2014). *Introduction to Wood and Natural Fiber Composites*. West Sussex, UK: John Wiley & Sons.
60 Yu, Y., Wang, H., Lu, F. et al. (2014). Bamboo fibers for composite applications: a mechanical and morphological investigation. *J. Mater. Sci.* 49: 2559–2566.
61 Satyanarayana, K.G., Sukumaran, K., Kulkarni, A.G. et al. (1986). Fabrication and properties of natural fibre-reinforced polyester composites. *Composites Part A* 17: 329–333.
62 Joseph, K., Filho, R.D.T., James, B. et al. (1999). A review on sisal fiber reinforced polymer composites. *Rev. Bras. Eng. Agric. Ambiental* 3: 367–379.
63 Mohanty, A.K., Misra, M., and Hinrichsen, G. (2000). Biofibres, biodegradable polymers and biocomposites: an overview. *Macromol. Mater. Eng.* 276 (277): 1–24.
64 Hill, C.A.S. and Abdul Khalil, H.P.S. (2000). Effect of fiber treatments on mechanical properties of coir or oil palm fiber reinforced polyester composites. *J. Appl. Polym. Sci.* 78: 1685–1697.
65 Gassan, J., Chate, A., and Bledzki, A.K. (2001). Calculation of elastic properties of natural fibers. *J. Mater. Sci.* 36: 3715–3720.
66 Mwaikambo, L.Y. and Ansell, M.P. (2002). Chemical modification of hemp, sisal, jute, and kapok fibers by alkalization. *J. Appl. Polym. Sci.* 84: 2222–2234.
67 Espert, A., Vilaplana, F., and Karlsson, S. (2004). Comparison of water absorption in natural cellulosic fibres from wood and one-year crops in polypropylene composites and its influence on their mechanical properties. *Composites Part A* 35: 1267–1276.
68 Swamy, R.P., Mohan Kumar, G.C., Vrushabhendrappa, Y., and Joseph, V. (2004). Study of areca-reinforced phenol formaldehyde composites. *J. Reinf. Plast. Compos.* 23: 1373–1382.

69 Ramakrishna, G. and Sundararajan, T. (2005). Studies on the durability of natural fibres and the effect of corroded fibres on the strength of mortar. *Cem. Concr. Compos.* 27: 575–582.

70 Tomczak, F., Sydenstricker, T.H.D., and Satyanarayana, K.G. (2007). Studies on lignocellulosic fibers of Brazil. Part II: morphology and properties of Brazilian coconut fibers. *Composites Part A* 38: 1710–1721.

71 Cheung, H.-Y., Ho, M.-P., Lau, K.-T. et al. (2009). Natural fibre-reinforced composites for bioengineering and environmental engineering applications. *Composites Part B* 40: 655–663.

72 Malkapuram, R., Kumar, V., and Negi, Y.S. (2009). Recent development in natural fiber reinforced polypropylene composites. *J. Reinf. Plast. Compos.* 28: 1169–1189.

73 Thiruchitrambalam, M., Athijayamani, A., Sathiyamurthy, S., and Thaheer, A.S.A. (2010). A review on the natural fiber-reinforced polymer composites for the development of roselle fiber-reinforced polyester composite. *J. Nat. Fibers* 7: 307–323.

74 Ku, H., Wang, H., Pattarachaiyakoop, N., and Trada, M. (2011). A review on the tensile properties of natural fiber reinforced polymer composites. *Composites Part B* 42: 856–873.

75 Saw, S.K., Sarkhel, G., and Choudhury, A. (2012). Preparation and characterization of chemically modified jute–coir hybrid fiber reinforced epoxy novolac composites. *J. Appl. Polym. Sci.* 125: 3038–3049.

76 Arifuzzaman Khan, G.M., Alam Shams, M.S., Kabir, M.R. et al. (2013). Influence of chemical treatment on the properties of banana stem fiber and banana stem fiber/coir hybrid fiber reinforced maleic anhydride grafted polypropylene/low-density polyethylene composites. *J. Appl. Polym. Sci.* 128: 1020–1029.

77 Zainudin, E.S., Yan, L.H., Haniffah, W.H. et al. (2014). Effect of coir fiber loading on mechanical and morphological properties of oil palm fibers reinforced polypropylene composites. *Polym. Compos.* 35: 1418–1425.

78 Zhang, L. and Hu, Y. (2014). Novel lignocellulosic hybrid particleboard composites made from rice straws and coir fibers. *Mater. Des.* 55: 19–26.

79 Vinayagamoorthy, R. and Rajeswari, N. (2014). Mechanical performance studies on *Vetiveria zizanioides*/jute/glass fiber-reinforced hybrid polymeric composites. *J. Reinf. Plast. Compos.* 33: 81–92.

80 Yan, L., Su, S., and Chouw, N. (2015). Microstructure, flexural properties and durability of coir fibre reinforced concrete beams externally strengthened with flax FRP composites. *Composites Part B* 80: 343–354.

81 Pérez-Fonseca, A.A., Arellano, M., Rodrigue, D. et al. (2016). Effect of coupling agent content and water absorption on the mechanical properties of coir-agave fibers-reinforced polyethylene hybrid composites. *Polym. Compos.* 37: 3015–3024.

82 Sanjay, M., Madhu, P., Jawaid, M. et al. (2018). Characterization and properties of natural fiber polymer composites: a comprehensive review. *J. Cleaner Prod.* 172: 566–581.

83 Al-Khanbashi, A., Al-Kaabi, K., and Hammami, A. (2005). Date palm fibers as polymeric matrix reinforcement: fiber characterization. *Polym. Compos.* 26: 486–497.

84 Kaddami, H., Dufresne, A., Khelifi, B. et al. (2006). Short palm tree fibers—thermoset matrices composites. *Composites Part A* 37: 1413–1422.

85 Bledzki, A.K., Reihmane, S., and Gassan, J. (1996). Properties and modification methods for vegetable fibers for natural fiber composites. *J. Appl. Polym. Sci.* 59: 1329–1336.

86 Kessler, R.W., Becker, U., Kohler, R., and Goth, B. (1998). Steam explosion of flax—a superior technique for upgrading fibre value. *Biomass Bioenergy* 14: 237–249.

87 Saheb, D.N. and Jog, J.P. (1999). Natural fiber polymer composites: a review. *Adv. Polym. Tech.* 18: 351–363.

88 Stamboulis, A., Baillie, C.A., and Peijs, T. (2001). Effects of environmental conditions on mechanical and physical properties of flax fibers. *Composites Part A* 32: 1105–1115.

89 Beckwith, S.W. (2008). Natural fibers: nature providing technology for composites. *SAMPE J.* 44: 64–65.

90 De Rosa, I.M., Santulli, C., and Sarasini, F. (2010). Mechanical and thermal characterization of epoxy composites reinforced with random and quasi-unidirectional untreated *Phormium tenax* leaf fibers. *Mater. Des.* 31: 2397–2405.

91 Barkoula, N.M., Garkhail, S.K., and Peijs, T. (2010). Effect of compounding and injection molding on the mechanical properties of flax fiber polypropylene composites. *J. Reinf. Plast. Compos.* 29: 1366–1385.

92 Demirbas, A. (1997). Calculation of higher heating values of biomass fuels. *Fuel* 76: 431–434.

93 Graupner, N., Herrmann, A.S., and Müssig, J. (2009). Natural and man-made cellulose fibre-reinforced poly(lactic acid) (PLA) composites: an overview about mechanical characteristics and application areas. *Composites Part A* 40: 810–821.

94 Satyanarayana, K.G., Sukumaran, K., Kulkarni, A.G. et al. (1983). Performance of banana fabric-polyester composites. *Compos. Struct.* 2: 535–538.

95 Martí-Ferrer, F., Vilaplana, F., Ribes-Greus, A. et al. (2006). Flour rice husk as filler in block copolymer polypropylene: effect of different coupling agents. *J. Appl. Polym. Sci.* 99: 1823–1831.

96 Lee, B.-H., Kim, H.-J., and Yu, W.R. (2009). Fabrication of long and discontinuous natural fiber reinforced polypropylene biocomposites and their mechanical properties. *Fibers Polym.* 10: 83–90.

97 Saha, P., Manna, S., Chowdhury, S.R. et al. (2010). Enhancement of tensile strength of lignocellulosic jute fibers by alkali-steam treatment. *Bioresour. Technol.* 101: 3182–3187.

98 Shibata, S., Cao, Y., and Fukumoto, I. (2005). Press forming of short natural fiber-reinforced biodegradable resin: effects of fiber volume and length on flexural properties. *Polym. Test.* 24: 1005–1011.

99 Yussuf, A.A., Massoumi, I., and Hassan, A. (2010). Comparison of polylactic acid/kenaf and polylactic acid/rise husk composites: the influence of the natural fibers on the mechanical, thermal and biodegradability properties. *J. Polym. Environ.* 18: 422–429.

100 Aji, I.S., Zainuddin, E.S., Abdan, K. et al. (2012). Mechanical properties and water absorption behavior of hybridized kenaf/pineapple leaf fibre-reinforced high-density polyethylene composite. *J. Compos. Mater.* 47: 979–990.

101 Mahjoub, R., Yatim, J.M., Mohd Sam, A.R., and Hashemi, S.H. (2014). Tensile properties of kenaf fiber due to various conditions of chemical fiber surface modifications. *Constr. Build. Mater.* 55: 103–113.

102 Satyanarayana, K.G., Sukumaran, K., Mukherjee, P.S., and Pillai, S.G.K. (1986). Materials science of some lignocellulosic fibers. *Metallography* 19: 389–400.

103 Hassan, M.L. and Nada, A.-A.M.A. (2003). Utilization of lignocellulosic fibers in molded polyester composites. *J. Appl. Polym. Sci.* 87: 653–660.

104 Athijayamani, A., Thiruchitrambalam, M., Natarajan, U., and Pazhanivel, B. (2009). Effect of moisture absorption on the mechanical properties of randomly oriented natural fibers/polyester hybrid composite. *Mater. Sci. Eng., A* 517: 344–353.

105 Venkata Reddy, G., Venkata Naidu, S., and Shobha Rani, T. (2009). A study on hardness and flexural properties of kapok/sisal composites. *J. Reinf. Plast. Compos.* 28: 2035–2044.

106 Sreekumar, P.A., Thomas, S.P., Saiter, J.M. et al. (2009). Effect of fiber surface modification on the mechanical and water absorption characteristics of sisal/polyester composites fabricated by resin transfer molding. *Composites Part A* 40: 1777–1784.

107 Gwon, J.G., Lee, S.Y., Doh, G.H., and Kim, J.H. (2010). Characterization of chemically modified wood fibers using FTIR spectroscopy for biocomposites. *J. Appl. Polym. Sci.* 116: 3212–3219.

108 Venkateshwaran, N. and Elayaperumal, A. (2010). Banana fiber reinforced polymer composites—a review. *J. Reinf. Plast. Compos.* 29: 2387–2396.

109 Milanese, A.C., Cioffi, M.O.H., and Voorwald, H.J.C. (2012). Thermal and mechanical behaviour of sisal/phenolic composites. *Composites Part B* 43: 2843–2850.

110 Lilholt, H. and Lawther, J.M. (2000). Natural organic fibers. In: *Comprehensive Composite Materials*. New York, USA: Elsevier.

111 Bisanda, E.T.N. and Ansell, M.P. (1992). Properties of sisal-CNSL composites. *J. Mater. Sci.* 27: 1690–1700.

112 Bledzki, A.K. and Gassan, J. (1999). Composites reinforced with cellulose based fibres. *Prog. Polym. Sci.* 24: 221–274.

113 Gupta, M.K. and Srivastava, R.K. (2016). Mechanical properties of hybrid fibers-reinforced polymer composite: a review. *Polym. Plast. Technol. Eng.* 55: 626–642.

114 J. Gassan, A.K. Bledzki, Influence of fiber surface characteristics on the flax/polypropylene interface. 7. Internationales Techtexil Symposium, Frankfurt, 20–22 June 1995.

115 Liu, L., Wang, Z., Yu, Y. et al. (2019). Engineering interfaces toward high-performance polypropylene/coir fiber biocomposites with enhanced friction and wear behavior. *ACS Sustainable Chem. Eng.* 7: 18453–18462.

116 Riccieri, J.E., Vázquez, A., and De Carvalho, L.H. (1999). Interfacial properties and initial step of the water sorption in unidirectional unsaturated polyester/vegetable fiber composites. *Polym. Compos.* 20: 29–37.

117 Lou, N. and Netravali, A. (1999). Mechanical and thermal properties of environment-friendly "green" composites made from pineapple leaf fibers and poly(hydroxylbutyrate-co-valerate) resin. *Polym. Compos.* 20: 367–378.

118 Ahmad, F., Choi, H.S., and Park, M.K. (2015). A review: natural fiber composites selection in view of mechanical, light weight, and economic properties. *Macromol. Mater. Eng.* 300: 10–24.

119 Milosevic, M., Valášek, P., and Ruggiero, A. (2020). Tribology of natural fibers composite materials: an overview. *Lubricants* 8: 42.

120 Syed, M.A. and Syed, A.A. (2012). Development of a new inexpensive green thermoplastic composite and evaluation of its physico-mechanical and wear properties. *Mater. Des.* 36: 421–427.

121 Raghavendra, G., Ojha, S., Acharya, S.K., and Pal, S.K. (2014). Mechanical properties of woven jute-glass hybrid-reinforced epoxy composite. *J. Compos. Mater.* 48: 3445–3455.

122 Satyanarayana, K.G., Arizaga, G.G.C., and Wypych, F. (2009). Biodegradable composites based on lignocellulosic fibers—an overview. *Prog. Polym. Sci.* 34: 982–1021.

123 Ruggiero, A., Valášek, P., and Müller, M. (2016). Exploitation of waste date seeds of Phoenix dactylifera in form of polymeric particle biocomposite: investigation on adhesion, cohesion and wear. *Composites Part B* 104: 9–16.

124 Elanchezhian, C., Ramnath, B.V., Ramakrishnan, G. et al. (2018). Review on mechanical properties of natural fiber composites. *Mater. Today Proc.* 5: 1785–1790.

125 Balla, V.K., Kate, K.H., Satyavolu, J. et al. (2019). Additive manufacturing of natural fiber reinforced polymer composites: processing and prospects. *Composites Part B* 174: 106956.

126 Nijjar, S., Sudhakara, P., Sharma, S. et al. (2021). Influence of alkali treatment and maleated polypropylene (MAPP) compatibilizer on the dry-sliding wear and frictional behavior of borassus fruit fine fiber (BFF)/polypropylene (PP) polymer composites for various engineering applications. *Adv. Mater. Sci. Eng.* 2021: 5822245.

127 Essabir, H., Bensalah, M.O., Rodrigue, D. et al. (2016). Structural, mechanical and thermal properties of bio-based hybrid composites from waste coir residues: fibers and shell particles. *Mech. Mater.* 93: 134–144.

128 Al-Oqla, F.M., Hayajneh, M.T., and Aldhirat, A. (2021). Tribological and mechanical fracture performance of Mediterranean lignocellulosic fiber reinforced polypropylene composites. *Polym. Compos.* 42: 5501–5511.

129 Kakroodi, A.R., Leduc, S., and Rodrigue, D. (2012). Effect of hybridization and compatibilization on the mechanical properties of recycled polypropylene-hemp composites. *J. Appl. Polym. Sci.* 124: 2494–2500.

130 Panthapulakkal, S. and Sain, M. (2007a). Injection-molded short hemp fiber/glass fiber-reinforced polypropylene hybrid composites—mechanical, water absorption and thermal properties. *J. Appl. Polym. Sci.* 103: 2432–2441.

131 Panthapulakkal, S. and Sain, M. (2007b). Studies on the water absorption properties of short hemp-glass fiber hybrid polypropylene composites. *J. Compos. Mater.* 41: 1871–1883.

132 Chand, N. and Dwivedi, U.K. (2006). Effect of coupling agent on abrasive wear behaviour of chopped jute fibre-reinforced polypropylene composites. *Wear* 261: 1057–1063.

133 Kumar, N.V., Priyanka, V.G., and Srikanth, K. (2014). Evaluation of wear properties of jute reinforced polypropylene composites. *Int. J. Eng. Res. Technol.* 3: 1212–1215.

134 Yallew, T.B., Kumar, P., and Singh, I. (2014). Sliding wear properties of jute fabric reinforced polypropylene composites. *Procedia Eng.* 97: 402–411.

135 Dwivedi, U.K. and Chand, N. (2009). Influence of MA-g-PP on abrasive wear behaviour of chopped sisal fibre reinforced polypropylene composites. *J. Mater. Process. Technol.* 209: 5371–5375.

136 Aslan, M., Tufan, M., and Küçükömeroğlu, T. (2018). Tribological and mechanical performance of sisal-filled waste carbon and glass fibre hybrid composites. *Composites Part B* 140: 241–249.

137 Ibrahim, M.A., Hirayama, T., and Khalafallah, D. (2020). An investigation into the tribological properties of wood flour reinforced polypropylene composites. *Mater. Res. Express* 7: 015313.

138 Sahoo, P. (2005). *Engineering Tribology*. India: Prentice Hall.

139 Vinayagamoorthy, R. (2020). Friction and wear characteristics of fibre-reinforced plastic composites. *J. Thermoplast. Compos. Mater.* 33: 828–850.

140 Nirmal, U., Hashim, J., and Lau, S.T.W. (2011). Testing methods in tribology of polymeric composites. *Int. J. Mech. Mater. Eng.* 6: 367–373.

141 Rajesh, J.J., Bijwe, J., and Tewari, U.S. (2002). Abrasive wear performance of various polyamides. *Wear* 252: 769–776.

142 Pıhtılı, H. and Tosun, N. (2002). Effect of load and speed on the wear behaviour of woven glass fabrics and aramid fibre-reinforced composites. *Wear* 252: 979–984.

143 Hummel, S.R. and Partlow, B. (2004). Comparison of threshold galling results from two testing methods. *Tribol. Int.* 37: 291–295.

144 Nusrat, Z. and Srivastava, A. (2019). Study on tribology natural fiber reinforced polymer composites: a review. *SVOA Mater. Sci. Technol.* 1: 4–10.

145 Omrani, E., Menezes, P.L., and Rohatgi, P.K. (2016). State of the art on tribological behavior of polymer matrix composites reinforced with natural fibers in the green materials world. *Eng. Sci. Technol.* 19: 1–26.

146 Moghadam, A.D., Omrani, E., Menezes, P.L., and Rohatgi, P.K. (2015). Mechanical and tribological properties of self-lubricating metal matrix nanocomposites reinforced by carbon nanotubes (CNTs) and graphene—a review. *Composites Part B* 77: 402–420.

147 Omrani, E., Moghadam, A.D., Menezes, P.L., and Rohatgi, P.K. (2016). Influences of graphite reinforcement on the tribological properties of self-lubricating aluminum matrix composites for green tribology, sustainability, and energy efficiency—a review. *Int. J. Adv. Manuf. Technol.* 83: 325–346.

148 Lv, M., Zheng, F., Wang, Q. et al. (2015). Friction and wear behaviors of carbon and aramid fibers reinforced polyimide composites in simulated space environment. *Tribol. Int.* 92: 246–254.

149 Tang, G., Hu, X., Sun, D.J. et al. (2016). The research on the friction and wear properties of ionic liquid-treated cellulose fibre-filled polyoxymethylene composites. *J. Thermoplast. Compos. Mater.* 29: 270–280.

150 Chegdani, F., El Mansori, M., Mezghani, S., and Montagne, A. (2017). Scale effect on tribo-mechanical behavior of vegetal fibers in reinforced bio-composite materials. *Compos. Sci. Technol.* 150: 87–94.

151 Nasir, R.M. and Ghazali, N.M. (2014). Tribological performance of paddy straw reinforced polypropylene (PSRP) and unidirectional glass-pultruded-kenaf (UGPK) composites. *J. Tribol.* 1: 1–17.

152 Chegdani, F. and El Mansori, M. (2019). Tribo-functional effects of double-crossed helix on surface finish, cutting friction and tool wear mechanisms during the milling process of natural fiber composites. *Wear* 426–427: 1507–1514.

153 Eyre, T.S. (1978). The mechanisms of wear. *Tribol. Int.* 11: 91–96.

154 Kato, K. (2002). Classification of wear mechanisms/models. *Proc. Inst. Mech. Eng. Part J J. Eng. Tribol.* 216: 349–355.

155 Dodiya, V.K. and Parmar, J.P. (2016). A study of various wear mechanism and its reduction method. *Int. J. Innovative Res. Sci. Eng. Technol.* 2: 242–248.

156 Arroyo, M., Zitzumbo, R., and Avalos, F. (2000). Composites based on PP/EPDM blends and aramid short fibres. Morphology/behaviour relationship. *Polymer* 41: 6351–6359.

157 Aurrekoetxea, J., Sarrionandia, M., and Gómez, X. (2008). Effects of microstructure on wear behaviour of wood reinforced polypropylene composite. *Wear* 265: 606–611.

7

Wear Behavior of the Natural Fiber-Reinforced Thermoplastic Composites

Ramu Sundaramoorthy[1], Vellaichamy Parthasarathy[2], Jeyanthi Subramanian[3], Lin Feng Ng[4], and Naveen Jesuarockiam[5]

[1]*Hindustan Institute of Technology and Science, Department of Aeronautical Engineering, Rajiv Gandhi Salai OMR, Kelambakkam Chennai 603103, Tamil Nadu, India*
[2]*Rajalakshmi Institute of Technology, Department of Physics, Poonamallee, Chennai 600124, Tamil Nadu, India*
[3]*School of Mechanical Engineering, VIT Chennai Campus, Vandalur-Kelambakkam Road, Chennai 600127, Tamil Nadu, India*
[4]*Universiti Teknologi Malaysia (UTM), School of Mechanical Engineering, Faculty of Engineering, 81310 Skudai, Johor Bahru, Johor, Malaysia*
[5]*Vellore Institute of Technology, School of Mechanical Engineering, Katpadi-Thiruvalam Road, Vellore 632014, Tamil Nadu, India*

7.1 Introduction

Tribology is associated with the relative motion of interfering surfaces. Wear and friction are the most predominant tribological parameters when two bodies are in relative motion. The durability of the tribomaterials depends mostly on these two parameters. Wear represents the loss of materials while solid surfaces are in contact during motion. The frictional force acts against the motion of two sliding bodies in contact.

The type of contact materials and lubricants that are utilized in a mechanical system have a direct bearing on the system's overall level of productivity and dependability. The point at which two different things come into contact is the principal focus of a tribologist's research. The absence of physical contact precludes the possibility of friction and wear. The opinion that normal touch is a necessary precondition for any and all other tribological occurrences as a result of their observations [1].

- The relationship that exists between normal body displacement and normal force is what ultimately determines dynamic and contact properties.
- In the region where the two surfaces come into contact, stresses are produced.

7.2 Wear Testing Methods

In order to evaluate the friction and wear qualities of various materials, tribo testing devices such as tribometers and tribo testers can be utilized. These devices make it

Tribological Properties, Performance, and Applications of Biocomposites, First Edition.
Edited by Chandrasekar Muthukumar, Senthilkumar Krishnasamy,
Senthil Muthu Kumar Thiagamani, and Ganesan Chinnachamy.

possible to apply a predetermined load and, if necessary, lubrication, and then obtain information regarding the friction and wear behavior of a sample. The tribometer is the most regularly used device to assess friction and wear. A tribometer must be carefully chosen and the experimental circumstances set up in order to eliminate measurement mistakes and imitate genuine systems. Tribo-contact tests, such as pin-on-flat, flat-on-flat, pin-on-rotating disc, ball-on-flat, and ball-on-disc, are the most commonly employed, as evidenced by a review of relevant literature. In the laboratory, tribometers almost often make use of specimens that have simple geometric shapes. This allows the test conditions to be determined in advance. It is customary for this to be one of the moving specimens whenever a rotating cylinder or disc is utilized to force another solid object with a different curvature into nominal point, line, or flat contact. Tribo-testing laboratory techniques can be used to simulate tribo-contacts in a specific machine, evaluate the influence of lubricants on a specific application, investigate the causes of friction, and more.

Sliding wear on various natural fiber reinforced composite (NFRC) materials occurs when the specimen comes into contact with the specimen's counterface. NFRC tribological behavior can be studied by using the pin/ball-on-disc arrangement shown in Figure 7.1a,b. The pin/ball is closed against a rotating disc and a normal load is applied to the pin/ball. The disc's flat surface or its perimeter can be the site of contact. The pin/ball and disc sizes can vary according to the intended mean contact pressure, depending on the test type. Figure 7.1c,d shows a reciprocating tribometer in which the pin/ball moves horizontally along a linear trajectory connected with a flat surface. Kinetic friction coefficient can be measured by setting parameters such as frequency (frequency), time (time interval), normal load (normal load), stroke (stroke length), lubricant type, and temperature, if applicable.

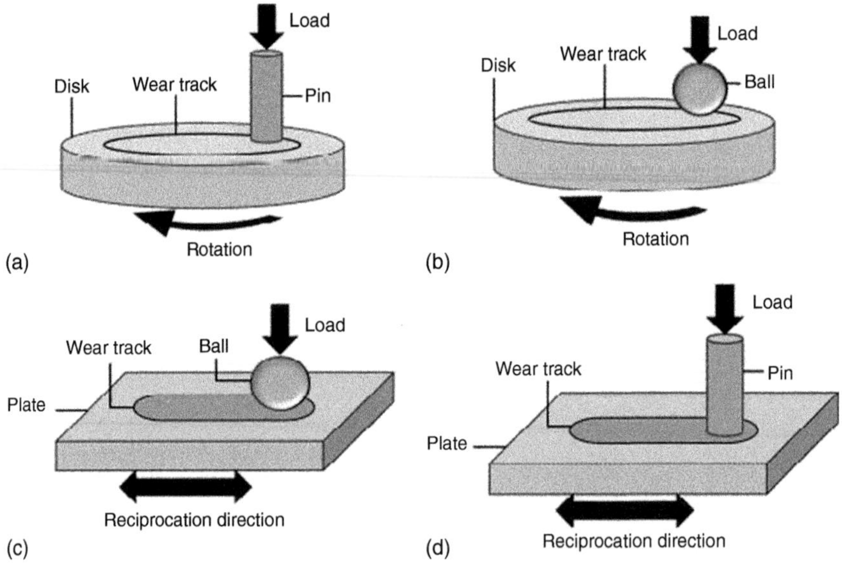

Figure 7.1 Tribological test setup (a) pin-on-disk rotating arrangement (b) ball-on-disk rotating arrangement (c) pin-on-disk reciprocating arrangement (d) BALL-on-disk reciprocating arrangement.

The use of sliding and rolling surfaces in modern technologies makes applied tribology an absolutely necessary field of study. In order to get the most out of a material's tribological qualities, its surface should be as smooth as possible while also being as hard as possible. This minimizes friction as well as wear, which results in an increase in efficiency. Friction is one of the two degrading phenomena that have a substantial impact on the efficiency of mechanical assemblies that include sliding surface contact. Friction is the more significant of the two. Additionally, it is accountable for wear, which is typically the aspect that determines the maximum useful life of a device. When developing new components, it is essential to take tribological loadings into consideration. The variability of the material, which can be seen on the exterior as well as the interior of composites, is the primary issue with these kinds of materials. Therefore, operations pertaining to surface finishing may be required in order to accomplish the desired level of smoothness in composite materials. In addition, an insufficient link between the matrix and the reinforcing fibers is often the root cause of excessive wear in sliding contact.

7.3 Factors Affecting Wear Behavior of the Composite

Operating variables for tribometers include the type of motion being measured, the load being applied, the velocity being measured, the ambient temperature being measured, and the length of the test being performed [2].

7.4 Motion Type

Sliding, rolling, slewing, and impact are the four primary types of motion, and any combination of these fundamental motions can be considered a motion type. The geometry of the testing system will determine the type of motion that is carried out.

7.5 Load

The test system can be loaded in a variety of ways, such as with a mass (dead load), a spring, hydraulics, or even electromagnetics. Strain gauges, inductive components, or pressure-sensitive detectors can be used to measure the load on a surface in the vicinity of the contact zone. Electronic force transducers have the advantage of allowing real-time calculation of a friction coefficient (μ) during a test by feeding the normal load signal into an electronic divider, and the actual value of the friction coefficient (μ = friction force/normal load) can be computed.

7.6 Velocity

For the purpose of controlling velocity, electronic or optoelectronic transducers are available in the commercial sector. Because velocity has an effect on lubrication as measured by the Stribeck curve, it is an essential factor to consider while carrying out tribo testing. This equation has an independent variable that refers to the power loss of the tribo testing system. This power loss is based on the friction force and sliding velocity.

7.7 Temperature

At the beginning stages of the test, the ambient temperature of the testing system is a critical operational variable because it determines the thermal state of the system. This makes the ambient temperature of the testing system an extra-critical operating variable. The approaches of thermocouples and infrared pyrometers are the ones that are utilized the most frequently for the purpose of temperature control.

7.8 Test Duration

The operational parameters, which include load, velocity, and temperature, have an effect on the tribo testing system as a function of time or duration. The duration of the investigation is another factor that ought to be selected in accordance with the kind of research being conducted. It is common practice in what are known as "accelerated tests" to make an effort to cut down on the number of time being spent testing by making the operating conditions significantly more difficult. On the other hand, this strategy is valid only under the condition that the tribological processes remain unchanged.

7.9 Performance Metrics From the Wear Test

The frictional force was used to estimate the coefficient of friction using the following mathematical expression [3];

$$\text{Coefficient of friction } (\mu) = E_f/F_n \tag{7.1}$$

where E_f and F_n are the developed friction force and the applied normal load, respectively.

The weight of the worn specimens was recorded after performing wear analysis on the counter disc for appropriate testing conditions by varying the sliding speed from 1 to 3 m s^{-1}, applied load from 10 to 30 N, and sliding distance from 1000 to 3000 m. The initial weight of the wear specimens was also recorded using an electronic balance to estimate the mass loss during the wear test. The specific wear (SW) rate of the specimens was computed using Eq. (7.2):

$$\text{Specific wear rate } K_s = \Delta m/\rho L F_n \,(\text{mm}^3) \tag{7.2}$$

where p—density of the wear specimen, Δm—mass loss during wear test (g), F_n—normal load (N) and sliding distance (m).

The abrasive wear (AW) rate was estimated using the following mathematical relation [4]:

$$W = V/\rho D$$

where, V—weight loss, D—sliding distance and ρ—density of the composite.

The dispersion of the material's fibers and the direction in which contact is made between the contact pairs are two of the elements that contribute to the wear resistance of a material.

7.10 Wear Studies on Natural Fiber-Reinforced Thermoplastic Composites

Biopolymers and synthetic thermoplastics can be reinforced with NFs to develop energy-efficient and economically viable tribomaterials. The tribological properties of polymer-based composites can also be enhanced by reducing wear and frictional force by reinforcing the polymers with NFs.

The laminated PLA composites were fabricated by incorporating three different NFs, such as *Grewia optiva* (GoF), sisal, and nettle into the PLA matrix to improve the tribological properties [3]. All the fabricated PLA/nettle, PLA/GoF, and PLAA/sisal composites were subjected to wear and friction analysis under dry contact conditions. An improvement in the tribological properties of NF-reinforced polymer composites significantly reduces failure in mechanical parts. Wear and friction characteristics were evaluated by tribometer and pin-on-disc rotating type friction, respectively. The morphology of the worn surface was evaluated using scanning electron microscope (SEM) to comprehend the wear mechanism in the NFs-reinforced PLA composites. The fabricated PLA composites exhibited a 70% reduction in SW rate and a 10–44% reduction in friction coefficient as compared with pristine PLA. An improvement in the wear behavior of PLA was due to the incorporation of natural fiber mats into PLA. The ASTM G99-95 standard was maintained to perform a wear test. All the fabricated PLA/GoF, PLA/nettle, and PLA/sisal composites were suitably cut with a dimension of $5 \times 5 \times 10\,\text{mm}^3$ to carry out the wear test. These wear specimens and counter-face disc were initially rubbed with silicon carbide abrasive paper of 1500 grits, and then they were rubbed with abrasive paper of 2000 grits to provide intimate contact between specimen surface and counter-face disc by providing constant initial roughness before each wear test. The wear specimens and counter-face disc were also cleaned properly before the wear analysis. Figure 7.2 illustrates the rubbing direction of the wear sample against the counter face.

The wear and friction monitor continuously displayed the tangential friction force after started rubbing the counter face and specimen pin. The friction force was recorded at different sliding speeds, normal loads and a sliding distance of 3000 m during the tribo-testing at every 60 s for the pristine PLA and the laminated composites. It was reported that the friction coefficient of the pristine PLA and its laminated fiber reinforced composites decreased with the increasing normal

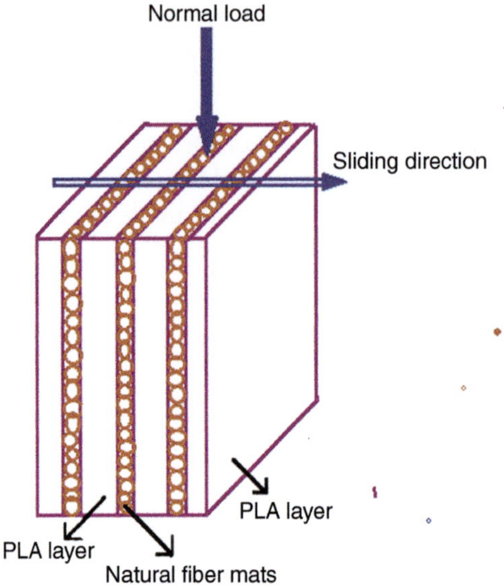

Figure 7.2 Schematic illustration of sliding direction of laminated PLA composite.

load from 10 N to 30 N at different sliding speed. The reason for the reduction in friction coefficient with the increasing load is due to the increase in temperature at the contact surface between the counterface and specimen during dry the sliding wear test. Hence, the specimen experiences thermal stress due to the developed thermal gradients, which reduce the bonding between the fiber matrix. The fibers easily shear upon the repeated axial thrust, resulting in the pullout of fibers during sliding [5–8]. However, the pristine PLA and its fiber-reinforced PLA composites did not show much reduction in friction coefficient. The dimensional stability of pristine PLA was greatly affected at higher loads (20 N and 30 N) and a sliding distance of 3000 m due to the repeated axial thrust. The weight loss of PLA was considerably high owing to the softening, removal, and film generation processes during sliding [9, 10]. However, the fiber- reinforced PLA composites exhibited a lower SW rate as compared with pristine PLA. It was concluded that the wear behavior of PLA was increased after incorporating the NFs into the PLA matrix as a reinforcing agent. In natural fiber-reinforced PLA composites, the PLA matrix layer was protected by the incorporated fibers due to the adhesion characteristics at the interface of the matrix and fibers.

Narish et al. [11] fabricated the kenaf fiber (KF)-reinforced polyurethane (PU) composite to investigate its frictional and adhesive wear characteristics using the block-on-disc method. The orientation of fiber like normal orientation (NO) anti-parallel orientation (AO) and parallel orientation (PO) was also examined with respect to sliding direction. The SEM micrographs of the worn surface of the PU/KF composite revealed different wear mechanisms such as pitting, detachment, micro-cracks ,and delaminating. The PU/KF composite with antiparallel orientation exhibited excellent wear resistance as compared to pristine PU. The thermoplastic PU is suitable for tribological applications owing to its chemical resistance, recyclability, flexibility, wear resistance, dry sliding ability and fast elastic recuperation [1, 12–15], The kenaf fiber with a length of 100 mm and 101 μm in diameter were arranged in a mat form. The PU/KF composite was prepared by placing 10 layers of fiber mat alternatively on 11 layers of the PU resin layer. The required dimension of $10 \times 10 \times 20$ mm PU/KF composite block was used to assess its tribological properties. PU/KF composite was subjected to a wear test by varying a sliding distance (0–2.7 km), different fiber orientations, applied loads (30–60 N), and at constant sliding velocity (2.8 m s^{-1}). There was an increase in SW rate while increasing the sliding distance from 0 to 1.5 km at all fiber orientations except the PU/KF composite with parallel fiber orientation. The initial SW rate of PU/KF composite with antiparallel fiber orientation was in the range of 30×10^{-8} and 75×10^{-8} mm^3 mN^{-1} for the different applied load (30 N–60 N), and the initial values of SW rate were decreased to 2×10^{-8} and 3.5×10^{-8} mm^3 mN^{-1} while increasing the sliding distance. However, there were a lot of fluctuations in the SW rate of PU/KF composite with normal fiber orientation for the same testing conditions. The pristine PU exhibited a poor SW rate due to its plastic nature. The friction coefficient was not significantly enhanced with the kenaf fiber- reinforced PU composite.

Chand and his research team [12] employed untreated jute fibers [UJF] and jute fibers treated with 1% maleic anhydride-g-polypropylene (MA-g-PP) solution as reinforcing agents to fabricate polypropylene (PP) composites. The main reason for treating jute fibers (JF) with a coupling agent of MA-g-PP was due to enhance the adhesion between the matrix and fiber through chemical bonds. The PP composite reinforced with untreated JF (UJF-PP) and PP composite reinforced with treated JF with MA-g-PP solution (TJF-PP) were prepared by a compression molding technique. The PP composite reinforced with MA-g-PP melt-mixed JF (MJF-PP) was fabricated by mixing 1% MA-g-PP with PP at 170 °C with the help of two roll mixer followed by compression molding. All the fabricated composites were cut in the form of rectangular slab of 45 mm × 35 mm × 5 mm dimension, and these wear specimens were kept on SUGA abrasion tester-NUS1 to slide against a rotating wheel to perform abrasive wear test under different sliding distances (6.4, 12.8, 19.2, and 25.6 m), different applied loads of 1, 3, 5, 7 N and a constant sliding speed (2.56 m min^{-1}). The highest wear loss was recorded by the UJF-PP composite among the fabricated composites, owing to the weak interfacial bonding of the PP matrix and untreated jute fibers. However, the TJF-PP composite showed a lower wear loss as compared to the UJF-PP composite owing to the improved matrix-fiber adhesion rusting with the incorporation of the coupling agent MA-g-PP onto JK fiber, which led to a considerable reduction in wear loss. It was also noticed that there was better wear resistance with the MJF-PP composite as compared to the TJF-PP composite. This was due to the incorporation of the MA-g-PP coupling agent into the PP matrix by melt mixing at 170 °C. The improvement in the wear resistance of the MJF-PP composite was due to its entangled nature as well as the strong interfacial adhesion upon melt mixing. The SEM micrograph of MJF-PP composite revealed strong fiber-matrix bonding and reduction in fiber pullout due to the addition of coupling agent into PP by melt mixing, resulting in enhanced wear properties. The wear of polymer composites generally occurs in four stages, such as fiber wear, matrix wear, fiber fracture, and fiber-matrix interfacial de-bonding. Wear and friction are two major tribological parameters resulting in loss of materials and dissipation of energy when there is any relative motion of solid surfaces [2, 16]. The tribological properties of polymers can be improved by embedding NFs into their matrix.

The polyamide/ZnO composite was reinforced with short jute fiber (6 μm) to study the effect of JF on the tribiological properties of the JF/PA/ZnO hybrid composite by Rajasekhar et al. [17]. A pin-on-disc machine was used to assess the wear rate and frictional coefficient of the JF/PA/ZnO hybrid composite under different input parameters. There was a reduction in the WR and FC after incorporating the short JF into polyamide (PA) as compared to pristine PA. The optimization of WR and FC was done by the developed second-order polynomial model. The optimized values of WR and FR were estimated as 0.6499 $mm^3\ Nm^{-1}$ and 0.1044 μ by response surface methodology (RSM) optimization techniques for a sliding velocity of 0.2970 $m\ s^{-1}$, and the normal force of 2 N for the fabricated JF/PA/ZnO composite. It was reported that an improvement in wear resistance of the JF/PA/ZnO composite was due to the formation of transfer film on the counterface during sliding with the applied load, which actually prevented the metal surface from direct contact with the composite, resulting in low plowing damage [18, 19]. A decrement in WR while increasing the sliding speed was understood from the SEM micrographs of the worn-out surface. This was due to the reduced shear strength of the composite while increasing the shear speed [20].

In another study conducted by Nishitani et al. [19], the various NFs such as jute fiber (JF), hemp fiber (HF), sisal fiber (SF), and ramie fiber (RF) were subjected to surface treatment by using ureidosilane coupling agents. These surface-treated NFs were used as reinforcing agents to improve the wear performance of biobased PA 1010. The tribological properties of PA1010 composites reinforced with four different NFs were evaluated using a ring-on-plate wear tester. The testing specimens were in the form of a plate with a dimension of 30 mm × 30 mm × 30 mm to carry out the sliding wear test. Carbon steel with a surface roughness of 0.4 μm was employed as a metal counterpart. The sliding wear test was performed under normal load (50 N), a sliding distance of 600 m and a sliding speed of 0.2 $m\ s^{-1}$ for the prepared NFs-reinforced PA1010 composites. The SW rate was estimated by measuring the sample weight before and after the wear test. The transfer film was formed on the counterface during the sliding wear test. The tribological properties of the PA1010 composites were also understood by analyzing the SEM micrographs of the metallic counterface (coated with transfer film during the sliding wear test) and the worn surface of the specimens. There was a variation in the thickness of the transfer films formed on the counterface due to the incorporation of different types of NFs into biobased PA1010. The thickness of the transfer film formed on the counterface against the RF/PA1010 composite was small as compared to other NFs/PA1010 composites (JF/PA1010, HF/PA1010, and SF/PA1010 composites). The incorporation of various NFs into biobased PA1010 considerably reduced the wear rate and frictional coefficient of PA1010 due to their self-lubricant properties. Ramie fiber was suggested as an efficient reinforcing agent to enhance the wear properties of biobased PA1010.

7.11 Conclusion

Researchers all over the world have been encouraged to experiment with locally available inexpensive fibers to determine whether they may match the requirements of a good reinforced polymer composite for tribological applications. This is due to the fact that NFs and their composites are easily available. It has been found that the use of NFs has a substantial impact on the tribological performance of composites. NFs have a number of downsides, including the fact that they absorb water, have lower fiber-to-matrix adhesion, and are found in a variety of geographic locations. In order to circumvent these limitations, hybrid composites and nanocomposites are now in focus. This chapter focused on the effect of operative conditions, fiber chemical treatments, and fiber orientation on the friction and wear behavior of the NFs-based thermoplastic composites. NFs-based composites have been found suitable for automotive and mechanical engineering applications due to their tribological characteristics, such as wear and friction resistance. Moreover, the environmental pollution associated with the disposal of NFs-based composites is also considerably reduced as these composites are reinforced with NFs due to their improved biodegradability. These composites also provide solutions to the growing environmental issues because they have the potential to replace synthetic fiber-reinforced composites in a wide range of industrial applications.

References

1 Elleuch, R., Elleuch, K., Salah, B., and Zahouani, H. (2007). Tribological behavior of thermoplastic polyurethane elastomers. *Mater. Des.* 28: 824–830.

2 Omrani, E., Moghadam, A.D., Menezes, P.L., and Rohatgi, P.K. (2016). Influences of graphite reinforcement on the tribological properties of self-lubricating aluminum matrix composites for green tribology, sustainability, and energy efficiency—a review. *Int. J. Adv. Manuf. Technol.* 83: 325–346.

3 Bajpai, P.K., Singh, I., and Madaan, J. (2013). Tribological behavior of natural fiber reinforced PLA composites. *Wear* 297: 829–840.

4 Chand, N. (2000). *An Introduction to Tribology of FRP Materials*. New Delhi: Allied Publishers.

5 Sumer, M., Unal, H., and Mimaroglu, A. (2008). Evaluation of tribological behaviour of PEEK and glass fibre reinforced PEEK composite under dry sliding and water lubricated conditions. *Wear* 265: 1061–1065.

6 Yousif, B.F. and El-Tayeb, N.S.M. (2008). Wear and friction characteristics of CGRP composite under wet contact condition using two different test techniques. *Wear* 265: 856–864.

7 Davim, J.P. and Cardoso, R. (2009). Effect of the reinforcement (carbon or glass fibres) on friction and wear behaviour of the PEEK against steel surface at long dry sliding. *Wear* 266: 795–799.

8 Chauhan, S.R., Kumar, A., and Singh, I. (2010). Sliding friction and wear behaviour of vinylester and its composites under dry and water lubricated sliding conditions. *Mater. Des.* 31: 2745–2751.

9 El-Tayeb, N.S.M., Yousif, B.F., and Yap, T.C. (2008). An investigation on worn surfaces of chopped glass fibre reinforced polyester through SEM observations. *Tribol. Int.* 41: 331–340.

10 Singh Gill, N. and Yousif, B.F. (2009). Wear and frictional performance of betelnut fibre-reinforced polyester composite. *Proc. Inst. Mech. Eng., Part J: J. Eng. Tribol.* 223: 183–194.

11 Narish, S., Yousif, B.F., and Rilling, D. (2011). Adhesive wear of thermoplastic composite based on kenaf fibres. *Proc. Inst. Mech. Eng. Part J: J. Eng. Tribol.* 225: 101–109.

12 Chand, N. and Dwivedi, U.K. (2006). Effect of coupling agent on abrasive wear behaviour of chopped jute fibre-reinforced polypropylene composites. *Wear* 261: 1057–1063.

13 Mohanty, A.K., Drzal, L.T., and Misra, M. (2002). Engineered natural fiber reinforced polypropylene composites: influence of surface modifications and novel powder impregnation processing. *J. Adhes. Sci. Technol.* 16: 999–1015.

14 Hashmi, S.A.R., Neogi, S., Pandey, A., and Chand, N. (2001). Sliding wear of PP/UHMWPE blends: effect of blend composition. *Wear* 247: 9–14.

15 Martínez, F.J., Canales, M., Bielsa, J.M., and Jiménez, M.A. (2010). Relationship between wear rate and mechanical fatigue in sliding TPU–metal contacts. *Wear* 268: 388–398.

16 Dorri Moghadam, A., Schultz, B.F., Ferguson, J.B. et al. (2014). Functional metal matrix composites: self-lubricating, self-healing, and nanocomposites—an outlook. *JOM* 66: 872–881.

17 Rajasekhar, P., Ganesan, G., and Senthilkumar, C. (2014). Studies on tribological behavior of polyamide filled jute fiber-nano-ZnO hybrid composites. *Proc. Eng.* 97: 2099–2109.

18 Chang, Q., Wang, K., and Zhao, H. (2013). The friction and wear properties of short carbon/CNT/PA6 hybrid composites under dry sliding conditions. *Int. J. Polym. Mater. Polym. Biomater.* 62: 540–543.

19 Nishitani, Y., Sugawara, N., Kawasaki, K., and Kajiyama, T. (2019). Tribological properties of ureidosilane treated natural fiber reinforced plant-derived polyamide 1010 biomass composites. In: *AIP Conf. Proc.*, vol. 2139, 120001. AIP Publishing LLC.

20 Wang, Q., Zhang, X., and Pei, X. (2010). A synergistic effect of graphite and nano-CuO on the tribological behavior of polyimide composites. *J. Macromol. Sci. Part B Phys.* 50: 213–224.

8

Tribological Characterization of the Natural Fiber-Reinforced Polyimide Composites

Anand Gobiraman[1], Santhosh Nagaraja[2], and Vishvanathperumal Sathiyamoorthi[3]

[1] *Achariya College of Engineering Technology, Department of Mechanical Engineering, Villiyanur, Pondicherry 560067, India*
[2] *MVJ College of Engineering, Department of Mechanical Engineering, Near ITPB Whitefield, Bengaluru 560067, Karnataka, India*
[3] *S.A. Engineering College, Department of Mechanical Engineering, Thiruverkadu 600077, Tamil Nadu, India*

8.1 Introduction: Background and Driving Forces

The constant examination of green composites for designing components for tribology-based applications has resulted in upgrades in the exploration of tribological conduct of natural fiber-built polymeric composites. Data on regular filaments, composites, and the wear attributes, for example, explicit wear rate and grinding, were effectively studied with regard to change of fiber weight or volume, expansion of nanofillers and fiber composition. Multiple working boundaries and wear components for different sorts of composites and their exhibition will be reviewed in this chapter.

Researchers across the globe have been involving natural filaments for tribological applications since days of yore, especially the composites supported with natural fiber strands, which acquired importance for their operational applications in tribological conditions. Composites built up with natural fiber strands have by and large yielded high strength composites for light weight and rough applications and have great mechanical properties and are likewise harmless to the ecosystem [1]. These composites possess predominant properties, and the properties of these composites rely essentially upon their constituents [2]. A few researchers are concentrating on natural filaments like sisal, jute and kenaf. Natural fiber composites have phenomenal mechanical properties and are improved when treated with epoxy polymers [3, 4]. They are likewise ecologically easy to be disposed and have excellent biodegradable properties [5]. Because of numerous unrivaled properties like high strength, low thickness with great tribological characteristics, eco-manageability and recyclability, and great mechanical attributes, natural fiber can be utilized in place of manufactured strands for fabrication of composites [6]. Different analysts

Tribological Properties, Performance, and Applications of Biocomposites, First Edition.
Edited by Chandrasekar Muthukumar, Senthilkumar Krishnasamy,
Senthil Muthu Kumar Thiagamani, and Ganesan Chinnachamy.

have involved natural strands in the manufacture of biodegradable composites and distinguished their unrivaled properties [7]. A few scientists have also finished their research on natural fiber-supported composites and examined their tribological, thermal, and mechanical properties [8, 9]. Natural fiber-supported polyamide composites have become popular in high-performance goods that must be lightweight while still being able to withstand dynamic stacking [10]. Further, the customary composites are not viable in the climate and easily degrade in humid conditions [11]. Notwithstanding, perhaps the greatest constraint to the utilization of this enormous group of composites is that there are two distinct parts present simultaneously [12]. This makes reuse and recyclability very troublesome toward the end of composites life cycles [13]. Therefore, it is frequently desirable to fabricate them straightforwardly so that when the life of such composites comes to an end, they can be disposed of in a landfill rather than burned [14]. Likewise, the manufacture of plastic calls for stupendous admission of nonrenewable energy source- based assets, which are not easily biodegradable [15]. Polymer composites provide a fantastic opportunity for an assortment of uses, but their impact on environmental factors restricts their level of usage [16]. Notwithstanding, with the expansion in the utilization of composites for tribological applications, the examination exercises on the utilization of regular filaments for epoxy composites [17]. Indeed, natural fiber-reinforced polyamide composites are widely employed in tribological applications such as brake liners, bearing cages, and other pilot drives that have been developed to build a new handling philosophy [18]. It is also critical to present a completely new process for the synthesis of polyamide natural fiber composites for a wide range of applications, as well as for life cycle assessment. Figure 8.1 illustrates some of the tribological applications of natural fiber reinforced polyamide composites.

In this aspect, natural fiber-reinforced polyamide composites have a lot of potential for usage in vehicle brake linings and friction clutches, vessels, submersibles, tribological structures, and other wear and friction components that face a lot of environmental problems. Therefore, natural fibers are generally considered to have higher durability and require little or no maintenance over long periods of time.

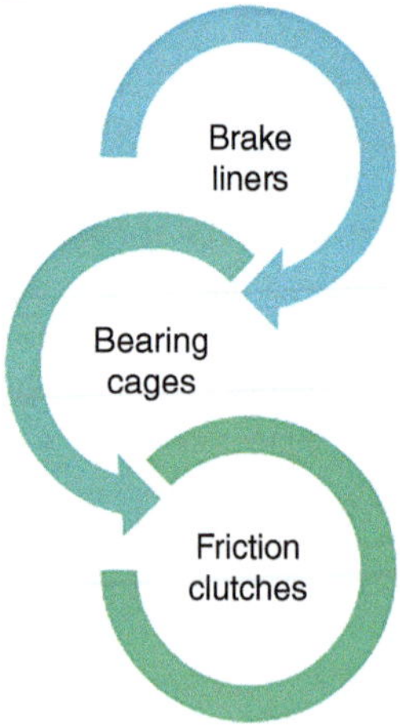

Figure 8.1 Natural fiber-reinforced polyamide composites have a variety of tribological uses.

Figure 8.2 Energy consuming sectors having major tribological concerns.

Additionally, light weight and wear resistance are a must for brakes and clutches to meet design requirements and operate at impeccable speed and reliability. The composites, with special reference to natural fiber-reinforced polymers, have found extensive use in the tribological industry, especially to overcome problems of friction and wear in steel, aluminum, and wood. Weight reduction is vital, especially for commercial vehicles, which have a higher weight. Natural fiber- reinforced composites have changed throughout time to fulfill the changing and expanding demands of the tribological industry, while still being useful for the production of consolidated elements. Composite materials are being used in all aspects of the tribological industry and in a wide range of components and structures, including brake liners, bearing components, friction clutches, and so on. In this regard, there is an urgent need to study the science and engineering of wear and friction.

Tribology is the study of how surfaces interact when they move relative to one another. It includes a review and application of erosion, friction, wear, and lubrication standards. Physical science, physics, materials science, and design are only a few of the scholastic subjects that are drawn upon in this field [19]. Following the publication of "The Jost Report" in 1966, the term tribology became widely accepted [20]. The cost of erosion, wear, and consumption to the UK economy was highlighted in this paper. Despite substantial research since then, the global impact of rubbing and wear on energy use and CO_2 emissions remains enormous. Researchers have attempted to quantify their impact on financial perspectives all over the world in this regard. They looked at the three major energy consumers: electricity, production, and transportation, and employed a variety of materials to reduce the negative impacts of wear and friction while causing minimal environmental damage [21] (Figure 8.2).

However, it is far more important to explore a totally new technique for each product and their manufacturability, including cutting-edge technologies for environmental evaluation, such as characterization of the mechanical, tribological properties, and lifecycle evaluation [22].

Thus, in this chapter, a new possibility of reinforcing the natural fiber in polyimide resin and synthesis of sustainable composites with better tribological characteristics are explored.

8.2 Materials and Methods

Natural fiber- reinforced composites can be made from a wide range of materials and processes. The use of natural fibers in composites has been researched by a number of researchers. According to De Paola et al. [23], a number of polymers are utilized mostly in composite structures (polyester, polyimide, epoxy, vinyl ester, etc.). Also, resins, such as polyethylene terephthalate (PET), polybutylene terephthalate (PBT), polypropylene, and polyamide, have lately been employed in the synthesis of composites. Composites for tribological applications have recently expanded to include more bio-based resins, which opens up a lot of possibilities for future research [24]. Bio-based resins have a wide range of uses, especially in terms of environmental sustainability and compatibility with the environment. The necessity of the hour is to reduce dangerous and toxic pollutants, with a focus on human health and the environment [25].

Environmental concerns are growing, and the demand for extra flexible polymer-based natural fiber-reinforced composites has an answer. As an end result of this reduction in environmental degradation, newer compositions of composites are being achieved as new options, which can reduce environmental pollution with substitutes that have a decreased environmental effect and therefore are frequently called "Eco composites" or "ecological compounds." These compounds may be considered if the polymer matrix is biodegradable (for example, if it is made entirely of herbal natural fillers), as well as if it is made from renewable resources. Fabrication and characterization of polymeric compounds based entirely on recyclable polymers (e.g. polyolefins) containing natural herbal fillers (e.g. plant-derived fibers and particles) is an exciting area of study and development. Simultaneously, fibers have been the focus of the majority of recent research. Instead of synthetic fibers such as glass, carbon, and aramid, which are already widely used in the automotive, aerospace, and construction industries, natural fibers such as flax, hemp, and cotton have been considered for more sustainable components [3]. While their overall performance and degree of sustainability are usually good research topics, especially to obtain better mechanical and tribological characteristics [4]. In this regard, the natural fibers and matrix materials utilized in composite synthesis are discussed in detail in this section.

8.2.1 Natural Fibers

Natural fibers are increasingly being used in composites due to their environmental sustainability. The natural fibers of hemicellulose, lignin, and cellulose composition can be sourced from seeds, leaves, basts, and fruits. The maximum advantage, the herbal fibers offer are associated with environmental aspects: they are biodegradable and carbon positive [22]. They are also nonabrasive, which means less wear and tear on industrial machinery. Increased ductility and rigidity, as well as improved tribological qualities, are some of the benefits of natural fiber reinforced composites. Furthermore, natural fibers can lower the likelihood of fiber debonding and delamination, while also increasing flexural strength [23]. Furthermore, due to the strong coherent connection between the fibers and the matrix, natural fibers can improve the properties of polymer composites [26]. Natural fibers have also been proven to improve the mechanical and tribological properties of composites when compared to glass fiber-reinforced composites [27].

Fiber from nature natural fibers of various sorts are used to make reinforced polymer composites. Most commonly used natural fibers among them are schematically illustrated in Figure 8.3.

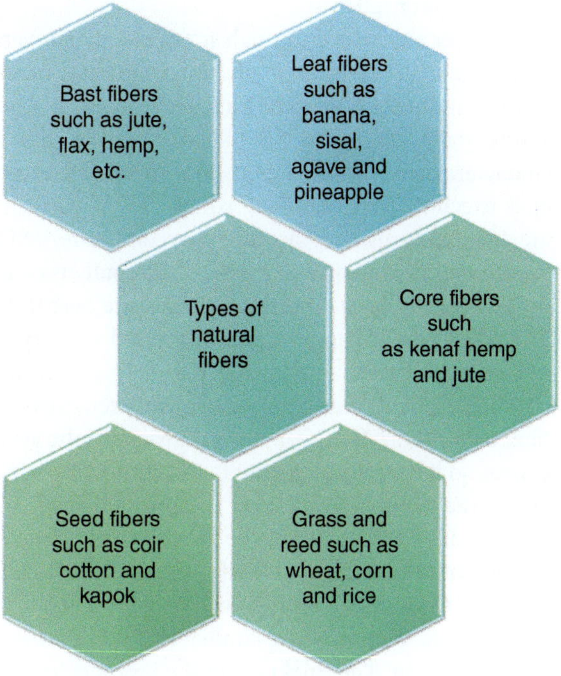

Figure 8.3 Some of the natural fibers that are used to make natural fiber-reinforced composites.

8.2.2 Methods of Production of Natural Fiber-Reinforced Polymer Composites

There are a variety of processes for making natural fiber-reinforced composites, and a few of them are covered in this chapter. Figure 8.4 depicts the various synthesis procedures used in the manufacturing of natural fiber reinforced polymer composites.

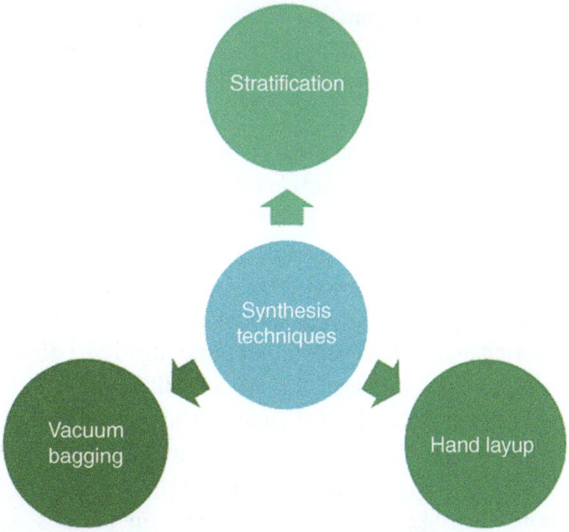

Figure 8.4 Some of the synthesis techniques for natural fiber-reinforced polymer composites.

8.2.2.1 Stratification

Composites are created through stratification of several layers, additionally referred to as lamina. The virgin- grade ply composites include fibers oriented in a single course, multiple courses, or multiple directions. The thickness of the layer varies substantially relying on the thickness of the fiber and resin layers, the desired characteristics, and the application. In several composite studies, along with the ones executed by Ghelli and Minak [13], natural fibers of 0.1–0.5 mm thickness are frequently used to acquire the composite lamina. Praveena et al. [28] and Ravichandran et al. [29–31] have additionally executed their studies on the thickness of the composites fabricated using the herbal fibers for low-cost applications. The properties of composites are assessed for various natural fiber compositions. According to Ravichandran et al. [29–31] and Santulli [32], when natural fibers are unidirectional, the composite has great strength and stiffness within the course of fiber orientation. However, maximum composite systems are not loaded in a single course, so it is very much vital for orienting fibers in a couple of directions. This is due to the fact that every layer is fabricated from unidirectional material, and the orientation of the fibers may be modified from one layer to another or through the use of a multidirectional material. Thus, a couple of layers are stacked collectively within the stratification approach to gain better tribological characteristics. This type of stack is known as a lamina, and the technique of stacking layers is known as stratification. The thickness of the layers determines the spectrum of laminates available, ensuring adequate tribological qualities while keeping the weight low. Generally, the composites are made of up to 42 layers of lamina and used for tribological applications [33]. Further, laminated composites can commonly be made using different lengths of natural fibers. However, the length of fibers is important to characterize the composites for tribological applications. The proper fiber length is extremely important when fabricating composites using the stratification method to achieve a high-strength low-weight composite.

8.2.2.2 Hand Layup

The technique is a labor-intensive, low-quantity technique that makes it especially appropriate for large components. In a manually opened chamber, a natural fiber reinforcing mat, fabric, or roving is placed, and resin is poured onto the fiber, which is then brushed or sprayed. To finalize the laminate construction, the trapped air is manually removed using a medical doctor blade or curler. The most commonly used matrix resins are room-temperature-cured polyester, polyimide, and epoxy. Curing is started by a catalyst built within the resin device, which efficiently treats the fiber-bolstered resin composite without the use of external heat. A colored gel coat is first applied on the die surface to make the composites' surfaces smoother. This coating makes it viable without difficulty to cost-effectively manufacture composites of any size and shape. It gives a realistic solution to generational limitations. To gain the artifact with a "stable laminate," a fiber layered to the ideal geometry, or a "sandwich," such as an outer composite layer, are appropriately employed. In addition to their important characteristics of wear loads, the internal and outer pores and skin have extra functions in phrases of aesthetics of the composite laminate. Ravichandran et al. [29–31] have reviewed the hand layup strategies for synthesizing natural fiber-bolstered composites, especially for tribological applications. The hand layup approach is the most viable method among all of the different strategies to manufacture composites of basic shape and size with relatively better characteristics, but it is restrained with respect to intricate profiles.

8.2.2.3 Vacuum Bagging Technique

Vacuum bag forming is a method employing vacuum pump and autoclave for curing the laminate at some stage in the curing cycle. Pressurization of composite laminates serves numerous functions. First, the technique eliminates the air trapped among the layers. Second, it densifies the fiber layer to allow for a green pressure changeover between fiber bundles and to prevent the fibers' orientation from shifting during the curing process. Third, it reduces air humidity. Finally, vacuum bagging composite items improves the fiber-to-resin ratio. These advantages help composites achieve their best tribological performance. It is worth noting that production methods other than vacuum packing and vacuum injection are favored when synthesizing composites. A comprehensive review of the techniques and methods employed for vacuum bag molding is stated in [15].

8.3 Polyimides

Polyimides are one of the important grades of polymers that have elevated thermal resistance and are broadly used within the manufacture of excessive-temperature plastics, dielectrics, adhesives, dielectrics, etc. Currently, polyimide resins are used as matrices to create strengthened composites primarily based on light-weight carbon fibers as an alternative for steel elements within the aerospace industry and airframe because of their excellent thermal, tribological, and mechanical resistance. Polyimide resins are broadly utilized in microelectronics, aerospace, automotive, and the fuel cells. They are used in the electrical industry for the fabrication of electrical insulators that have excessive thermal resistance, appropriate dielectric properties, and elasticity. Depending on their chemical composition and structure, polyimides (PI) can be aliphatic or alicyclic. Polyimides can be linear or three-dimensional, depending on the chemical chain structure [34]. There are polyimides with aliphatic hyperlinks within the primary chain of the macromolecule. Polypyromellitimides primarily based totally on aliphatic diamines containing much less than seven carbon atoms within the molecule have excessive melting factors, which can be better than the decomposition temperatures (above 360 °C). Polypyromellitimides primarily based on aliphatic diamines contain seven carbon atoms in the hydrocarbon chain, in addition to polyimides of diverse aliphatic diamines that melt at temperatures of 310 °C.

8.4 Natural Fibers/Polyimides Composites

For the manufacture of natural fibers/Polyimide composites, compatibility between the fiber and matrix phases is critical. The drawbacks related to lignocellulosic natural fibers should be appropriately solved before using them in polymer composites. The largest challenge is their hydrophilic nature due to strongly polarized hydroxyl groups, which affords their swelling and, ultimately, rotting degradation by means of microbiological marketers, which include fungi. This hydrophilicity confers incompatibility with hydrophobic thermoplastic polymers, which include polyolefins, inflicting bad miscibility [35]. A viable answer to enhance interplay among the polymer–fiber combination is by using coupling and adhesion promoters [36]. In the case of polyamides, the compatibility is not always a sizable hassle due to their low hydrophobicity. The most difficult task in these composites is to use polyamide 6.6 as the polymer matrix in composites reinforced with natural fibers while maintaining the cellulose filler's thermal balance. The processing temperature in this situation is around 280–300 °C, so the physical–chemical remedies accomplished on the fiber intention are majorly focused at maximizing their thermal balance. In fact, untreated cellulosic fibers no longer support such temperatures and are susceptible to weight loss, resulting in a worsening of the composites' mechanical and tribological properties. In one of the few research projects defined within the literature [37], it becomes viable to combine Poly Amide 6.6 resin with unique kinds of cellulosic fibers, together with timber fiber and jute. The thermal balance and tribological characteristics of the fibers after remedy have been analyzed by means of thermogravimetric studies, indicating that once appropriate treatments are carried out, the fibers have little or no weight reduction even at temperatures as high as 310 °C. Thus, making it as a potential candidate for synthesis of the composites.

8.5 Tribological Applications of Natural Fibers/Polyimides Composites

There are many components within the automotive domain where hybrid bushings have to possess the essential tribological characteristics for their operation for an extended duration of time. Windshield wiper motors, suspension systems, shafts and joints, door hinges, and other components are included. A low-friction matrix that combines advantageous tribological qualities with decreased friction and wear loss between the metals is created by inserting a plastic fabric directly into a porous bronze sinter layer attached to a metallic back (hybrid structure). The use of powerful lubricants within the plastic matrix creates a lubrication film between the sliding surfaces, allowing such bearings to be used in packages without the need for oil or grease. It also contributes to low-noise operation at consistent speeds and improved coefficients of friction throughout the component's wear and tear [38]. In the last few years, the necessity of improving the overall performance of such bushings has been accelerated in the direction of better temperature packages and better operational conditions. Such bushings' unique technology is related to their employment as bearings in high-pressure diesel gas injection pumps. One option to achieve these unusually high demands is to manufacture and use a polymer matrix combined with unique micro- and nano-scaled fillers, which improves its tribological properties [39]. Furthermore, the composites' compounding and extrusion into flat tapes must be investigated on a laboratory scale. The necessity for tape fabric to be included in sinter bronze metallic substrates, as well as the utilization of hot press procedures when inserting these polymer composites in between the metals, have reduced wear and friction rates thus far.

The tribological industry has made clear technological developments, and today the manufacturing of brake liners, clutches, and bearing bushes is based on the latest generation production systems involving natural fiber composites. Natural fiber composites have constituted a watershed moment in the tribological sector. Composite technology has allowed manufacturers to improve the quality of their products by maintaining a rigid and lightweight structure that has advantages in terms of performance and durability. With the ability to automate and accelerate the manufacturing process, the introduction of composites has reduced overall costs and made the final product affordable to more people. The natural fiber-reinforced composite is primarily used in the development of adhesive films that reduce surface wear and friction. The composites reduce the wear on the outer surface of the component and have a soft and thick core, which avoids the rubbing action between the mating components [40]. Natural fibers have a hard skin that provides excellent wear resistance, while the core sustains shear and compressive stresses, stabilizes the matrix, and avoids global and local instability [41]. The weight reduction and increased strength improve the load-bearing capacity of the tribo-structures, improve fuel economy, reduce inertia, and improve stability and buoyancy [42]. Natural fibers have a hard skin that provides excellent wear resistance, while the core sustains shear and compressive stresses, stabilizes the matrix, and avoids global and local instability [41]. Tribological sandwich structures typically use natural fibers and filler materials as the covering layer instead of metal, thereby reducing the friction between the surfaces. Polyimide resins and honeycombs are the main core materials used. With the use of aramid natural fibers to support the high-quality resin and ultra-lightweight core, a bearing bush with unrivaled tribo-mechanical qualities and dimensions can be created. The composite sandwich construction, on the other hand, has greater costs than standard materials, and the production process is more labor-intensive, which limits its broad application. The advantages of employing natural fiber-reinforced polyimide composites for tribological applications are shown in Figure 8.5.

The material composition and filler materials, application requirements, counter materials, and wear mechanisms are all investigated in the research on the tribological performance of natural fiber-reinforced polyimide composites.

The tribological performance of polymeric matrix composites and their blends is influenced by the composition of the reinforcements, filler weight percent, compatibilizing agents for interfacial lubrication, and other additives.

The load, operational regime, contact circumstances, temperature conditions, dependability, and durability are all important considerations when choosing such composites for real-time operations.

The hardness, surface texture, nature, and shape of the composite specimens are deciding elements in appraising the tribological performance of the composite materials, as are the counter materials and wear mechanisms. Also, the wear mechanisms, viz., plastic deformation, adhesion, abrasion, fretting, and tribochemical

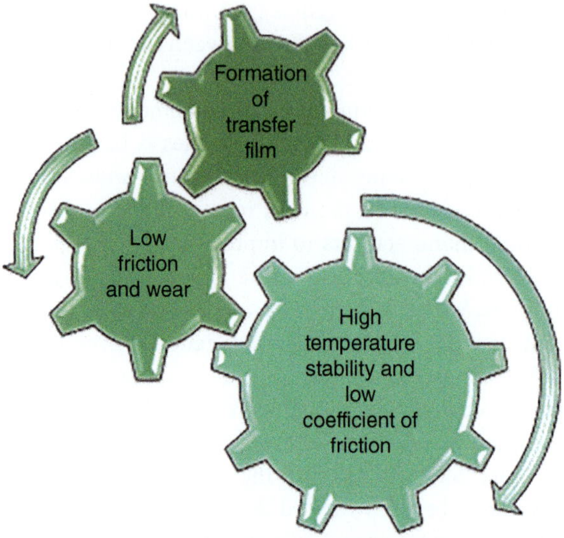

Figure 8.5 Advantages of using natural fiber-reinforced polyimide composites for tribological applications.

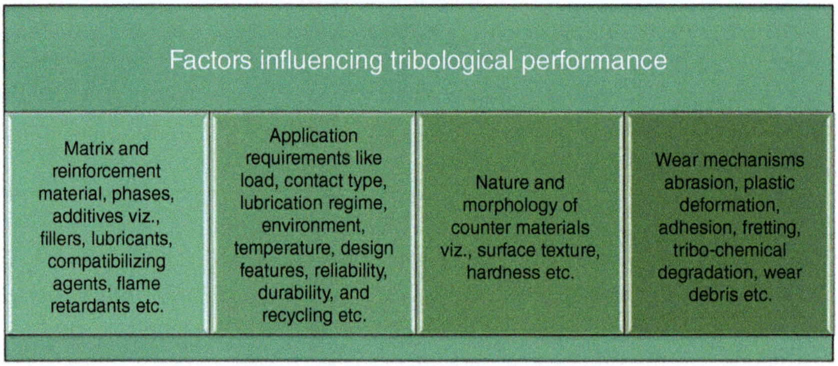

Figure 8.6 Factors influencing tribological performance of natural fiber-reinforced polyimide composites.

degradations, apart from the nature of wear debris, are deciding parameters for evaluating the performance attributes of the composites [43].

The elements influencing the tribological behavior of natural fiber-reinforced polyimide composites are graphically depicted in Figure 8.6.

8.6 Conclusions

For better mechanical and tribological performance, there is a growing demand for improved tribological features of composite constructions in today's environment. The use of natural fiber composites to improve surface characteristics is one prospective development path. The introduction of natural fibers in the polyimide can induce a further improvement of the tribological properties. Among the various reinforcements for synthesis of the polyimide composites, natural fibers are finding prime importance. The earlier studies on the tribological characteristics of natural fiber-reinforced composites were reviewed in this chapter. The tribological performance of composites is influenced by the matrix and reinforcement composition, synthesis techniques and chemical composition of natural fiber-reinforced composites, additives and filler materials, lubrication regimes and operating conditions, and wear mechanisms, according to research. Further, there are several studies, which have described the application of natural fibers for reinforced polyimide composites for multi-functional operations, especially tribological operations. The detailed studies have brought out the importance of natural fibers in the synthesis of composites, especially for the tribological components, which eventually lead to reduced wear and friction and better lubrication between the contact surfaces attributed to the formation of transfer films, reducing the wear and tear in case of bearings, brake liners, friction clutches, etc. The elevated thermal and tribo-chemical protection offered by the polyimides and the lubrication characteristics of the natural fibers have together offered the prospectus for the synthesis of composites to be effectively used for tribological applications.

References

1 Bakkal, M. and Savas, M. (2013). Development of natural fiber reinforced laminated hybrid composites. *Adv. Mater. Res.* 628: 15–20.

2 Barberio, G., Buttol, P., Righi, S., and Andretta, M. (2010). Combined approach of risk assessment and life cycle assessment for the environmental evaluations: an overview. In: *Proceedings of the Ecomondo Congress*, Rimini (3–6 November) (ed. Maggioli), 1262–1268.

3 Camargo, F.V., Guilherme, C.E.M., Fragassa, C., and Pavlovic, A. (2016). Cyclic stress analysis of polyester, aramid, polyethylene and liquid crystal polymer yarns. *Acta Polytech.* 56 (5): 402–408. https://doi.org/10.14311/AP.2016.56.0402.

4 Sahari, J. and Sapuan, S.M. (2011). Natural fiber reinforced biodegradable polymer composites. *Rev. Adv. Mater. Sci.* 30: 166–174.

5 Hyseni, A., De Paola, S., Minak, G., and Fragassa, C. (2013). Mechanical characterization of ecocomposites. In: *Proceedings of the 30th Danubia Adria Symposium on Advanced Mechanics*, Primosten, Croatia (25–28 September 2013) (ed. Croatian Society of Mechanics), 175–176.

6 Fragassa, C. (2016). Effect of natural fibers and bio-resins on mechanical properties in hybrid and non-hybrid composites. In: *Proceedings of the 8th conference on times of polymers & composites: from aerospace to nanotechnology*, (19–23 June 2016). Ischia, Italy: American Institute of Physics (AIP) AIP Conference Proceedings, vol 1736, No. 4949693. doi:https://doi.org/10.1063/1. 4949693.

7 Fragassa, C., Pavlovic, A., and Santulli, C. (2017). Mechanical and impact characterisation of flax and basalt fiber vinylester composites and their hybrids. *Composites, Part B* https://doi.org/10.1016/j.compositesb.2017.01.004.

8 Hardaker, K.M. and Richardson, M.O.W. (1980). Trends in hybrid composite technology. *Polym. Plast. Technol. Eng.* 15 (2): 169–182.
9 Friedrich, K., Lu, Z., and Häger, A.M. (1995). Recent advances in polymer composites' tribology. *Wear* 190: 139–144.
10 Khalil, H.A., Bhat, A.H., and Yusra, A.I. (2012). Green composites from sustainable cellulose nanofibrils: a review. *Carbohydr. Polym.* 87 (2): 963–979.
11 Mishra, S. and Mohanty, A.K. (2003). Studies on mechanical performance of biofiber/glass reinforced polyester hybrid composites. *Compos. Sci. Technol.* 63 (10): 1377–1385.
12 Hall, B.J., Schaffer, G.B., Ning et al. (2003). Al/AlN layered composites by direct nitridation of aluminium. *J. Mater. Sci. Lett.* 22: 1627–1630.
13 Ghelli, D. and Minak, G. (2011). Low velocity impact and compression after impact tests on thin carbon/epoxy laminates. *Composites, Part B* 42 (7): 2067–2079.
14 Duval, A., Bourmaud, A., Augier, L., and Baley, C. (2011). Influence of the sampling area of the stem on the mechanical properties of hemp fibers. *Mater Lett* 65: 797–800.
15 LC, B. (2006). *Composites for Construction: Structural Design With FRP Materials.* London: Wiley.
16 Schwartz, C.J. and Bahadur, S. (2000). Studies on the tribological behavior and transfer film-counterface bond strength for polyphenylenesulfide filled with nanoscale alumina particles. *Wear* 237: 261–273.
17 Boria, S., Pavlovic, A., Fragassa, C., and Santulli, C. (2016). Modeling of falling weight impact behavior of hybrid basalt/flax vinylester composites. *Proc. Eng.* 167: 223–230. https://doi.org/10.1016/j. proeng.2016.11.691.
18 Abdelbary, A. (2014). *Wear of Polymers and Composites*, 256. Cambridge: Woodhead Publishing https://doi.org/10.1016/C2014-0-03367-9.
19 Fox, M. (2016). *Polymer Tribology*, vol. 135, 32–37. Lube Magazine.
20 Jost, H.P. (1966). *Lubrication (Tribology)—A Report on the Present Position and Industry's Needs, Department of Education and Science.* London, UK: H.M. Stationary Office.
21 Holmberg, K. and Erdemir, A. (2017). Influence of tribology on global energy consumption, costs and emissions. *Friction* 5: 263–284.
22 Bogoeva-Gaceva, G., Avella, M., and Errico, M. (2007). Natural fiber eco-composites. *Polym. Compos.* 28: 98–107.
23 De Paola, S., Minak, G., Fragassa, C., and Pavlovic, A. (2013). Green composites: a review of state of art. In: *Proceedings of the 30th Danubia Adria Symposium on Advanced mechanics*, Primosten, Croatia (25–28 September 2013) (ed. Croatian Society of Mechanics), 77–78.
24 Pavlovic, A., De Paola, S., Fragassa, C., and Minak, G. (2013). Mechanical characterization of green composites: earliest experimental evidences by flexural tests. In: *Proceedings of the ICCE 21 Composite/Nano Engineering Conference.* Tenerife, Spain.
25 Zivkovic, I., Pavlovic, A., Fragassa, C., and Brugo, T. (2017). Influence of moisture absorption on the impact properties of flax, basalt and hybrid flax/basalt

fiber reinforced green composites. *Composites, Part B* 111: 148–164. https://doi.org/10.1016/j.compositesb.2016.12.018.

26 Zivkovic, I., Pavlovic, A., and Fragassa, C. (2016). Improvements in wood thermoplastic composite materials properties by physical and chemical treatments. *Int. J. Q. Res.* 10 (1): 205–218.

27 Lopresto, V., Leone, C., and De Iorio, I. (2011). Mechanical characterization of basalt fiber reinforced plastic. *Composites, Part B* 42: 717–723.

28 Praveena, B.A., Buradi, A., Santhosh, N. et al. (2021). Study on characterization of mechanical, thermal properties, machinability and biodegenerability of natural fiber reinforced polymer composites and its applications, recent developments and future potentials: a comprehensive review. *Mater. Today Proc.* 52 (Part 3): 1255–1259. https://doi.org/10.1016/j.matpr.2021.11.049.

29 Ravichandran, G., Rathnakar, G., Santhosh, N. et al. (2019). Enhancement of mechanical properties of epoxy/halloysite nanotube (HNT) nanocomposites. *SN Appl. Sci.* 1: 296. https://doi.org/10.1007/s42452-019-0323-9.

30 Ravichandran, G., Rathnakar, G., and Santhosh, N. (2019). Effect of heat treated HNT on physico-mechanical properties of epoxy nanocomposites. *Compos. Commun.* 13 (2): 42–46.

31 Ravichandran, G., Rathnakar, G., Santhosh, N., and Thejaraju, R. (2019). Antiwear performance evaluation of halloysite nanotube (HNT) filled polymer nanocomposites. *Int. J. Eng. Adv. Technol.* 9 (1): 3314–3321.

32 Santulli, C. (2007). Impact properties of glass/plant fiber hybrid laminates. *J. Mater. Sci.* 42 (11): 3699–3707.

33 Poodts, E., Minak, G., Mazzocchetti, L., and Giorgini, L. (2014). Fabrication, process simulation and testing of a thick CFRP component using the RTM process. *Composites, Part B* 56: 673–680.

34 Yegorov, A., Bogdanovskaya, M., Ivanov, V. et al. (2018). Nanocomposite polyimide materials. In: *Characterizations of Some Composite Materials* (ed. H.E.M. Saleh and M. Koller). IntechOpen https://doi.org/10.5772/intechopen.79889.

35 Veazie, D.R., Robinson, K.R., and Shivakumar, K. (2004). Effects of the tribological environment on the interfacial fracture toughness of PVC core sandwich composites. *Composites, Part B* 35 (461): 466.

36 Sridhar, M.K., Basavarajappa, G., Kasturi, S.G., and Balasubramanian, N. (1982). Evaluation of jute as a reinforcement in composites. *Indian J. Fiber Text Res.* 7: 87–92. T800H Technical Data Sheet CFA-007, Carbon Fibers America Inc Twaron Technical Data Sheet, Teijin Ltd.

37 Werner, P., Altstädt, V., Jaskulka, R. et al. (2004). Tribological behavior of carbon-nanofiber-reinforced poly(ether ether ketone). *Wear* 257: 1006–1014.

38 Aldousiri, B., Shalwan, A., and Chin, C.W. (2013). A review on tribological behavior of polymeric composites and future reinforcements. *Adv. Mater. Sci. Eng.* 2013: 8.

39 Deleanu, L., Botan, M., and Georgescu, C. (2020). Tribological behavior of polymers and polymer composites. In: *Tribology in Materials and Manufacturing—Wear, Friction and Lubrication* (ed. A. Patnaik, T. Singh, and V. Kukshal). IntechOpen https://doi.org/10.5772/intechopen.94264.

40 Friedrich, K. (1997). Wear performance of high temperature polymers and their composites. In: *Application of High Temperature Polymers* (ed. R.R. Luise), 221–246. Boca Raton, USA: CRC Press.

41 Bismarck, A., Baltazar-Y-Jimenez, A., and Sarikakis, K. (2006). Green composites as panacea Socio-economic aspects of green materials. *Environ. Dev. Sustain.* 8 (3): 445–463.

42 Savoia, M., Stefanovic, M., and Fragassa, C. (2016). Merging technical competences and human resources with the aim at contributing to transform the Adriatic area in a stable hub for a sustainable technological development. *Int. J. Q. Res.* 10 (1): 1–16.

43 Ravichandran, G., Rathnakar, G., Santhosh, N., and Suresh, R. (2020). Wear characterization of HNT filled glass-epoxy composites using Taguchi's design of experiments and study of wear morphology. *Compos. Theory Pract.* 20 (2): 85–91.

9

Investigations of the Friction and Wear Resistance of the Natural Fiber-Reinforced Polyamide Composites

İbrahim Can Kaymaz[1], Alperen Doğru[2,3], Miray Batıkan Kandemir[4], and Mehmet Özgür Seydibeyoğlu[4,5]

[1] *ThermoPlastic composites Application Center, Saxion University of Applied Sciences, 7513 AB Enschede, the Netherlands*
[2] *Ege University, Aviation Higher Vocational School, 35414, Izmir, Turkey*
[3] *Muğla Sıtkı Koçman University, Department of Metallurgical and Materials Engineering, 48000 Muğla, Türkiye*
[4] *İzmir Katip Çelebi University, Department of Metallurgical and Materials Engineering, 35620 İzmir, Türkiye*
[5] *University of Maine, Advanced Structures and Composites Center, Orono, ME, USA*

9.1 Introduction

The fact is that most technological developments are achieved in parallel with developments in materials science. Regardless of the industry field, the usage limits of the products depend on the materials rather than the design geometry. Polymeric materials, which have been the subject of extensive research in recent years, find use in many new fields and play a role in determining product performances [1]. Polymer materials have a wide range of applications, from simple parts in our daily lives to complex industrial products [2]. So that it is difficult to imagine today's world without polymers. Polymer materials have unique properties not found in conventional materials, such as lightness, flexibility, corrosion resistance, transparency, and easy processing [3].

Polymer materials were one of the materials first introduced by JJ Berzelius in 1833 [4]. The structures of polymers consist of long molecules and macromolecules in the form of chains [2]. Polymers are familiar plastic and rubber materials. They are mostly organic compounds based on hydrogen, carbon, and other non-metallic elements [5]. Typically, polymers contain various additives. Additives are explained in the following categories: antistatic agents, binding agents, fillers, extenders, flame retardants, lubricants, pigments, plasticizers, supplements, and stabilizers [3]. Polymer materials are divided into four main groups depending on the origin of source, structure, molecular forces, and mode of polymerization [2]. Polymers can be categorized for better understanding, as shown in Figure 9.1.

Although polymers are divided into four main groups, one of these "based on molecular forces" is more important in terms of industrial application. This group

Tribological Properties, Performance, and Applications of Biocomposites, First Edition.
Edited by Chandrasekar Muthukumar, Senthilkumar Krishnasamy,
Senthil Muthu Kumar Thiagamani, and Ganesan Chinnachamy.

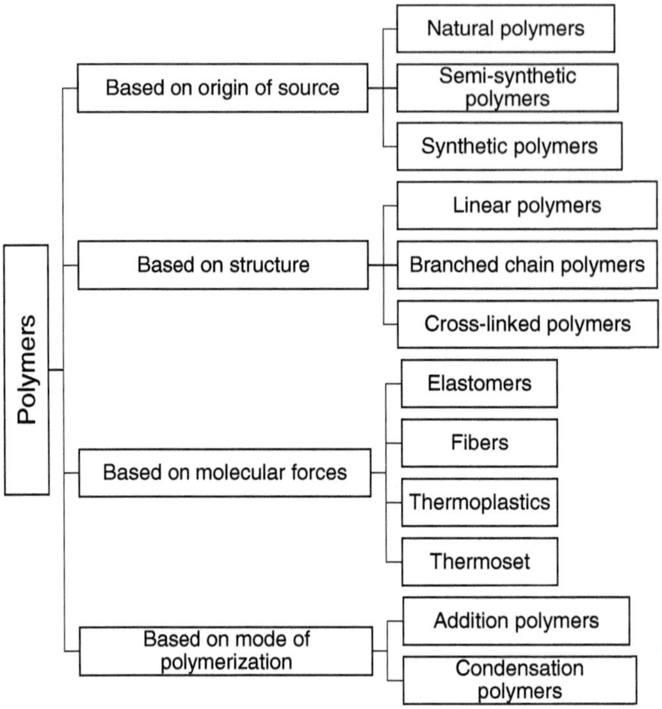

Figure 9.1 Classification of polymers [2, 6].

consists of four different categories: elastomers, fibers, thermoplastics, and thermosetting plastics. Thermoplastic or thermosetting materials can be processed with different methods. Table 9.2 contains some properties of thermoplastic and thermosetting materials and examples of different polymer matrix materials used in industry.

9.1.1 Thermosetting

Thermosetting polymers are formed by chemical reactions in two steps. They are first produced as chains of macromolecules, such as reactive thermoplastics. In the second stage, these macromolecular chains form cross-links with the effect of temperature and pressure [3]. The cross-linking in thermosetting polymers strengthens the molecular bonds and makes the polymer durable. Thermosetting polymers do not soften when heated, due to the cross-linking molecular bonds they have. These polymers do not become fluid as they are reheated, but rather decompose with an increase in temperature [7, 8]. Thermosetting polymers, which are widely used in the polymer industry today, have recycling problems [9].

9.1.2 Thermoplastics

Polymers that soften or melt when heated are called thermoplastic polymers. These polymers are suitable to form flow with temperature increases [10]. The macromolecules that make up the thermoplastic polymers are bonded to each other by weak van der Waals forces. When thermoplastic polymers are heated, the intermolecular forces are greatly reduced, so the material begins to soften. With the increase in temperature, the material becomes flexible and becomes a viscous liquid at high temperatures. It becomes solid again when allowed to cool [3]. This cycle can be repeated many times, which is an advantage for recycling. However, of course, with more than one heating-cooling cycle, the properties of thermoplastic polymers deteriorate [11].

Thermoplastic materials are divided into two groups according to the arrangement of their macromolecules. These are called crystalline and amorphous structures. In the crystalline structure, macromolecules are characterized in an ordered array, while in the amorphous structure, the macromolecules are randomly arranged [12, 13]. Polymers such as polyamide can have a high degree of crystallinity. However, it is not possible to make a perfect crystalline thermoplastic because of the complex structures. So it can be called semi-crystalline or partially crystalline [14]. The crystallization of polymers is largely dependent on the thermal processes during their manufacture and results from the denser aggregation of macromolecules. Crystal structure affects the mechanical properties of polymers. [15]. Compared to amorphous polymers, the superior properties of crystalline polymers are given in Table 9.1.

9.1.3 Thermoplastic Composites

Both thermoplastics and thermosetting polymers can be reinforced with different reinforcing elements, thereby developing new materials with unique properties [16, 17]. The properties and sample polymer types of thermosetting and thermoplastic materials are given in Table 9.2. The structures, called composite materials, are not new inventions. There are many examples of composite structures within the scope of biomimetics. The natural nanocomposite of seashells with a laminated structure is an example of this [18]. Another example is that the coconut leaf consists of cellulose fibers in a lignin matrix. It is a leaf fibrous compound; cellulose fibers are contained in the lignin matrix. Cellulose fibers have high-tensile strength but are very flexible (i.e. low stiffness); the lignin matrix binds the fibers together and provides stiffness [3]. Composites are at many points in our lives, from mud

Table 9.1 Properties of crystalline polymers polyamide.

Properties	Value
Hardness	High
Coefficient of friction	Low
Impact resistance	High
Ability to add reinforcements	High
Creep resistance	High

Source: Guedes [15]/with permission of Elsevier.

Table 9.2 Polymer matrix and properties.

	Thermosetting	Thermoplastic
Properties	High stiffness and strength	Better impact resistance
	Adhesion properties	Higher fracture toughness Moderate stiffness and strength
Examples	Epoxy	Polyamide (PA) Polycarbonates (PC) Polyethylene (PE)
	Polyester	Polyethylene Terephthalate (PET)
	Bismaleimide	Polyether ketone (PEK)
	Vinyl ester	Polyether ketone ketone (PEKK)

and straw used in the construction of adobe houses to the carbon fiber-reinforced epoxy materials used in the construction of today's airplanes [19].

The use of polymers as matrix materials in composite materials consisting of two different components, matrix and reinforcement, and creates polymeric composites [20]. Polymeric composites are greatly affected by the applied force, deformation, temperature, humidity, and time. The main feature of this behavior is the viscoelastic response. The crystal structure of the macromolecules of thermoplastic polymers provides higher impact strength compared to thermosetting [12]. The wide range of fracture stresses in thermoplastics is due to large variations in the amount of crystallinity. The crystal structure of the thermoplastic matrix changes greatly with the addition of reinforcing materials [21]. To ensure the flow of a thermoplastic matrix during manufacture, the matrix must be heated to a temperature above its melting point [22]. Increasing temperature decreases the viscosity of the polymer, but degradation is inevitable at higher temperatures. For this reason, performing the process at the appropriate temperature is critical for thermoplastics.

9.1.4 Thermoplastic Polymer Matrix

Thermoplastic composites are the focus of new-generation research and development studies on sustainable materials, green composites, and circular economy concepts [23]. Thermoplastic matrices can be recyclable to some extent [11]. Heat and pressure are required to produce and shape composites with a thermoplastic matrix. The disadvantages of thermoplastics are that they have very high expansion and contain high viscosity [24].

In Figure 9.2, thermoplastics used industrially are shown with their properties and classifications. Among these polymers, polyamide, which is included in the classification of engineering polymers, has found wide application in many different industries in recent years. Polyamide matrix composites have thermal and chemical stability [22]. In addition, thanks to their mechanical properties, they have been used in different applications in many vehicles, such as marine, aviation, and

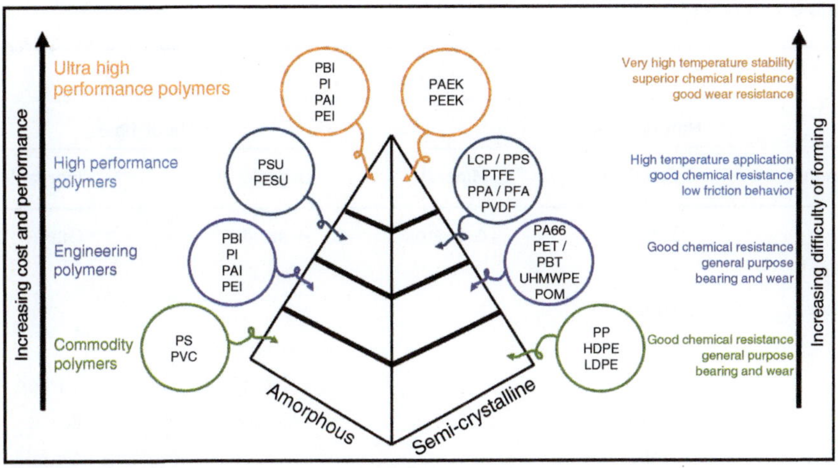

Figure 9.2 Thermoplastic polymers. Source: Adapted from TRI-MACK [25].

automotive. Polyamide is among the most durable engineering polymers thanks to its high fatigue strength and temperature resistance [26].

Although all these properties of thermoplastic polymers have their limits, the application limits are only up to the imagination of the designer. In addition, reinforcements to improve strength properties can be realized as fiber, particle, and structural unity [27].

9.1.5 Fibers

In composite structures where recyclable thermoplastic materials are used as matrix materials, it is possible to produce functional products by changing many properties with fibers or particle reinforcement, as shown in Figure 9.3. Reinforcement of thermoplastic polymers with fibers is a frequently applied method. The fibers can be in short and continuous forms. Continuous fibers can be given different forms by weaving. Short fiber reinforcements are used extensively with the thermoplastic matrix, but continuous fiber-reinforced thermoplastics have also been produced since the 1980s [15].

Fibers used for reinforcement are divided into two categories: natural and synthetic according to their sources (Table 9.3). Natural fibers are increasingly

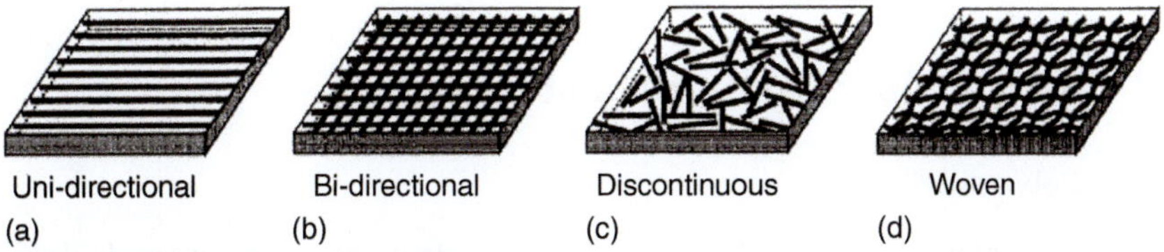

Figure 9.3 Composite material reinforcement types (a) Unidirectional, (b) bi-directional, (c) discontinuous, and (d) woven. Source: Tawfik et al. [28]/with permission of Elsevier.

Table 9.3 Fiber types.

Fibers				
Natural fibers			Synthetic fibers	
Animal	**Cellulose**	**Mineral**	**Organic**	**Inorganic**
Silk	Jute	Asbestos	Aramid	Glass
Wool	Flax			
Hair	Hemp			
	Kenaf			
	Wood		Polyethylene	Carbon
	Cotton		Polyester	Boron
	Stalk			Aramid

used in many different engineering applications. Natural fibers are produced from renewable resources and are the next-generation supplements to replace petroleum-based materials and synthetic fibers. Natural fibers are named after their origin, regardless of whether they are derived from plants, animals, or minerals [29]. Compared to synthetic fibers, natural fibers are low-cost, lightweight, biodegradable, and have high thermal and acoustic insulation. It also has high specific strength and hardness values [30].

Also, synthetic fibers have some serious disadvantages, such as high cost, high energy consumption in processing and manufacturing processes, recycling problems and nonrenewability, CO_2 emissions, and health hazards when inhaled [30]. To eliminate these problems, natural fibers have significant advantages. However, disadvantages such as low thermal degradation resistance to moisture absorption and exposure to weather conditions, lower durability, and poor interfacial adhesion limit the use of natural fibers [31].

As a result, functional composite products can be produced with appropriate design and material selection. Due to the adoption of natural fibers by automotive manufacturers due to their low cost, reasonably superior mechanical properties, abrasion resistance, high specific strength, environmental friendliness, and biodegradability, they are used to replace traditional fibers such as glass, aramid, and carbon [30].

9.2 Natural Fiber-Reinforcement Polyamide

9.2.1 Polyamide

Polyamide, which has a crystalline structure, is made up of recurring amide links, such as —CO—NH—. Polyamides can be found in nature as polypeptides. Also, polyamides can be seen commercially as nylon and aramids. The term nylon has

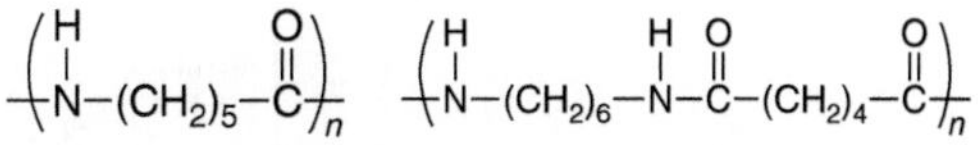

Figure 9.4 Molecular structure of PA6 (left) and PA66 (right).

been discovered by Dupont in 1928 [32]. But today, nylons are categorized into the aliphatic PAs while aramids are classified in PAs containing aromatic diamines and aromatic dicarboxylic acids [33].

The most widely used PA types are PA 6 and PA66, whose molecular structures are shown in Figure 9.4. The monomer of PA 6 is caprolactam while the monomer of PA66 is hexamethylene diamine/adipic acid. Polyamide 12, Polyamide 69, Polyamide 6-10, Polyamide 6-12, Polyamide 46, and Polyamide 1212 are some of the polyamide types found in literature [32].

Polyamides are excellent engineering materials that have high strength, good heat resistance (up to 230° [33]), good fatigue resistance, and good chemical resistance. These materials have good wear-resistant properties. That is why they are commonly used in the clothing industry, airbags, and bullet-proof vests [34]. On the other hand, they are also used in the manufacturing of automotive parts, electrical electronics devices, packaging, leisure (ski fixations and in-line skate frames), lawn, and furniture (chair bases and other chair components) [33].

9.2.2 Natural Fibers

Plants, animals, and minerals may all be used to make natural fibers. Composite materials with a wide variety of properties can be produced by combining these raw materials, which can be turned into filaments, nonwoven fabrics, paper, or thread, with a suitable matrix. Natural fibers are generally categorized into three types: animal fibers, plant fibers, and mineral fibers [35]. The comparison of the mechanical properties of natural fibers with glass fibers is shown in Table 9.4.

9.2.2.1 Animal Fiber

These fibers, which mostly contain protein, are divided into three categories: animal hair, silk fiber, and avian fiber. Wool from animals such as sheep, goats, alpacas, and horses, as well as hairy mammals, is referred to as animal hair. The fibers derived from the dried saliva released by insects (bugs) during cocoon construction are known as silk fibers. This type of fiber includes silk from the silkworm. Avian fiber is a type of fiber derived from the feathers of birds [35].

9.2.2.2 Plant Fiber

Cellulose is the major component of these fibers. They may be grouped into the following categories based on where they came from [35].

- Seed fiber: cotton and kapok
- Leaf fiber: sisal and agave
- Bast fiber or skin fiber: flax, jute, kenaf, industrial hemp, ramie, rattan, soybean fiber, vine fibers, and banana fibers
- Fruit fiber: coconut (coir) fiber.
- Stalk fiber: straws of wheat, rice, barley, and other crops including bamboo, grass, and tree wood

Table 9.4 Mechanical properties of natural fibers as compared to e-glass fiber.

Fiber	Density ($g\,cm^{-3}$)	Length (mm)	Diameter (μm)	Tensile strength (MPa)	Tensile modulus (GPa)	Elongation (%)	Cellulose (wt%)	Hemi-cellulose (wt%)	Lignin (wt%)	Pectin (wt%)	Waxes (wt%)	Moisture content (%)
Cotton	1.5–1.6	10–60	10–45	287–800	5.5–12.6	3–10	82.7–90	5.7	<2	0–1	0.6	7.85–8.5
Jute	1.3–1.49	1.5–120	20–200	320–800	8–78	1–1.8	59–71.5	13.6–20.4	11.8–13	0.2–0.4	0.5	12.5–13.7
Flax	1.4–1.5	5–900	12–600	343–2000	27.6–103	1.2–3.3	62–72	18.6–20.6	2–5	2.3	1.5–1.7	8–12
Hemp	1.4–1.5	5–55	25–500	270–900	23.5–90	1–3.5	68–74.4	15–22.4	3.7–10	0.9	0.8	6.2–12
Kenaf	1.4	—	—	223–930	14.5–53	1.5–2.7	31–72	20.3–21.5	8–19	3–5	—	—
Bamboo	0.6–1.1	1.5–4	25–40	140–800	11–32	2.5–3.7	26–65	30	5–31	—	—	—
E-glass	2.5–2.59		<17	2000–3500	70–76	1.8–4.8	—	—	—	—	—	—

Source: Adapted from Dittenber and Gangarao [36].

Cotton Gossypium, a tropical and subtropical cotton plant belonging to the Malvaceae family, is a plant species that thrives in tropical and subtropical climates [37]. This plant, which is mostly grown in China, the United States, Russia, and India, grows to a height of approximately half a meter but can grow to two meters.

Flower buds begin to develop once the seedlings have grown to a certain size (see Figure 9.5a). Following pollination, the cocoon grows into cotton seeds, which are subsequently transformed into flowers (Figure 9.5b). The seeds undergo cellularization/fiber development, and the cotton fiber in the cocoon thickens, while the blossoms fall off (Figure 9.5c) [40]. The method of turning raw cotton bolls into fiber is extremely efficient, with a maximum weight loss of one-tenth after the procedure. Long (lint) and short fibers are two types of fibers that are created (linter). Cotton is a rather weak fiber with little extensibility, although it is known to improve in strength and stiffness as the fibers lengthen (Figure 9.5d) [41]. Cotton is a biodegradable fiber that can be made into fiber, yarn, fabric, and apparel. It may also be combined with a variety of other fibers, such as polyester, wool, and silk [42].

Cotton is suitable for the production of filters, pads, quilts, pillows, cushions, and seats [43], and polymer/cotton fiber composites are used for surgical purposes in the medical industry to prevent the leakage of wounds due to their high water absorption ability [44]. The major problem is that cotton growing requires a lot of chemical fertilizers, insecticides, and water [42]. Especially polyester/cotton composites are preferred because polyester is cheap, liquid, readily available, easy to process, durable, and has good mechanical properties when reinforced with cotton. Because polyester is inexpensive, liquid, easily accessible, quick to manufacture and cure, and has strong mechanical qualities when reinforced with cotton, polyester/cotton composites are chosen. Furthermore, these composites with thermo-acoustic insulation qualities can comprise up to 80% cotton by weight, making them appealing for automotive applications that need lightweight materials. However, good fiber-matrix interfacial adhesion is required for the resulting composite to have acceptable mechanical strength; otherwise, chemical modification or compatibilizer can be applied [42].

Figure 9.5 The life cycle of the cotton plant Watch Me [38]/YouTube and the cotton fiber. Source: Cotton Fibre [39], Saiful Islam/Textile Craft.

Figure 9.6 Jute fiber [45] Biswarup Ganguly/Wikimedia Commons/CC BY 3.0, and jute composite door. Source: Gon et al. [46]/Scientific & Academic Publishing.

Jute Jute, derived from the Corchorus genus of the Tiliaceae family, is one of the cheapest and most significant natural fibers after cotton. Jute is a plant that is cultivated purely to produce fiber. This fiber, as shown in Figure 9.6, which is available in a wide variety of colors from off-white to brown, is soft and shiny. To thrive, the jute plant needs a warm, humid atmosphere. Jute crops may grow to be two to three and a half meters tall. There are two types of jute: white jute (*Corchorus capsularis*) and black jute (*C. olitorius*). India, Bangladesh, Brazil, Thailand, and China are among the countries that cultivate it. It is also critical to the economies of these countries [47].

Jute, which has biodegradable and antistatic properties, is resistant to microorganisms. It also has good insulation properties and heat resistance. Having high specific strength makes it attractive. However, it is resistant to chemicals and moisture. In addition, it has low extensibility. Because they are one of the fibers that absorb moisture the fastest, they should be kept dry. Its tensile strength is weaker than that of flax and hemp. As the amount of lignin in the fiber increases, the number of the aromatic rings and rigid backbone structures will increase, increasing brittleness [48, 49]. Jute fiber's strength is damaged by exposure to direct sunlight [50].

Jute fiber is used in the production of fashion, travel, luggage, furnishing, carpet, packing, doors, floor coverings, and wall decoration. Today, more ecologically friendly and lasting interior constructions in the aviation and automotive sectors are being attempted by blending epoxy and polyester resins with jute fiber [50, 51].

Flax Flax, belonging to the Linaceae family, is one of the oldest known textile products. The production of linen from flax dates back 5000 years and is depicted on tomb and temple walls in the ancient Egyptian city of Thebes [52]. The flax plant, which belongs to the dicotyledonous group, is mostly grown and produced in countries with temperate climates such as China, France, and Belarus. This plant can be grown for the production of fiber, linseed, and linseed oil. Harvesting for fiber production is expected for about 100 days, or until the base of the plant turns yellow. While the flax plant has a size between 80 and 150 cm, the flax fibers obtained have a length of 60–140 cm and a diameter of 40–80 μm [37].

Figure 9.7 Porsche flax fiber composite panels. Source: Eckhardt [54]/nova-Institut GmbH.

Flax fiber has different colors, from light yellow to gray and is lustrous [51]. It is both flexible and durable (tensile strength can be up to 100 MPa [53]). It has inextensible property. It shows elasticity only at small amounts of elongation. It also has good wear resistance. It is biodegradable and has high toughness [53]. It absorbs moisture quickly and dries quickly. It decomposes above 200° and is not resistant to sunlight. It can withstand alkaline solutions and mild acids, but it cannot withstand hot dilute acids or cold concentrated acids. It is costly to produce because it requires high-level labor [50]. Flax is used in thread, twine, sewing thread, suture, tablecloths, decorative fabric, apparel, composite helmets, the suitcase of instruments, and interior automotive components, as shown in Figure 9.7 [51, 55]

Hemp This plant, which belongs to the *Cannabis* family, is an annual herb that grows in temperate climates. There are two common types: *Cannabis sativa* (*C. sativa*) for fiber production and *Cannabis indica* (*C. indica*) for recreational and medicinal drugs use. It is known to be the oldest cultivated fiber plant in the world and is native to Central Asia [37]. One-third of the total cannabis production belongs to China [56]. Hemp plants may grow up to 4.5 m in length in around 145 days [57]. The hemp fiber as shown in Figure 9.8 is roughly 25 mm in length [60]. *Cannabis* cultivation is gaining popularity again today, as it grows very quickly and has strong fibers [52]. The most essential characteristics that impact the cost of hemp cultivation are color, brightness, and weight.

Hemp fiber has the feature of not passing harmful UV rays from the sun [51]. It is also nonallergenic [61] and rot-resistant. However, it breaks without warning as it shows decay from the inside out [52]. Ship rope, rugs, yarns, paper, carpet, home furnishings, building, interior design, and garment materials are all made from hemp [51, 62].

Kenaf Kenaf, which belongs to the Malvaceae family, is an annual herb that grows in temperate climates. The kenaf plant, which has a woody base, has leaves of various shapes. This plant, which is native to Asia and Africa, grows to 6 meters in length in 150 to 180 days [63].

Figure 9.8 Hemp fiber [58] Natrij/Wikimedia Commons/Public Domain and hemp fiber composite chair. Source: Inhabitat – Sustainable Design Innovation [59], Werner Aisslinger/MH Sub I, LLC dba Inhabitat.

Figure 9.9 Kenaf fiber [69] Textile Value Chain, and the inner space of Toyota LQ which is made of PU-kenaf fiber. Source: Rothbarth [70], Punekar News by BIMS India.

Kenaf fibers are obtained by retting the thin bark layer under the stem. The average length of kenaf fiber is around 5 mm [52]. The quality of kenaf fiber is affected by factors including the growth season, sowing date, and sunshine [64]. Low density, better durability, high aspect ratio, and biodegradability are just a few of the benefits of kenaf fiber [51, 65]. Kenaf fiber is an important reinforcing candidate in thermoplastic composites since it can have a tensile strength of 11.9 GPa and an elastic modulus of 60 GPA. Furthermore, kenaf is used to make rope, sackcloth [66] paper, animal feeds, manufacturing oil, medication, food additives, mushroom cultivation platforms, automotive industries, textiles, construction materials, oil, and chemical absorbents [67, 68]. Kenaf fibers show up in Figure 9.9.

Bamboo Bamboo is a perennial woody plant that grows in tropical and subtropical regions. It has the distinction of being one of the world's fastest-growing plants, capable of growing 3 cm every hour [71]. This plant, which reaches maturity in 6 to 8 months, can reach 40 m in height [51]. The average bamboo fiber length is approximately 2.7 mm [60]. Bamboo is widely grown, particularly in Malaysia, and is utilized as a building material in low-cost traditional bridge and home construction [72] China and Taiwan are two other major bamboo producers [73].

Figure 9.10 Bamboo fiber [74] Stéphane Schröder/Guadua Bamboo, and the Phoenix bamboo concept car. Source: [75] Ben Coxworth/New Atlas.

Bamboo fiber, as shown in Figure 9.10, is hard and brittle due to its high lignin content [52, 76]. Because of the longitudinal alignment of fibers, bamboo has high tensile strength, and it can be combined with matrix resins such as polystyrene, polyester, and epoxy to produce more environmentally friendly nonhazardous composites with better mechanical properties [76]. Other uses for bamboo include textiles, home paneling, furniture, automobile (Figure 9.10) and airplane interiors, helmets, surfboards, boat decks and floors, garden fences, and outdoor garbage cans [73, 76].

9.2.3 Mineral Fiber

These fibers can be found in nature or can be created by slightly altering the mineral. Asbestos is the only mineral fiber found naturally in the environment. Asbestos comes in two varieties: serpentine (chrysotile) and amphiboles (amosite, crocidolite, tremolite, actinolite, and anthophyllite) [35].

9.2.4 Production

Adding natural fibers to oil-based polyamides is often a problematic process. Although polyamide has a high melting temperature (over 220 °C), natural fibers begin to degrade at these temperatures due to the thermally sensitive components they contain (cellulose 60–70 wt%, hemicelluloses 10–20 wt%, lignin 2–10 wt% [77]) [78]. Furthermore, the fibers' high moisture absorption limits the compounding process in the twin-screw extruder, as shown in Figure 9.11. Humid raw ingredients in the extruder preclude homogenous mixing. The hollow structure creates a discontinuity in the line and causes a twisted exit from the die. As a result, both the polymer raw material and the fibers must be completely dried before processing (below 1–2% moisture by mass).

However, technologies like the resin drier are not ideal for drying fibers like wood fiber because tiny wood particles might ignite. In addition, high temperatures cause the closure of small spaces in the cellulosic structure of the fiber, reducing plasticity and causing a more brittle structure [80]. The depolymerization with the effect of high temperature reduces the tensile strength by almost half [81]. Natural fibers with a large volume (low apparent density) are difficult to process because they cannot pass through small extruder openings.

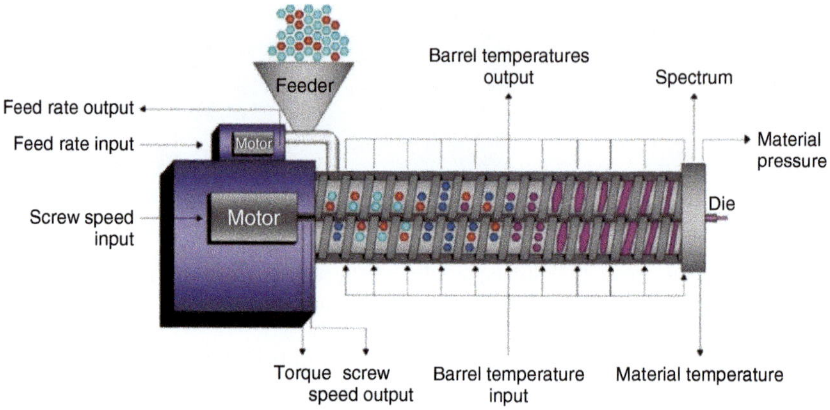

Figure 9.11 Twin-screw extruder. Source: [79]/Engineering Learn.

To avoid harming or breaking the extruder screw, it is vital to choose the proper screw pitch design, zone temperatures, screw speed, and fiber size. According to studies, composites with natural fiber additions can only be made by utilizing certain additives and compounding techniques. For example, it has been observed that fibers treated with N2 plasma [82] or caustic soda (NaOH) tend to exhibit higher mechanical properties due to better fiber/matrix adhesion and no gaps between the fiber and matrix, but these processes are costly, and effluent formation has been observed after NaOH chemical treatment. In addition, the mechanical properties of this new composite vary with moisture content. The very fast water absorption of the composite is one of the factors that should be considered in the design of the part. Insufficient surface treatment of the fibers causes an unbonded structure around the fiber. For this reason, surface treatment of natural fibers is critical [83].

9.3 Friction and Wear Resistance at Natural Fiber-Reinforcement Polyamide

9.3.1 Friction

The simplest definition of friction can be explained by energy dissipation. When two substances or materials have relative motion, the energy needed to start or maintain this motion. In solid phases, we can face this energy in the connection as heat rises or deforms. The same phenomenon can be observed with a solid body that tries to move on the liquid. This time, we can see the same energy in the waves [84]. The friction was discovered in the manuscripts of Leonardo da Vinci. Afterward, Guillaume Amontons showed that this friction force depends on the applied load while being independent of the contact area, and this proof is accepted as the two first friction laws [85]. Nowadays, we know that the phenomena of friction are based on principles of physics "There is always a reaction force that has the same magnitude and the opposite direction for any forces." [86].

These explorations also show that there would be a relationship between the friction force and the normal force. The relation can be expressed as the ratio of the friction force to the normal force, which is named the friction coefficient [87]. Today, there are two types of friction coefficients. One is known as the static friction coefficient (μ_s), which creates the challenge when the motion starts, and the other is the kinetic friction coefficient (μ_k) which opposes the relative motion. These coefficients can be expressed (in the case of solid–solid configuration) as shown below in the formulas [88]. The developed formulas can be illustrated as shown below (Figure 9.12) [89].

$$\mu_s = {}^{F_s}/_{F_N} \quad \mu_k = {}^{F_k}/_{F_N}$$

Both of the coefficients are related and determine the characteristics of the systems. Static friction (F_s) is usually accepted as the first maximum signal in the system. Afterward, system coefficient of friction is descending through an average value (F_k) [90]. Static and kinetic coefficients of friction have a relationship that is always considered in design projects, as shown in Figure 9.13 [91]. Table 9.5 shows some of the coefficients that belong to the common materials [92].

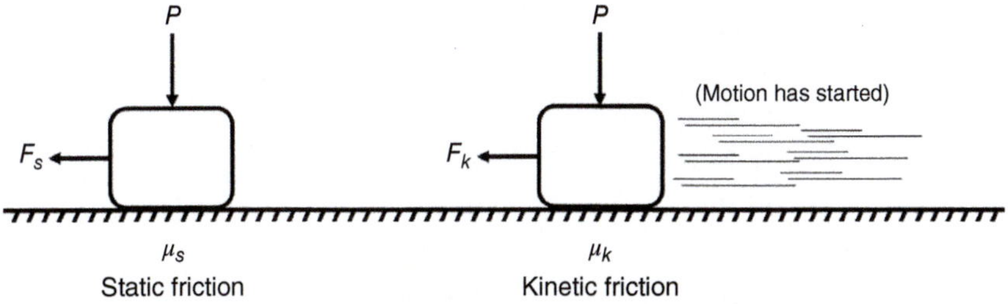

Figure 9.12 Static and kinetic friction. Source: Jones and Ashby [89]/with permission of Elsevier.

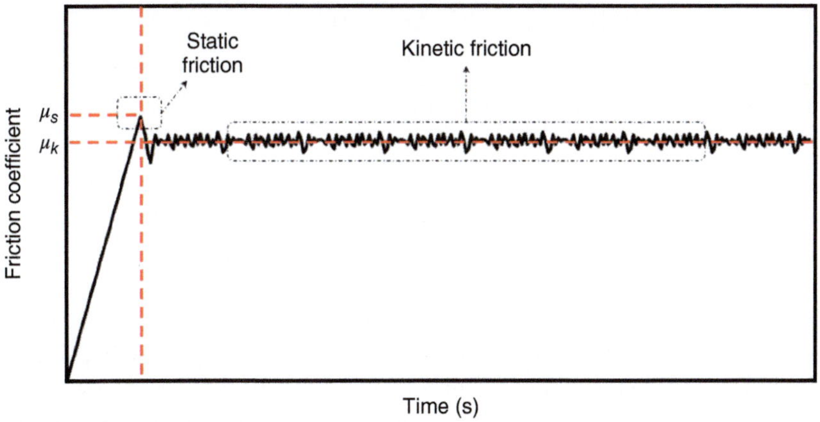

Figure 9.13 The relation between static and kinetic friction. Source: Adapted from Cho et al. [91].

Table 9.5 Coefficient of friction of common materials.

Material 1	Material 2	Static coefficient of friction	Dynamic coefficient of friction
ABS	Steel	0.21–0.28	
ABS	Aluminum	0.40	
Cellulose	Cellulose	0.25	0.2
HDPE	Steel	0.17	
HDPE	Aluminum	0.27–0.33	
LDPE	LDPE		0.6
PA4.6	PA4.6	0.08–0.38	
PA 6	PA 6	0.32–0.43	0.16–0.8
PA 6	Steel	0.26	
PA 6	Aluminum	0.44–0.47	
PA 6.6	PA 6.6	0.35–0.5	
PA 11	PA 11	0.1–0.3	
PC	Steel	0.21	
PC	Aluminum	0.41–0.54	
PE	PE	0.6	
Peek (dry)	Peek (dry)	0.25–0.3	
Peek (wet)	Peek (wet)	0.08–0.09	
POM	POM	0.10–0.38	0.11
POM	Steel	0.18–0.41	
PP	Steel	0.27–0.29	
PP	Aluminum	0.35–0.36	
PTFE	PTFE	0.08	0.06
PTFE	Glass fiber	0.24–0.31	
PTFE	Carbon fiber	0.20–0.24	
PVC	Steel	0.35–0.8	0.72–0.93

Source: Wypych [92]/MDPI/Public Domain CC BY 3.0.

Besides the calculation of the friction coefficient, the determination of the friction mechanism is one of the critical criteria. In this manner, four different friction mechanisms have been found from past to present in literature.

A mechanism known as sliding friction is a system that includes a solid block that slides on the other surface. This motion creates a resistance force and this force is proportional to the normal load being applied to a solid block [93]. Sliding friction theory is based on the roughness of connecting surfaces, the proportion between the resistance that occurs in the real connection area, and external forces [94]. In cases of round materials moving on planar surfaces, a small resistance occurs between the contact regions. When compared to sliding friction, rolling friction is very small. This resistance cannot be eliminated, even on surfaces that have improved roughness [95]. Fretting friction is one of the friction mechanisms. A small- amplitude external vibration affects the system through this mechanism. The load can be considered a cyclic load, which might cause fatigue failure as a result [96]. The last friction mechanism is impact friction. Besides the other mechanisms, impact friction has no continuous contact surface in the system. While the relation between friction and energy dissipation can be linked in systems that have a continuous contact surface, in impact friction mechanisms, energy dissipation can be defined as the energy loss that occurred in impact [97].

9.3.2 Wear

The wear has always been searched for as a mysterious thing in our lives. The term was introduced in 1973 by famous scientist Nam P. Suh. Researchers have always been looking for a description of the phenomenon of energy loss. According to the founding scientist, this phenomenon can only be understood by asking the right questions, which are "Where does all energy go?", "Why and how does the friction affect the wear issue?" [98].

Wear is the process of losing, transferring, or moving material from a surface. Some sources explain this phenomenon as a result of relative motion [99]. Progressive loss can be observed on the operating surface while it is exposed to relative motion, and this causes the reduction of the tolerances or direct surface failure [100, 101]. Furthermore, the wear phenomenon also provides a ground that can lead to fatigue failure [101].

Wear is the response of a system that depends on different parameters. Contact conditions (match material, load, sliding velocity, shape, stiffness, environment, and lubrication system) are key characteristics of this response. Regarding these conditions, the rate can change in the range of 10^{-3} to 10^{-10} $mm^3\ Nm^{-1}$ [99]. Scientists are only succeeded if they classify their mechanisms and modes of study. Table 9.6 shows the categorization of the wear and its parameters [102].

Wear does not always occur in one mode. Sometimes observed more than one mode affecting together with the related systems [103]. Rolling, sliding, fretting, and impact wear are the most common wear terms in literature (Figure 9.14) [101, 102]. They usually illustrate the type of friction that creates wear as a result. Wear mechanisms include three sections: mechanic, chemical, and thermal wear [104].

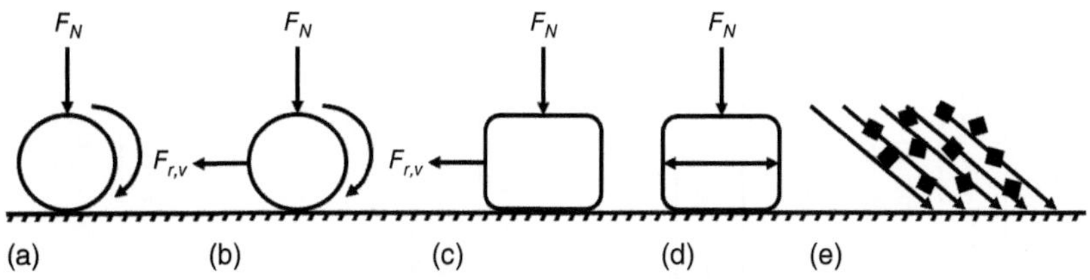

Figure 9.14 Types of friction-related wear process. (a) Rolling, (b) rolling–sliding, (c) sliding, (d) fretting and (e) impact [101, 102].

Table 9.6 Classification of wear parameters.

Class	Parameter					
Friction type	**Rolling**	**Rolling sliding**	**Sliding**		**Fretting**	**Impact**
Contact shape	**Sphere–sphere**	**Cylinder–cylinder**	**Flat–flat**	**Sphere–flat**	**Cylinder–flat**	**Punch–flat**
Contact pressure level	Elastic		Elasto-plastic		Plastic	
Sliding speed or loading speed	Low		Medium		High	
Flash temperature	Low		Medium		High	
Mating contact material	Same	Harder	Softer		Compatible	Incompatible
Environment	Vacuum	Gas		Liquid		Slurry
Contacty cycle	Low (single)		Medium		High	
Contact distance	Short		Medium		Long	
Phase wear	Solid	Plastic	Liquid	Gas	Atom	Ion
Structure of wear particle	Original		Mechanically mixed		Tribochemical	
Freedom of wear particle	Free	Trapped		Embedded		Agglomerated
Unit size of wear	mm scale		µm scale		nm scale	
Elemental physics and chemistry in wear	Physical adsorption, chemical adsorption, tribochemical activation and tribofilm formation, oxidation and delamination, oxidation and dissolution, oxidation and gas formation, phase transition, recrystallization, crack nucleation and propagation, adhesive transfer and retransfer					
Elemental system dynamics related to wear	Vertical vibration	Horizontal vibration	Self-excited vibration		Harmonic vibration	Stick-slip motion
Dominant wear process	Fracture (ductile or brittle)	Plastic flow	Liquid flow	Dissolution	Oxidation	Evaporation
Wear mode	Abrasive	Adhesive	Flow	Fatigue	Corrosive	Melt Diffusive
Wear type	Mechanical		Chemical		Thermal	

Source: Kato [102]/John Wiley & Sons.

Mechanical wear describes material defects under mechanical loads. This wear type is understood by results like deformation (in ductile materials) and cracking (in brittle materials). Basic loads that cause mechanical wear are rolling, sliding, and cyclic impacts. Adhesive, abrasive, and fatigue wear modes are the key modes of this wear type [105].

Chemical wear is defined by the growth rate of chemical reaction films. This effect increases with the growth of the exposed friction. This type is known as tribochemical wear due to the effects of chemical and mechanical loads [102]. Thermal wear mechanisms are systems affected by local heating or surface temperature conditions [105].

Abrasive wear mode can be observed in between frictional pair surfaces where one of them is harder and rougher than the other [84]. This situation can occur when a rough steel surface tries to mount on a softer metal [106]. Abrasive wear mode can be seen in Figure 9.15 [107]. On the other hand, adhesive wear is one of the most common types of wear modes with abrasive wear in the literature. This mode can occur in contacts that have strong bonding; as a result, some particles can be detached from the connecting surface and create a wear mark on the surface [108]. K. Kato and his colleague performed a study in 1987 that showed thin film-like tissues can be formed as debris as a result of the plastic flow on smooth surfaces. This debris was named flow wear by Kato and Akagaki [109]. Fatigue wear can occur as a result of cyclically loaded system configurations [102]. The diffusive wear mode can be defined by the dissolution of material on one of the contact surfaces [110]. The melt-wear mode can occur as particles that have different surface morphologies. This mode can be observed as tiny droplets on the connecting surfaces [102]. The corrosive mode is the wear mode that can be faced as the result of interactions with environmental effects. This mode can be categorized into two classes, which are dry and wet, and environmental conditions are one of the most critical factors for preventing this failure. Thus, material selections must be done carefully by considering these conditions [111].

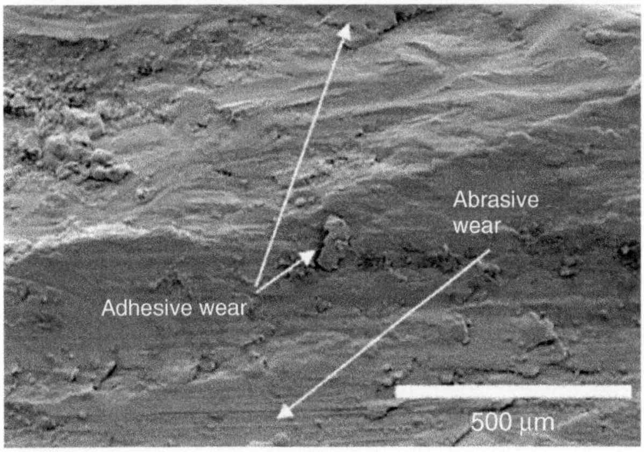

Figure 9.15 Adhesive and abrasive wear modes. Source: Fabbro et al. [107]/Elsevier/CC BY 4.0.

9.3.3 Testing and Measurement

Wear and friction are the key properties that must be considered in any system. These properties need to be calculated carefully and taken into account during the design process. There are several methods available for measuring the wear rate and friction.

9.3.3.1 Friction Test Methodologies

The frictional properties can be measured and reported by following the ASTM G115 standard. Amonton's formula can be used as a valid equation except for the web wrapped around a cylinder system. The capstan formula is the proper methodology to calculate web wrapped around a cylinder system. Frictionometer is usually used for the friction coefficient calculation. This device follows the ASTM D3028 standard [84].

9.3.3.2 Wear Test Methodologies

Wear calculations are quite hard to measure correctly due to the complexities of the system, like the connecting surface materials, roughness, and contact types. To overcome these challenges and measure the correct wear specifications, a field test must be done for each system. These tests have disadvantages in time and cost domains. That is why wear measurements need to be done using a simplified and well-designed methodology [101]. To avoid any deviation effects, environmental factors (moisture, humidity, and temperature) should be controlled before the test starts. After the standardization of these factors, the most suitable wear mode is selected for the system, and tests are carried out with the methods commonly used in the literature specified in Table 9.7 [84].

Table 9.7 Common wear tests.

Test	Features	Drawbacks
Pin-on-disc	- Continuous motion - Test specimens (basic) - Easy to apply	- Different disc radii may affect the results - Does not simulate real tribosystems
Ball-on-plane	- Reciprocating motion - Multiple tests can be applied with a small specimen - Easy to apply	- Only reflect reciprocating tribosystems - Limited frequency (<10 Hz)
Block-on-ring	- Available for bushing/shaft applications - High forces - It is easy to produce a specimen	- Sometimes alignment can be a problem - Rings are expensive

Source: Budinski [84]/ASTM.

9.3.4 Applications

Nowadays, many systems are exposed to friction and wear. Some systems are using this exposure as an advantage. Braking systems, safety systems, and clutch mechanisms are examples of friction-needed systems. Some systems are struggling with the effects of friction and wear. In some cases, these effects can cause a major failure.

Energy usage and environmental concerns are growing day by day. These concerns cannot be met with conventional materials [112]. Fiber-reinforced polymeric composites have been aimed for being an alternative to conventional materials. There are two types of these composites as mentioned in Section 9.2 depending on the fiber type (synthetic and natural) [113, 114]. Due to their lightweight and good performance, there is huge interest in fiber-reinforced polymer composites [115]. These materials are generally used in the aircraft industry (landing gear gates, cabins, and jet engines), automotive, shipbuilding, and military applications. Composite materials that contain synthetic fibers (aramid, carbon, glass, etc.) are highly preferred due to their long service life, adaptedness, and good mechanical specifications [116]. Fiber-reinforcement studies of polyamides are showing that fibers are decreasing the contact between matrix and mating face. Moreover, they are also able to support applied loads. These specifications put forward fiber reinforcement studies on friction and wear-related applications [117]. J. JyotiKalita and K. K. Singh studied the tribological properties of different synthetic fiber-reinforced polymer composites and reported that these properties can be affected directly by the fiber orientation, sliding speed, and matrix material [118]. On the other hand, D.X. Li et al. investigated the friction and wear characteristics of glass fiber-reinforced polyamide 6 combined with polytetrafluoroethylene (PTFE), ultra-high molecular weight polyethylene (UHMWPE) in different configurations. They also examined the effects of different configurations on friction and wear specifications [117]. Furthermore, J. Li and Y. C. Xia studied friction and wear characteristics of various combinations of carbon fiber-reinforced polyamide 6 in dry sliding conditions. They have reported the comparative results of these combinations. Pure polyamide 6's main wear mechanism in dry sliding conditions is adhesive wear. The results showed that the specific amount of carbon fiber addition would positively affect resistance to this wear mechanism [119]. Although synthetic fibers have good mechanical strength, it is a fact that they are struggling in recycling processes. This creates serious disadvantages in industrial applications of synthetic fibers. Natural fibers are one of the alternative solutions to synthetic fibers via their proven quality in different applications [116]. Nishitani et al. investigated the tribological properties of Polyamide 1010 material reinforced with four different natural fibers (hemp, jute, ramie, and sisal), which aimed to determine its availability for use in mechanical sliding parts. While they were doing that, they also looked at the effects of surface treatments to improve tribological properties. They also reported the relation between friction and wear rates, which clearly shows that the best results can be achieved by using ramie (Figure 9.16) [120].

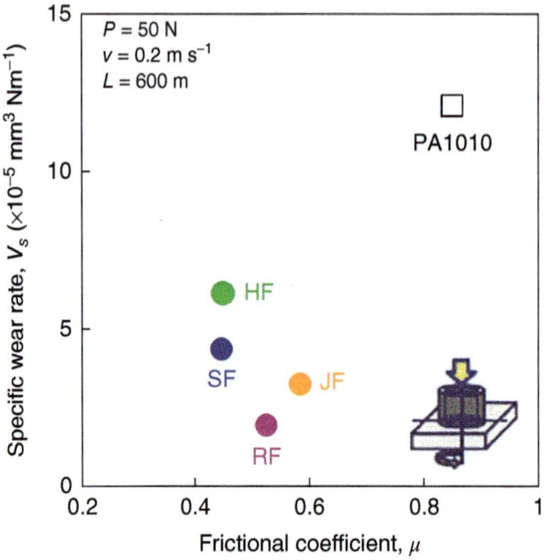

Figure 9.16 Relation between wear rate and friction coefficient of different natural fiber reinforced PA 1010 composites surface treated by ureidosilane. Source: Nishitani et al. [120]/AIP Publishing.

References

1 Gouzman, I., Grossman, E., Verker, R. et al. (2019). Advances in polyimide-based materials for space applications. *Adv. Mater.* 31 (18): 1–15. https://doi.org/10.1002/adma.201807738.

2 Vendan, S.A., Natesh, M., Garg, A., and Gao, L. (2019). *Confluence of Multidisciplinary Sciences for Polymer Joining*. Singapore: Springer Singapore.

3 Chawla, K. (2012). *Composite Materials Science and Engineering*. Springer.

4 Batten, S.R., Champness, N.R., Chen, X.-M. et al. (2012). Coordination polymers, metal–organic frameworks and the need for terminology guidelines. *CrystEngComm* 14 (9): 3001–3004. https://doi.org/10.1039/C2CE06488J.

5 Callister, W.D. and Rethwisch, D.G. *Materials Science and Engineering: An Introduction*. Wiley.

6 Brinson, H.F. and Brinson, L.C. (2015). *Polymer Engineering Science and Viscoelasticity: An Introduction*, 2ee. US: Springer.

7 Utekar, S., Suriya, V.K., More, N., and Rao, A. (2021). Comprehensive study of recycling of thermosetting polymer composites—driving force, challenges and methods. *Composites, Part B* 207: 108596. https://doi.org/10.1016/J.COMPOSITESB.2020.108596.

8 Abali, B.E., Yardımcı, M.Y., Zecchini, M. et al. (2021). Experimental investigation for modeling the hardening of thermosetting polymers during curing. *Polym. Test.* 102: 107310. https://doi.org/10.1016/J.POLYMERTESTING.2021.107310.

9 Zhao, P., Rao, C., Gu, F. et al. (2018). Close-looped recycling of polylactic acid used in 3D printing: an experimental investigation and life cycle assessment. *J. Clean. Prod.* 197: 1046–1055. https://doi.org/10.1016/j.jclepro.2018.06.275.

10 Ebewele, R.O., Raton, B., and York, N. (2000). *Polymer Science and Technology*. CRC Press.

11 Grigore, M.E. (2017). Methods of recycling, properties and applications of recycled thermoplastic polymers. *Recycling* 2: 24. https://doi.org/10.3390/RECYCLING2040024.

12 Coppola, S., Acierno, S., Grizzuti, N., and Vlassopoulos, D. (2006). Viscoelastic behavior of semicrystalline thermoplastic polymers during the early stages of crystallization. *Macromolecules* 39 (4): 1507–1514. https://doi.org/10.1021/MA0518510.

13 Hotta, A., Cochran, E., Ruokolainen, J. et al. (2006). Semicrystalline thermoplastic elastomeric polyolefins: advances through catalyst development and macromolecular design. *Proc. Natl. Acad. Sci. U.S.A.* 103 (42): 15327–15332. https://doi.org/10.1073/PNAS.0602894103/ASSET/F6A85671-8E5C-4346-8B53-E3B933D4A18B/ASSETS/GRAPHIC/ZPQ0330632090006.JPEG.

14 Karsli, N.G. and Aytac, A. (2013). Composites: Part B Tensile and thermomechanical properties of short carbon fiber reinforced polyamide 6 composites. *Composites, Part B* 51: 270–275. https://doi.org/10.1016/j.compositesb.2013.03.023.

15 Guedes, R.M. (2019). *Creep and Fatigue in Polymer Matrix Composites*. Woodhead Publishing.

16 Dhinakaran, V., Surendar, K.V., Riyaz, M.S.H., and Ravichandran, M. (2020). Materials today: proceedings review on study of thermosetting and thermoplastic materials in the automated fiber placement process. *Mater. Today Proc.* 27: 10–13. https://doi.org/10.1016/j.matpr.2019.12.355.

17 Chang, I.Y., Lees, J.K. (1988). *Recent Development in Thermoplastic Composites: A Review of Matrix Systems and Processing Methods. Journal of Thermoplastic Composite Materials.* 1 (3), 277–296. doi:10.1177/089270578800100305.

18 Bhushan, B. (2009). Biomimetics: lessons from naturean overview. *Phil. Trans. R. Soc. A* 367 (1893): 1445–1486. https://doi.org/10.1098/RSTA.2009.0011.

19 Seydibeyoğlu, M.Ö., Doğru, A., Kandemir, M.B., and Aksoy, Ö. (2020). Lightweight composite materials in transport structures. In: *Lightweight Polymer Composite Structures* (ed. J.P.S.K.S.M. Rangappa), 103–130. CRC Press.

20 Q. T. H. Shubhra and A. Alam, *Mechanical Properties of Polypropylene Composites: A Review*, 2011, doi: https://doi.org/10.1177/0892705711428659.

21 Friedrich, K. (2018). Polymer composites for tribological applications. *Adv. Ind. Eng. Polym. Res.* 1 (1): 3–39. https://doi.org/10.1016/j.aiepr.2018.05.001.

22 Pan, Z., Han, S., Wang, J. et al. (2020). Polyimide fabric-reinforced polyimide matrix composites with excellent thermal, mechanical, and dielectric properties. *High Perform. Polym.* 32 (10): 1085–1093. https://doi.org/10.1177/0954008320928387.

23 Yao, S.S., Jin, F.L., Rhee, K.Y. et al. (2018). Recent advances in carbon-fiber-reinforced thermoplastic composites: a review. *Composites, Part B Eng.* 142: 241–250. https://doi.org/10.1016/J.COMPOSITESB.2017.12.007.

24 T. Rajmohan and R. Vinayagamoorthy, *Review on Effect Machining Parameters on Performance of Natural Fibre—Reinforced Composites (NFRCs)*, 2019, doi:https://doi.org/10.1177/0892705718796541.

25 TRI-MACK. 2022 *Materials | Tri-Mack Custom Plastic Parts Manufacturing.* http://www.trimack.com/materials .

26 Jiang, Y., Chen, L., Xiao, C. et al. (2021). Friction and wear behaviors of steel ball against polyimide-PTFE composite under rolling-sliding motion. *Tribol. Lett.* 69 (3): 1–12. https://doi.org/10.1007/s11249-021-01476-0.

27 Friedrich, K., Zhang, Z., and Schlarb, A.K. (2005). Effects of various fillers on the sliding wear of polymer composites. *Compos. Sci. Technol.* 65 (15–16 Spec. Iss): 2329–2343. https://doi.org/10.1016/j.compscitech.2005.05.028.

28 Tawfik, B.E., Leheta, H., Elhewy, A., and Elsayed, T. (2017). Weight reduction and strengthening of marine hatch covers by using composite materials. *Int. J. Nav. Archit. Ocean Eng.* 9 (2): 185–198. https://doi.org/10.1016/J.IJNAOE.2016.09.005.

29 George, M., Chae, M., and Bressler, D.C. (2016). Composite materials with bast fibres: structural, technical, and environmental properties. *Prog. Mater. Sci.* 83: 1–23. https://doi.org/10.1016/j.pmatsci.2016.04.002.

30 Holbery, J. and Houston, D. (2006). Natural-fiber-reinforced polymer composites in automotive applications. *JOM* 58 (11): 80–86. https://doi.org/10.1007/S11837-006-0234-2.

31 A. Lotfi, H. Li, D. V. Dao, and G. Prusty, *Natural Fiber—Reinforced Composites: A Review on Material, Manufacturing, and Machinability*, 2019, https://doi.org/10.1177/0892705719844546.

32 PA Plastic 2022 *Polyamide/Nylon (PA Plastic): Uses & Properties [Updated 2022].*

33 Rulkens, R. and Koning, C. (2012). *Chemistry and Technology of Polyamides*, vol. 5. Elsevier B.V.

34 Gong, Y. and Yang, G. (2009). Manufacturing and physical properties of all-polyamide composites. *J. Mater. Sci.* 44 (17): 4639–4644. https://doi.org/10.1007/s10853-009-3708-0.

35 Chandramohan, D. and Marimuthu (2011). A review on natural fibers. *Int. J. Res. Rev. Appl. Sci.* 8 (2): 194–206.

36 D. B. Dittenber and H. V. S. Gangarao, Critical review of recent publications on use of natural composites in infrastructure, *Composites Part A. Applied Science and Manufacturing*, vol. 43, no. 8. Elsevier, pp. 1419–1429, Aug. 2012, doi: 10.1016/j.compositesa.2011.11.019.

37 Bismarck, A., Mishra, S., and Lampke, T. (2005). Plant fibers as reinforcement for green composites. In: *Natural Fibers, Biopolymers, and Biocomposites*, 37–108. CRC Press.

38 Watch Me (2019). *Life Cycle of Cotton Plant | How Cotton is Produced Full Video | How to Grow Cotton—YouTube.* Youtube.

39 Cotton Fibre (2013). *Physical Properties and Chemical Properties of Cotton.* Textile Craft.

40 DeJoode, D.R. and Wendel, J.F. (1992). Genetic diversity and origin of the Hawaiian Islands cotton, Gossypium tomentosum. *Am. J. Bot.* 79 (11): 1311. https://doi.org/10.2307/2445059.

41 Hearle, J.W.S. (2006). Physical structure and properties of cotton. In: *Cotton: Science and Technology*, 35–67. Elsevier Inc.

42 Wankhede, B. (2008). Cotton reinforced polymer composites. In: *Tribology of Natural Fiber Polymer Composites*, 129–161. New Delhi: Woodhead Publishing.

43 M. M. A. Mangat, *Structure and Properties of Cotton Fiber: A literature Review*, 2009.

44 Bajpai, S.K., Mary, G., and Chand, N. (2015). The use of cotton fibers as reinforcements in composites. In: *Biofiber Reinforcements in Composite Materials*, 320–341. Elsevier Inc.

45 Jute—Wikipedia *Jute— Wikipedia*. 2023.

46 Gon, D., Das, K., Paul, P., and Maity, S. (2013). Jute composites as wood substitute. *Int. J. Text. Sci.* 1 (6): 84–93. https://doi.org/10.5923/j.textile.20120106.05.

47 Ali, M. (2012). Natural fibres as construction materials. *J. Civ. Eng. Constr. Technol.* 3 (3): 80–89. https://doi.org/10.5897/jcect11.100.

48 Yang, S., Yuan, T.-Q., Shi, Q., and Sun, R.-C. (2018). Application of lignin in thermoplastic materials. In: *Encyclopedia of Sustainability Science and Technology*, 1–22. New York: Springer.

49 Mohanty, A.D. and Bae, C. (2015). Transition metal-catalyzed functionalization of polyolefins containing C-C, C=C, and C-H bonds. In: *Advances in Organometallic Chemistry*, vol. 64, 1–39. Academic Press Inc.

50 Mohanty, A.K., Misra, M., and Drzal, L.T. (ed.) (2005). *Natural Fibers, Biopolymers, and Biocomposites*, 1ee. CRC Press.

51 Keya, K.N., Kona, N.A., Koly, F.A. et al. (2019). Natural fiber reinforced polymer composites: history, types, advantages, and applications. *Mater. Eng. Res.* 1 (2): 69–87. https://doi.org/10.25082/mer.2019.02.006.

52 Pickering, K.L. (2008). *Institute of Materials, Properties and Performance of Natural-Fibre Composites*. Boca Raton, FL: CRC Press.

53 Zhu, J., Zhu, H., Njuguna, J., and Abhyankar, H. (2013). Recent development of flax fibres and their reinforced composites based on different polymeric matrices. *Materials (Basel).* 6 (11): 5171–5198. https://doi.org/10.3390/ma6115171.

54 Eckhardt, H. (2020). *Racing cars with body parts made from renewable raw materials*, vol. 49, 1–2. Renewable Carbon News.

55 Foulk, J., Akin, D., Dodd, R., and McAlister, D. III, (2002). Flax fiber: potential for a new crop in the Southeast. *Trends New Crop. New Uses* 2015: 361–370.

56 Hu, R. and Lim, J.K. (2007). Fabrication and mechanical properties of completely biodegradable hemp fiber reinforced polylactic acid composites. *J. Compos. Mater.* 41 (13): 1655–1669. https://doi.org/10.1177/0021998306069878.

57 S. Batra, *Other Long Vegetable Fibers*, 2006.

58 Marialma Blog *Hemp 101: Key Facts You Need To Know*. 2023.

59 Inhabitat – Sustainable Design Innovation *The Best Green Designs at Fiera Milano During the 2013 Milan Furniture Fair*.

60 Ilvessalo-Pfäffli, M.-S. (1995). *Fiber Atlas*. Berlin, Heidelberg: Springer Berlin Heidelberg.

61 Kostic, M., Pejic, B., and Skundric, P. (2008). Quality of chemically modified hemp fibers. *Bioresour. Technol.* 99 (1): 94–99. https://doi.org/10.1016/j.biortech.2006.11.050.

62 Indran, S., Edwin Raj, R., and Sreenivasan, V.S. (2014). Characterization of new natural cellulosic fiber from Cissus quadrangularis root. *Carbohydr. Polym.* 110: 423–429. https://doi.org/10.1016/j.carbpol.2014.04.051.

63 Chandramohan, D. and Marimuthu, K. *Tensile and Hardness Tests on Natural Fiber Reinforced Polymer Composite Material*, no. 6, 97–104.

64 Webber, C.L.I. and Bledsoe, V.K. (2002). Kenaf yield components and plant composition. *Trends New Crop New Uses* 348–357.

65 Karnani, R., Krishnan, M., and Narayan, R. (1997). Biofiber-reinforced polypropylene composites. *Polym. Eng. Sci.* 37 (2): 476–483. https://doi.org/10.1002/pen.11691.

66 Abilash, N. and Sivapragash, M. (2013). Environmental benefits of eco-friendly natural fiber reinforced polymeric composites materials. *Int. J. Appl. Innov. Eng. Manag.* 2 (1): 2319–4847.

67 Akil, H.M., Omar, M.F., Mazuki, A.A.M. et al. (2011). Kenaf fiber reinforced composites: a review. *Mater. Des.* 32 (8–9): 4107–4121. https://doi.org/10.1016/j.matdes.2011.04.008.

68 Bongarde, U.S. and Shinde, V.D. (2014). Review on natural fiber reinforcement polymer composites. *Int. J. Eng. Sci. Innov. Technol.* 3 (2): 431–436.

69 Clothing From Kenaf Fibres *Textile News, Apparel News, Fashion News*, 2019.

70 Frank Rothbarth, *Covestro provides sustainable solution for new concept car Toyota 'LQ,'* 2020.

71 Lobovikov, M., Paudel, S., Piazza, M. et al. (2007). World bamboo resources: a thematic study prepared in the framework of the Global Forest Resources, assessment 2005. *FAO Tech. Pap.* 1–74.

72 Bahari, S.A. and Krause, A. (2016). Utilizing Malaysian bamboo for use in thermoplastic composites. *J. Clean. Prod.* 110: 16–24. https://doi.org/10.1016/j.jclepro.2015.03.052.

73 Md Shah, A.U., Sultan, M.T., Jawaid, M. et al. (2016). Bamboo composites review. *BioResources* 11 (4): 10654–10676.

74 Guadua Bamboo, *Bamboo Textiles And Fabrics: What You Need To Know Before Buying*, 2021.

75 Biodegradable car made from rattan and bamboo.

76 Abdul Khalil, H.P.S., Bhat, I.U.H., Jawaid, M. et al. (2012). Bamboo fibre reinforced biocomposites: a review. *Mater. Des.* 42: 353–368. https://doi.org/10.1016/j.matdes.2012.06.015.

77 Gassan, J. and Bledzki, A.K. (2001). Thermal degradation of flax and jute fibers. *J. Appl. Polym. Sci.* 82 (6): 1417–1422. https://doi.org/10.1002/app.1979.

78 Xu, S., Sun, L., He, J. et al. (2018). Effects of LiCl on crystallization, thermal, and mechanical properties of polyamide 6/wood fiber composites. *Polym. Compos.* 39: E1574–E1580. https://doi.org/10.1002/pc.24507.

79 *Engineering Learn Types of Extruders: Application & Working Principle Types of Extruders.*

80 Bos, H.L., Müssig, J., and van den Oever, M.J.A. (2006). Mechanical properties of short-flax-fibre reinforced compounds. *Compos. Part A Appl. Sci. Manuf.* 37 (10): 1591–1604. https://doi.org/10.1016/j.compositesa.2005.10.011.

81 Baley, C., Le Duigou, A., Bourmaud, A., and Davies, P. (2012). Influence of drying on the mechanical behaviour of flax fibres and their unidirectional composites. *Compos. Part A Appl. Sci. Manuf.* 43 (8): 1226–1233. https://doi.org/10.1016/j.compositesa.2012.03.005.

82 Yuan, X., Jayaraman, K., and Bhattacharyya, D. (2002). Plasma treatment of sisal fibres and its effects on tensile strength and interfacial bonding. *J. Adhes. Sci. Technol.* 16 (6): 703–727. https://doi.org/10.1163/156856102760099898.

83 Clemons, C. (2002). Wood-plastic composites in the United States: the interfacing of two industries. *For. Prod. J.* 52 (6): 10–18.

84 Budinski, K.G. (2007). *Guide to friction, wear and erosion testing*. ASTM International.

85 Desplanques, Y. (2015). Amontons-Coulomb friction laws, a review of the original manuscript. *Int. J. Mater. Manuf.* 8 (1): 98–103. https://doi.org/10.2307/26268696.

86 Taylor, J.R. and Taylor, C.M.U.S. (2005). *John R. Taylor—Classical Mechanics-University S*, 2005. Class. Mech. Univ. Sci.

87 Bartenev, G.M. and Lavrent′ev, V.V. (1981). *Friction and Wear of Polymers*, vol. Vol. 6. Elsevier.

88 P. J. Blau, *The Significance and Use of the Friction Coefficient*, 2001.

89 Jones, D.R.H. and Ashby, M.F. (2019). Friction and wear. In: *Engineering Materials 1*, 469–484. Elsevier.

90 Wypych, G. (2018). The coefficient of friction. In: *Handbook of Surface Improvement and Modification*, 165–193. Elsevier.

91 Cho, D.H., Bhushan, B., and Dyess, J. (2016). Mechanisms of static and kinetic friction of polypropylene, polyethylene terephthalate, and high-density polyethylene pairs during sliding. *Tribol. Int.* 94: 165–175. https://doi.org/10.1016/j.triboint.2015.08.027.

92 Wypych, G. (2012). *Handbook of Polymers*. ChemTec Pub.

93 B. N. J. Persson, *Theory and Simulation of Sliding Friction*, 1993.

94 Bowden, F.P. and Tabor, D. (1968). *Friction and Lubrication*. London: Methuen and Co. Ltd.

95 Ma, L., Cai, C., Tan, Y. et al. (2019). Theoretical model of transverse and longitudinal surface roughness and study on brittle-ductile transition mechanism for turning fluorophlogopite ceramic. *Int. J. Mech. Sci.* 150: 715–726. https://doi.org/10.1016/J.IJMECSCI.2018.10.059.

96 Zhou, Z.R. and Vincent, L. (1999). Lubrication in fretting—a review. *Wear* 225–229, no. PART II: 962–967. https://doi.org/10.1016/S0043-1648(99)00038-1.

97 Ratner, S.B. and Styller, E.E. (1981). Characteristics of impact friction and wear of polymeric materials. *Wear* 73 (2): 213–234. https://doi.org/10.1016/0043-1648(81)90292-1.

98 Suh, N.P. (1977). *An Overview of the del~ina~on Theory of Wear*. Elsevier Sequoia S.A.

99 G. W. Stachowiak, *Wear: Materials, Mechanisms and Practice*. 2006.

100 Bayer, R.G. (ed.) (1976). *Selection and Use of Wear Tests for Metals*, vol. 615. ASTM International.

101 G. Straffelini, *Springer Tracts in Mechanical Engineering Friction and Wear Methodologies for Design and Control.*

102 Kato, K. (2005). *Classification of Wear Mechanisms/Models*, vol. 216, 349–356. https://doi.org/10.1243/135065002762355280.

103 Eyre, T.S. (1981). Wear mechanisms. *Powder Metall.* 24 (2): 57–63. https://doi.org/10.1179/pom.1981.24.2.57.

104 Bhushan, B. (2000). *Modern Tribology Handbook, Two Volume Set*. CRC Press.

105 Bayer, R.J. (2004). *Mechanical Wear Fundamentals and Testing, Revised and Expanded*. CRC Press.

106 Khruschov, M.M. (1974). Principles of abrasive wear. *Wear* 28 (1): 69–88. https://doi.org/10.1016/0043-1648(74)90102-1.

107 Fabbro, S., Araujo, L.M., Engel, J. et al. (2021). Abrasive and adhesive wear behaviour of metallic bonds in a synthetic slurry test for wear prediction in reinforced concrete. *Wear* 476: https://doi.org/10.1016/j.wear.2021.203690.

108 Brink, T., Frérot, L., and Molinari, J.F. (2021). A parameter-free mechanistic model of the adhesive wear process of rough surfaces in sliding contact. *J. Mech. Phys. Solids* 147: https://doi.org/10.1016/j.jmps.2020.104238.

109 Akagaki, T. and Kato, K. (1987). Plastic flow process of surface layers in flow wear under boundary lubricated conditions. *Wear* .

110 Saketi, S., Bexell, U., Östby, J., and Olsson, M. (2019). On the diffusion wear of cemented carbides in the turning of AISI 316L stainless steel. *Wear* 430–431: 202–213. https://doi.org/10.1016/j.wear.2019.05.010.

111 Upadhyay, R.K. and Kumaraswamidhas, L.A. (2018). Bearing failure issues and corrective measures through surface engineering. In: *Handbook of Materials Failure Analysis*, 209–233. Elsevier.

112 Fuchs, E.R.H., Field, F.R., Roth, R., and Kirchain, R.E. (2008). Strategic materials selection in the automobile body: economic opportunities for polymer composite design. *Compos. Sci. Technol.* 68 (9): 1989–2002. https://doi.org/10.1016/j.compscitech.2008.01.015.

113 Billady, R.K. and Mudradi, S. Influence of filler incorporation on the mechanical and wear behaviour of synthetic fiber reinforced polymer matrix composites—a review. In: *AIP Conference Proceedings*. (May 2020), vol. 2236, https://doi.org/10.1063/5.0006881.

114 Mahir, F.I., Keya, K.N., Sarker, B. et al. (2019). A brief review on natural fiber used as a replacement of synthetic fiber in polymer composites. *Mater. Eng. Res.* 1 (2): 88–99. https://doi.org/10.25082/MER.2019.02.007.

115 Rahman, R. and Putra, S.Z.F.S. (2018). Tensile properties of natural and synthetic fiber-reinforced polymer composites. In: *Mechanical and Physical Testing of Biocomposites, Fibre-Reinforced Composites and Hybrid Composites*, 81–102. Elsevier.

116 Begum, K. and Islam, M.A. (2013). "Natural Fiber as a substitute to Synthetic Fiber in Polymer Composites: A Review," *Res. J. Eng. Sci.* vol. 2, no. 3, pp. 46–53.

117 Li, D.X., You, Y.L., Deng, X. et al. (2013). Tribological properties of solid lubricants filled glass fiber reinforced polyamide 6 composites. *Mater. Des.* 46: 809–815. https://doi.org/10.1016/j.matdes.2012.11.011.

118 Jyotikalita, J. and Kumar Singh, K. (2018). Tribological properties of different synthetic fiber reinforced polymer matrix composites—a review. *IOP Conf. Ser. Mater. Sci. Eng.* 455 (1): https://doi.org/10.1088/1757-899X/455/1/012134.

119 Li, J. and Xia, Y.C. (2010). The friction and wear properties of thermoplastic PA6 composites filled with carbon fiber. *J. Thermoplast. Compos. Mater.* 23 (3): 337–349. https://doi.org/10.1177/0892705708103397.

120 Nishitani, Y., Sugawara, N., Kawasaki, K., and Kajiyama, T. (2019). Tribological properties of ureidosilane treated natural fiber reinforced plant-derived polyamide1010 biomass composites. *AIP Conf. Proc.* 2139: https://doi.org/10.1063/1.5121683.

10

Friction and Wear Resistance of the Natural Fiber-Reinforced Polymer Composites With Metal Oxide Fillers

Niket Suresh Powar[1] *and Mariyappan Shanmugam*[2]

[1] *Daegu Gyeongbuk Institute of Science and Technology (DGIST), Department of Energy Science & Engineering, 333 Techno Jungang-daero, Hyeonpung-eup, Dalseong-gun, Daegu 42988, Republic of Korea*
[2] *Hindustan Institute of Technology and Science, Department of Physics, Centre of Excellence in Underwater Robotics and Communication, Chennai 603103, Tamil Nadu, India*

10.1 Introduction

Natural fiber/polymer-based composites (NFP) are potential alternative to synthetic fiber/polymer-based composites because of their lower environmental and pollution consequences [1]. Natural fibers are being held to the same standards as their synthetic analogues more and more frequently, such as when it comes to the materials used in brake pads for automobiles. In order to withstand the tremendous temperatures caused by braking, high-temperature-resistant fibers must be adequately distributed throughout the manufacturing process [2]. The matrix and reinforcements utilized in producing NFP materials allow further classification of the resulting materials. In most cases, the materials used for matrices fall into three categories: thermosets, thermoplastics, or elastomers. The characteristics of the matrix polymer that is utilized impact the processes utilized while dealing with NFP composites. Most of the time, thermosets are simple to cast or mold, but thermoplastics can be handled at higher temperatures, and elastomer-based NFP materials can be vulcanized [3–6]. The ability of the fiber to reinforce is directly proportional to the structure of the fiber itself. Several types of reinforcement are available, including particle, short-fiber, long-fiber, and laminar reinforcement [7–9].

The large surface area-to-volume ratio of the nanoparticles is responsible for the superior wettability between the nanoparticles and the polymer matrix, as reported by many industry professionals. This increased bonding is what leads to better transfer film strength. A thinner and smoother transfer layer generated during dry sliding wear led to a lower friction coefficient and a lower wear rate on the composites. In addition, it was emphasized that the wear resistance of the composites would be significantly increased the smaller the particles were. To determine friction and wear, a wide variety of tribometers were utilized. There are a few different varieties of

Tribological Properties, Performance, and Applications of Biocomposites, First Edition.
Edited by Chandrasekar Muthukumar, Senthilkumar Krishnasamy,
Senthil Muthu Kumar Thiagamani, and Ganesan Chinnachamy.

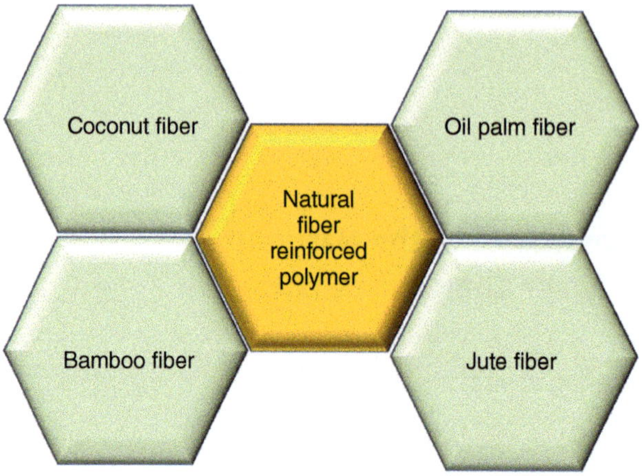

Figure 10.1 Representation of discussed natural fiber for reinforcement concerning the metal oxides.

tribometers that can be used to forecast friction and wear [10–14]. Most components used in industrial and manufacturing settings are subjected to tribological loadings such as adhesives, abrasives, and other substances. As a result, the tribological performance of materials has developed into an issue that is becoming an increasingly significant concern in the design of mechanical parts. As a result, the tribological performance of materials has turned into an issue that is becoming an increasingly significant concern in the manufacture of mechanical parts. In other words, the tribological behavior of natural fiber/polymer composites needs to be accorded the same weight as mechanical attributes [15–17].

This discussion leads us to the conclusion that reinforcement is required to maximize wear and friction in the natural fiber. Filler particles in the micro/nano range that are distributed evenly and linked appropriately with the matrix material can improve the material's mechanical characteristics. Particles with lower rigidity than the polymer matrix can add ductility to the material, although at the expense of the modulus. In addition, adding elastomer raises the viscosity of the resin mixture, making it more challenging to include a more significant weight percentage of fiber, and incorporating the stiff metal oxide particles into the matrix increases the biocomposites' toughness. In this chapter, we have discussed natural fiber, primarily used in applying wear and friction, as shown in Figure 10.1 [18–21].

10.2 Oil Palm Fiber

In several countries, oil palm generates major revenue, and plantation zones have spread across the globe. Malaysia's fourth-largest contribution to its gross domestic product is oil palm, a major agricultural crop in the country. The porous surface structure of oil palm fiber permits mechanical interlocking with the matrix resin, making it an effective reinforcement for composites. Due to their high cellulose content, mechanical strength, and crystalline structure, oil palm fibers are beneficial for composite applications. The cellulose and hemicellulose that make up natural fiber are loaded with hydroxyl groups. Hydrophilic fibers have poor interfacial adhesion with hydrophobic polymers and suffer from inferior mechanical properties in composites. As an added downside, natural fibers have many drawbacks. Their poor compatibility characterizes oil palm fiber for composite production with polymers, low thermal degradation, and high-water absorption capabilities. Thus, it is necessary to alter the fiber's surface. By eliminating hemicellulose, the hydrophilicity of fibers is reduced. The removal of hemicellulose through surface treatment of natural fibers can be used to enhance fiber-polymer matrix adhesion [22]. Ahmad Fahad Ahmad et al. studied the use of NiO in the form of a hybrid composite with oil palm empty fruit bunches/polycaprolactone as a new absorption material. They improved the thermal and dielectric properties of the resulting hybrid composite at X-band frequency. The fabrication and preparation of composite substrates were accomplished using the hot-blending approach [23]. In another study, Devaraju et al. reported that palm fiber composites (PFC) were made by the hand lay method by reinforcing a palm fiber with epoxy glue. The ceramic palm fiber composite (CPFC) was then formed by incorporating 0.3% wt% Al_2O_3 ceramic particles into the PFC (CPFC). Their study demonstrated that ceramic palm fiber has superior mechanical qualities to those of plain palm fiber composite, including hardness, tensile strength, impact strength, and flexural strength [24].

10.3 Jute Fiber

The polymer contains various essential characteristics, including simplicity of processing and economic suitability, which make it viable for number of applications [25] Nevertheless, single polymers lack strength, hardness, and other qualities, necessitating modification. Due to their superior strength and rigidity, glass and carbon fibers have supplanted lignocellulosic natural fibers such as jute. However, increasing environmental, health, safety, and biodegradability concerns about synthetic fibers compelled the designer to consider NFP. In addition, natural fibers are inexpensive and lighter in weight. Epoxy composites made from jute and glass fibers see extensive use in many different fields, including building, automobiles, aeroplanes, and wind turbines, as investigated by Abu Hena Md Shamim Hasan et al. Epoxy composites of jute and glass fibers have been examined, with and without fillers, such as Al_2O_3 and SiO_2 particles. The results showed that composites with higher composition of Al_2O_3 and SiO_2 have a maximum viscosity of 4172.95 $mm^2\ s^{-1}$ and at a weight of 15 g, maximum viscosity of 4977.15 $mm^2\ s^{-1}$ was achieved. The maximum tensile strength was determined to be 113.438 MPa without filler and 80.1202 MPa with the addition of Al_2O_3 and SiO_2 to the composite [26].

The wear performance of jute fiber composites may be significantly enhanced by adding appropriate wear resistant filler materials like SiC and Al_2O_3. Ahmed et al. indicated that 15 wt% Al_2O_3-filled jute-epoxy composites' resulted in lowest coefficient of friction, indicating superior wear resistance. The friction and wear loss coefficients of Al_2O_3-filled jute epoxy composites are significantly lower than those of SiC-filled composites. Fiber breakage and plastic deformation dominated the wear mechanism of unfilled jute-epoxy composites, whereas microcracking, pit and debris development dominated the wear mechanism of filled composites [27]. Mechanical parameters such as micro-hardness, flexural strength and impact strength of jute fiber with Al_2O_3 filler-based polymer composites were examined by Swain et al. in a separate study conducted under both dry and wet circumstances. Composites having 30% fiber content and 10 wt% Al_2O_3 content achieved flexural strength of 72.94 MPa and impact strength of 1.902 J. Increases in fiber and filler loading, such as 40 wt% fiber content and 10 wt% of Al_2O_3, improve the composite's hardness.

The higher fiber content in jute-reinforced polymer composites increased their weight at room temperature due to moisture absorption [28]. Improving the performance of a composite by treating the fiber surface and adding a nanofiller may be a realistic option. To this end, Adnan Amjad et al. developed a vacuum bagging approach to produce epoxy bio-nanocomposites reinforced with jute fiber (both alkali-treated and untreated) and nanofiller (Al_2O_3 and MgO). Tensile strength and modulus are similar in jute epoxy BNC with Al_2O_3 or MgO nanofillers. Because sodium hydroxide (NaOH) affects the cellulose structure of jute fibers, treated reinforcement performs better than untreated reinforcement. The alkali treatment removes oil, wax, lignin, and pectin from fibers. Using functional moieties to remove dirt enhances fiber attachment to the matrix [16, 21]. The Young's modulus of composites with treated reinforcement was 1% higher than untreated (41% higher than 0% Al_2O_3) and 83% higher than untreated. When the nanofiller concentration was increased to 4%, ultimate tensile strength and Young's modulus declined by 24% and 20%, respectively. Agglomeration of the nanofiller reduces surface area and fiber-matrix connectivity.

Treated jute had a higher tensile strength than untreated jute when MgO was present, by 21%, 44%, and 60%, respectively. Treated reinforced laminate's modulus climbed 29%, 77%, and 137%, while untreated jute's expanded 27%, 72%, and 125%. Increasing the nanofiller to 4% lowered the tensile strength and modulus by 31% and 73%, respectively. The nanofiller is agglomerated rather than dispersed weakening fiber/matrix bonding and surface area [29].

10.4 Bamboo Fiber

Bamboo fiber is prevalent in India, several Asian nations, and South America. It is strong and has a rapid growth rate. Due to its mechanical qualities, however, bamboo fiber is also known as natural glass fiber. Several investigations have demonstrated that bamboo fiber composites have excellent tensile, static, and bending strengths. However, with the rapid development of the bamboo processing sector, many processing residues and wastes have been formed, which, if not sensibly handled, would lead to severe environmental damage and cost increases.

Therefore, it is essential to develop efficient methods for maximizing the use of these wastes and residues from bamboo processing. To get rid of this problem, Guo et al. [30] stated that commercial nano-TiO_2 particles dispersed in a silane solution and sprayed to the surface of bamboo fibers were referred to as solid phase grafting (by creating O—H bonds). Injection molding was used to make the composites out of modified bamboo fiber-reinforced polypropylene. This paper's primary purpose was to investigate the compatibility of nano-TiO_2-modified bamboo fibers with a polypropylene matrix.

Incorporating TiO_2 nanoparticles into polypropylene/bamboo fiber composites resulted in a marginal improvement in the composites' mechanical strength. The tensile strength and impact strength of the composite treated with 0.4 wt% TiO_2 nanoparticles improved considerably. The composites' enhanced crystallinity and thermal stability suggest that the nanoeffect has the potential to modify the crystallization behavior of polypropylene matrix [30]. Microparticles are used to fill the spaces between macrofilter particles to improve wear resistance and polishing qualities. The material's physical attributes have enhanced appreciation of the filler component, particularly its toughness and durability. Particles with a submicron size have a greater specific surface area, which might facilitate stress transmission from the matrix to the particles. Submicron particles added to the matrix may enhance transfer films generated on the counter-surface through tribochemical processes. Therefore, Latha et al. looked into the impact of inorganic fillers on the mechanical and tribological characteristics of hybrid composites by studying the development of lightweight hybrid composites supplemented with TiO_2 and ZrO_2. They have found that ZrO_2 filler outperforms TiO_2 filler in terms of wear performance. The hybrid composite with 6 wt% ZrO_2 filler exhibits the least amount of wear compared to neat polymer composites [31]. Wear can be categorized into sliding wear, abrasive wear, and erosive wear, depending on the specific application. Erosion occurs when solid particles are subjected to the repetitive impact of gases and liquids, resulting in the removal of surface materials. When preparing composites using bamboo fibers as the reinforcement phase, adding inorganic fillers with high hardness values can further enhance the mechanical properties and wear resistance of bamboo fiber-reinforced polymer composites. Nanoscale Al_2O_3 has a high specific surface area and specific surface energy, making it a valuable additive for bamboo fiber-reinforced polymer composites since it can effectively compensate for inferior hardness of the bamboo fiber. This was achieved by Jiang et al. who reported superior wear resistance for the Al_2O_3/bamboo fiber-reinforced high-density polyethylene composite. In contrast, deficient Al_2O_3 levels (2 wt%) and very high Al_2O_3 contents (8 wt%) reduce composites' impact strength and hardness, respectively. Possible explanations for this behavior include the island effect of Al_2O_3 particles at low levels and the aggregation effect of Al_2O_3 particles at high content. Hardness and abrasion resistance tend to change with Al_2O_3 content, with the inflexion point occurring at 6 wt% Al_2O_3. The abrasion and abrasion resistance increased initially, then decreased with increasing Al_2O_3 content. They found the maximum erosive wear rate at a firing distance of 0.5 cm and an erosive time of 180 s [32].

10.5 Coconut Fiber

The coconut (*Cocos nucifera*) is the sole member of the genus Cocos. The shell and the coir from coconut fruit can be employed as reinforcement in biocomposite production. In tropical nations like Malaysia, Indonesia, Thailand, Sri Lanka, and India, the coconut shell is a common reinforcement used in biocomposites. This is due to its high modulus and specific strength. Coconut particles appeal to the automobile sector because of their durability, high hardness, good acoustic resistance, nontoxicity, resistance to microbial and fungal degradation, and inflammability. Because of their high compressive strength, matrix composites reinforced with coconut shells can be employed in various load-carrying applications. This classifies them as structural biocomposites, which are defined as those that require pressure to carry weight in use [33]. Friction and wear properties of TiO_2-Al-matrix-coconut-shell-ash were reported by Kumar et al. By using stir casting, they were able to create composites using an aluminum 6061 matrix and reinforcing materials of titanium dioxide and coconut shell ash. Ashes from coconut shells were set at a constant 3 wt%, while titanium dioxide amounts ranged from 2 to 6 wt%. Microstructural analysis of the hybrid composites confirmed that both types of reinforcing particles were evenly dispersed inside the matrix. The hardness of hybrid composites steadily rises as CSA increases, and TiO_2 particles are included as reinforcements. The friction coefficients of materials made from Al6061 and its hybrid compounds vary with the applied force. They found that as the load increased from 20 to 100 N, the coefficient of friction for both Al6061 alloy and its composites decreased. The study found that the Al6061-3%CSA–6%TiO_2 composite has the lowest friction coefficient. In addition, the high weight percentage of TiO_2 and the 3% coconut shell ash, which reduces wear by over 42%, account for excellent wear resistance. The hybrid composite of Al6061, 3% copper silicate alumina, and 6% TiO_2 has been found to have the maximum hardness [34]. In another study, Ashish Kumar et al. demonstrated that AA7075 hybrid metal–matrix composites are influenced in mechanical, microstructural, and tribological ways by reinforced particles. They concluded that when Al_2O_3 and CSA-reinforced particles were added to the Al matrix, the hybrid composite's hardness rose. There was a drop in wear rate and coefficient of friction compared to the base metal. The base alloy's highest corrosion rate decreased with each hybrid composite specimen [35].

10.6 Conclusion

Natural fibers offer a desirable substitute for synthetic fiber. On the other hand, reinforcement is needed to improve its mechanical qualities. Thus, the incorporation of metal oxide into the natural fibers is promising for the future of these products. However, it is not easy to further improve the mechanical properties of these materials. As a result, there is a need to discover cutting-edge metal oxides and chemical processing to achieve long-term stability in outdoor applications subjected to severe climatic conditions, including high temperatures, humidity, and UV radiation. In order to reduce pollution in the environment through the use of eco-friendly materials in various sectors, society needs to invest in research on producing NFRPCs.

References

1 Mohammed, L., Ansari, M.N.M., Pua, G. et al. (2015). *Int. J. Polym. Sci.* 2015.

2 Sinha, R.K., Sridhar, K., Purohit, R., and Malviya, R.K. (2020). *Mater. Today Proc.* 26: 3183–3186.

3 Petroudy, S.R.D. (2017). *Advanced High Strength Natural Fibre Composites in Construction*, 59–83. Elsevier.

4 Saheb, D.N. and Jog, J.P. (1999). *Adv. Polym. Technol.* 18: 351–363.

5 Ouarhim, W., Zari, N., and Bouhfid, R. (2019). *Mechanical and Physical Testing of Biocomposites, Fibre-Reinforced Composites and Hybrid Composites*, 43–60. Elsevier.

6 Thomason, J.L. and Rudeiros-Fernández, J.L. (2018). *Front. Mater.* 5: 60.

7 Gohal, H., Kumar, V., and Jena, H. (2020). *Mater. Today Proc.* 26: 1368–1372.

8 Laure, P., Silva, L., and M. (2011). Vincent, in Composite Reinforcements for Optimum Performance. *Elsevier* 616–650.

9 Vu-Khanh, T. and Denault, J. (1994). *J. Mater. Sci.* 29: 5732–5738.

10 Mishra, T., Mandal, P., Rout, A.K., and Sahoo, D. (2022). *Composites, Part C: Open Access* 100298.

11 Ahsan, Q., Mustafa, Z., and Chang, S.Y. (2021). *Tribological Applications of Composite Materials*, 347–381. Springer.

12 Rosamah, E., Hossain, M.S., Abdul Khalil, H.P.S. et al. (2017). *Adv. Compos. Mater* 26: 259–272.

13 Castellanos, L.J., Blanco-Tirado, C., Hinestroza, J.P., and Combariza, M.Y. (2012). *Cellulose* 19: 1933–1943.

14 Ferreira, D.P., Cruz, J., and Fangueiro, R. (2019). *Green Composites for Automotive Applications*, 3–41. Elsevier.

15 Shalwan, A. and Yousif, B.F. (2013). *Mater. Des.* 48: 14–24.

16 Baydoun, S., Arnaud, P., and Fouvry, S. (2020). *Tribol. Int.* 151: 106490.

17 Czichos, H. (1983). *Wear* 88: 27–43.

18 Evora, V.M.F. and Shukla, A. (2003). *Mater. Sci. Eng., A* 361: 358–366.

19 Zhang, M. and Singh, R.P. (2004). *Mater. Lett.* 58: 408.

20 Peng, G., Li, Q., Yang, Y. et al. (2008). *Polym. Adv. Technol.* 19: 1629–1634.

21 Adhikari, J., Biswas, B., Chabri, S. et al. (2018). *Polym. Compos.* 39: E101–E112.

22 Asyraf, M.R.M., Ishak, M.R., Syamsir, A. et al. (2021). *J. Mater. Res. Technol.* .

23 Fahad, A., Abbas, Z., Zainuddin, M.F. et al. (2015). *Procedia Environ. Sci.* 30: 273–278.

24 Devaraju, A. and Murali, P. (2020). *Mater. Today Proc.* 22: 1161–1166.

25 Wirawan, R. and Sapuan, S.M. (2018). Natural Fibre Reinforced Vinyl Ester and Vinyl. *Polym. Compos.* 157–168.

26 Shamim Hasan, A.H.M., Chowdhury, M.A., Almahri, A. et al. (2022). *Polym. Compos.*

27 Ahmed, K.S., Khalid, S.S., Mallinatha, V., and Kumar, S.J.A. (2012). *Mater. Des.* (1980–2015) 36: 306–315.

28 Swain, P.T.R. and Biswas, S. (2017). *IOP Conf Ser Mater Sci Eng*, 012010. IOP Publishing.

29 Amjad, A., Awais, H., Abidin, M., and Rahman, A.A.A. (2022). *Biomass Convers. Biorefin.* 1–14.

30 Guo, J., Cao, M., Ren, W. et al. (2021). *J. Mater. Sci.* 56: 12643–12659.

31 Latha, P.S. and Rao, M.V. (2018). *Silicon* 10: 1543–1550.

32 Jiang, L., Yang, Y., Fu, J. et al. (2022). *Polym. Compos.* .

33 Shahidan, N., Bakar, M.B.A., Mohamed, M. et al. (2020). *IOP Conference Series: Materials Science and Engineering*, 012039. IOP Publishing.

34 Hemanth Kumar, K.S., Siddeswarappa, B., Kulkarni, P.P., and Rokhade, K.K. (2022). *J. Inst. Eng. (India): Ser. D* 1–12.

35 Kumar, A., Singh, R.C., and Chaudhary, R. (2022). *Arab. J. Sci. Eng.* 1–15.

11

Investigation of Sliding Wear Properties of Nanofiller-Based Biocomposites

Ajish Babu[1], Anusree Thilak[2], Harikrishnan Pulikkalparambil[3], Sandhya Alice Varghese[4], Sanjay Mavinkere Rangappa[4,5], Kuruvilla Joseph[6], and Suchart Siengchin[4,5]

[1] *Indian Institute of Technology Patna, Department of Metallurgical and Materials Engineering, Bihta, Patna 801106, India*
[2] *Cochin University of Science and Technology (CUSAT), Department of Polymer Science and Rubber Technology, Kochi, Kerala 682022, India*
[3] *School of Nanoscience and Nanotechnology, Mahatma Gandhi University, Priyadarshini Hills, Kottayam, Kerala 686560, India*
[4] *King Mongkut's University of Technology North Bangkok, Materials and Production Engineering, The Sirindhorn International Thai-German Graduate School of Engineering (TGGS), Bangkok 10800, Thailand*
[5] *King Mongkut's University of Technology North Bangkok, Natural Composites Research Group Lab, The Sirindhorn International Thai-German Graduate School of Engineering (TGGS), Bangkok 10800, Thailand*
[6] *Indian Institute of Space Science and Technology, Department of Chemistry, Valiyamala PO, Trivandrum 695547, Kerala, India*

11.1 Introduction

Ever since their inception, polymers have continued to be fascinating materials and have evolved in numerous ways to replace traditional materials in many applications [1, 2]. In this technological era, in order to satisfy the ever-growing demand of customers, intensive research and development have been ongoing in both academia and industrial divisions on modification of polymeric materials. Especially, several research are going on in the direction of adding reinforcing fillers to polymer matrix in order to achieve synergistic effects of both polymer as well as filler. Such modifications in polymeric materials with the addition of particulate or fibrous fillers are known as polymer composites. Generally speaking, polymer matrix composites (PMCs) consist of two parts: the matrix phase (thermoset or thermoplastic) and the reinforcement phase (organic or inorganic). The matrix encases the filler particles and distributes the load evenly, while the incorporated fillers, serve as the primary load-bearing component with unique physical and chemical properties. These materials possess the advantage of easy tailoring when compared to metals and ceramics. For instance, the wear behavior of high temperature epoxy composites can be improved by incorporating self-lubricating filler like graphite [3]. In addition to being lightweight, polymer composites can be used to replace heavy metals without sacrificing strength.

Tribological Properties, Performance, and Applications of Biocomposites, First Edition.
Edited by Chandrasekar Muthukumar, Senthilkumar Krishnasamy,
Senthil Muthu Kumar Thiagamani, and Ganesan Chinnachamy.

Over the past few decades, considerable attention has been paid to sustainable engineering in order to minimize the threat to environment [4]. In this regard, extensive research is conducted on biopolymers due to their reliability, renewability, and biodegradability [5, 6]. This led to the development of sustainable biocomposites to replace conventional polymer composites. Polymer biocomposites consist of a polymer matrix and a reinforcement phase, with either one or both phases being of bio-origin or biodegradability [7, 8]. Biocomposites are gradually emerging, and macroscale tailoring to achieve the desired properties of biocomposites is no easier than it is for conventional composites [9].

Wear is one of the most significant causes of energy consumption worldwide. Several classifications of wear can be defined based on the type of materials involved in the frictional force. Among them, the sliding wear of materials has more significance, especially for products like gears in power transmission and automotive applications. Minimal wear and frictional loss are among the highly desired properties of polymeric materials. The wear behavior of self-lubricating polymers, such as nylon [10], polytetrafluoroethylene (PTFE) [11], and high density polyethylene (HDPE) [12], has been widely investigated. However, along with their low wear properties, their load-carrying capacity and strength are also significant during the design phase. Consequently, the addition of fillers to these polymer matrixes results in improved composite performance [13]. Fibers and other biofillers added to biopolymers and synthetic polymers have been reported to display intriguing wear behavior in biocomposites. Numerous studies have been conducted to determine the tribology and wear behavior of biocomposites with natural fiber reinforcement [14]. However, reviews based on the sliding wear behavior of biocomposites are seldom reported. In this context, this chapter focuses primarily on providing a comprehensive overview of the sliding wear behavior of polymer biocomposites reinforced with biofillers apart from natural fibers against their polished metallic counterparts.

11.2 Wear General Aspects

Friction and wear are an inevitable part of our daily activities, such as brushing our teeth, writing with a pencil, eroding the written words, tire rolling on the road, and so on. In brief, *wear* is a process involving the mechanical action of two contacting bodies in relative motion, resulting in the removal and deformation of materials from the surface [15]. It is noteworthy that wear and friction are both interrelated parameters. Wear does not occur in the absence of friction, and increased frictional force does not always result in wear of a body. However, in most cases, the loss of mass due to wear causes deformation, and changes in the subjected components actually result in crack-related failures. Furthermore, there can be nonuniform wear of materials due to the design of components. For instance, unbalanced tires (weight of tire components or wheel balance) will abrade their service life in-vehicle, leading to severe wobbling and accidents. The loss caused by wearing is determined by a number of factors (Figure 11.1), such as

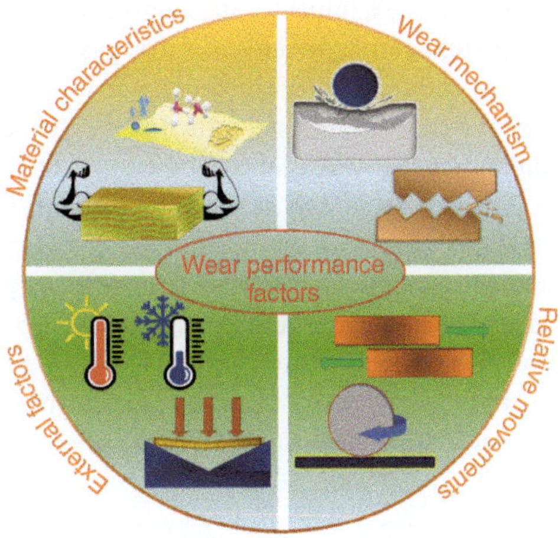

Figure 11.1 Factors influencing the wear of materials.

(a) wear mechanism,
(b) bulk and topographical characteristics of the mating materials,
(c) external conditions of the system applied,
(d) relative movements.

Recognition of the above factors aids in selecting suitable materials and engineering precisely according to the desired applications.

Based on the mechanisms and involved surface properties of the wear process, wear can be classified as adhesive or sliding, abrasive, corrosive, erosive, fatigued, or fretting wear. Sliding wear occurs when metallic or nonmetallic surfaces of two bodies slide against each other in wet (lubricated) or dry (air present) conditions [16]. Generally, wear occurring due to sliding against hard particles is considered abrasive wear [16]. While sliding happens between two surfaces, wear can occur due to surface fatigue, abrasion, adhesion, and chemical reactions like oxidation of surfaces. In the presence of temperature, accelerated wear may occur due to a faster oxidation rate, resulting in premature failure. Factors, such as sliding speed, movement direction (unidirectional or reciprocating), and applied load on body, have significant impact on the sliding wear rate [17]. Similar to sliding wear, in abrasive wear, hard particles are forced to slide against soft or comparable rough surfaces. Surface destruction, cracking, cutting, and scratches on the mating surfaces could happen due to these hard obtrusions, leading to failure of the component [18]. Furthermore, the wear loss of material is also dependent on the number of passes (repeated sliding or rolling through the same path) subjected to the mating surfaces. Usually, single-pass abrasive wear conditions are more prone to material loss than multiple-pass abrasive wear conditions [19].

Erosive wear occurs when material is lost or degrades from the constituent body as a result of the repeated impact of angular and hard particles. In comparison to abrasive wear, the impairment of components subjected to erosive wear will be relatively high due to the ready expulsion of materials caused by the action of impinging particles. The rate of material erosion is highly dependent on various factors such as impact velocity, impact angle, shape, size, and hardness of eroded materials [20]. In most engineering systems, such as gas and sand slurries carrying pipelines, steam and jet turbines, and fly ash-exposed boiler tubes, erosive wear negatively impacts the service life of exposed components. Conversely, they are also beneficial for applications such as water jet cutting and sandblasting [21].

Corrosive wear occurs when two bodies slide against each other in a corrosive or oxidizing environment. Components susceptible to friction and wear in marine and gas pipeline applications are primarily prone to wear failure due to corrosion. In such cases, hydrophobic coatings over the materials or high-corrosion-resistant polymer composites are typically used [22, 23]. Apart from these, other mechanisms are also significant for designing engineering components undergoing wear failure. However, a detailed study of the wear mechanism is beyond the scope of this chapter.

11.3 Methods to Measure Wear

Wear is affected by a variety of factors, as aforementioned. A variety of tribo-testers are available to address the volume or mass loss caused by material wear. However, the test conditions entirely depend on the desired application [24]. For instance, to test the wear loss of bicycle brake pads, the sliding wear test can be performed using polished metal counterparts. While abrasion wears measurements are taken against hard and rough surfaces to understand the wear behavior of the same bicycle tire. Although the temperature is not a key factor in both cases, the load would be according to the designed safety factor of the tire and brake pad. As a result, the wear evaluations of these components would be carried out while keeping practical aspects in mind.

Sliding wear tests for polymer composites can be either unidirectional or reciprocating. For estimating wear loss and coefficient of friction (COF) in unidirectional wear testing, the pin-on-disc method is commonly used. The basic design of the apparatus consists of a pin sliding against a rotating counter disc. In a laboratory testing scale, this apparatus setup can be modified with abrasive or rough counterpart surfaces, load, temperature, lubrication, air or vacuum, pin shape, and speed so as to meet the specifications for testing components [25–27]. A unidirectional sliding wear test cannot be used to estimate wear behavior for components that are subjected to periodic motion that includes both forward and reverse motion. In such cases, equipment such as pin-on-plate is set up to investigate the reciprocating wear characteristics of the parts [28]. Typically, a radius-tipped, or plain, test sample in pin shape is forced against a flat counterpart and moved back and forth to measure the wear loss associated with the movement. Alternative methods, in which the flat counterparts move while the test specimen remains fixed, are also commonly used.

In power transmission applications, the role of polymer materials is inevitable for designing components like gears, cams, and rollers, which undergo wear due to rolling motion. Understanding the wear behavior under this motion is quite complex, as both sliding and rolling are associated with the movement. Testing methods like ball-on-plate, where a steel ball is rolled against a polymer plate with

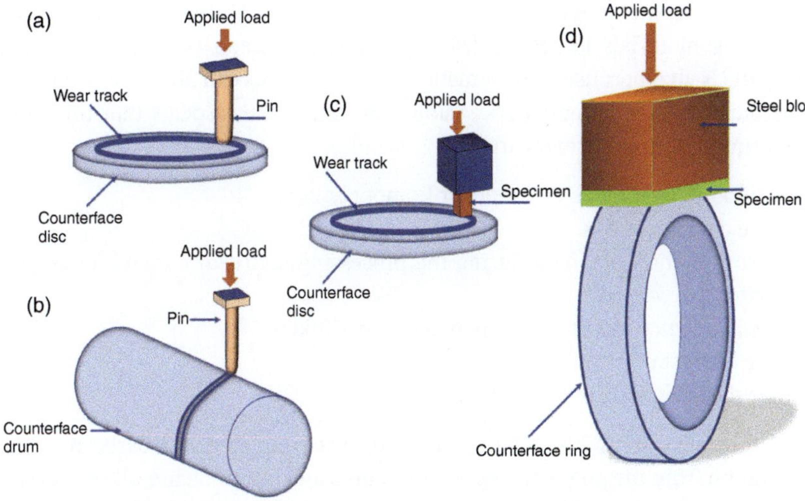

Figure 11.2 Illustrations showing some commonly used sliding wear test methods (a) pin-on-disc, (b) pin-on-drum, (c) block-on-disc, and (d) block-on-ring.

an orbital circuit, are commonly used to investigate the rolling wear of polymer composites. Here, the wear induced by both rotation of the ball and translational motion along the circular path is obtained. Furthermore, two disc testing methods have been reported to study the wear due to rolling motions, in which two polymer discs are arranged on parallel shafts and roll with opposite rotations under load on their circumferences [29]. Other methods, like scratch tests, where a rigid material is scratched against the specimen at a specified speed, geometry, and load, are also widely adapted. Similarly, for abrasive wear resistance analysis, test equipment such as hard counterpart surfaces or a dry sand rubber wheel is used [30, 31]. For erosive wear analysis, test rigs with gas supplies containing abrasive particles are provided in controlled conditions [32]. In short, the testing conditions for each wear mechanism are designed in accordance with the criteria for the desired applications. Figure 11.2 shows the schematic diagrams of some of the sliding wear techniques.

11.4 Sliding Wear in Polymer Composites

Applications of polymeric materials for tribological purposes have grown drastically in recent years, owing to their self-lubricating properties and ease of processing when compared to metals and ceramics [33, 34]. Polymers demonstrate less frictional force when rubbing against a surface due to their low stiffness and strength, but they also have a higher wear loss. Because of their soft and viscoelastic properties, polymeric materials are involved in mechanisms such as deformation and adhesion, where sliding wear occurs. The testing conditions, such as sliding speed, temperature, and applied load, have a significant impact on changes in the microstructure, physical and chemical properties, and surface and bulk properties of the polymeric materials involved with the process. Unlike metals, the use of lubricating materials to reduce frictional forces between polymers and their counterparts is vital because the interaction with lubricants can affect the structural and chemical changes of polymers. Aside from that, other aspects that influence the wear properties of polymers are:

(a) type of polymers (block polymers or homopolymers),
(b) surface characteristics,
(c) chain orientation obtained during the processing (extrusion, melt blending or shearing),
(d) molecular structure (linear, branched or crosslinked),
(e) degree of crystallinity,
(f) molecular weight distribution [35].

When polymers are subjected to sliding against counter-metal parts, material fragmentation from the polymeric surface and interlocking at the metal counterpart surface result in the formation of transfer films. Understanding specific factors that could affect transfer film formation is difficult, as many parameters and interactions are involved in wear process that affect the deformation and adhesion behavior of the polymer to the counterpart metal surfaces [36]. However, it is widely believed that transfer film formation over metal surfaces reduces the COF and provides a lubricant effect between the polymer and the counterpart metallic surface [37]. It is worth noting that the transfer film formations and thickness of the formed film in polymer composites can either increase or decrease the wear resistance depending on the fillers used. For fillers with good bonding with the polymer matrix, it is hard to form thin films, while for fillers that undergo easy degradation or discontinuity in the matrix, there may be a decrease in the wear resistance [37]. Aside from the effect of fillers, the surface finish of the counter surface also influences film formation, with a coated surface being able to retain a more stable transfer film than an uncoated one [38].

Polymer materials, as discussed previously, are likely to result in significant wear loss even under low loads due to their low strength and stiffness. Polymer reinforcement with appropriate fillers can alleviate this problem and improve overall wear resistance. Many studies have been conducted on the sliding wear properties of polymer composites, where both the polymer and reinforcement material types play a key role in determining the wear behavior [39–41]. To enhance the wear behavior of a polymer composite, it is essential to possess improved stiffness, strength, and reduced friction. Many efforts have been made to determine the cause of the increase in wear resistance for composites. The enhancement has been observed to be dependent on the properties of the fillers used. Some fillers, such as PTFE and graphite, reduce ploughing and adhesion between sliding parts, form thin, stable transfer films, and inhibit the harmful oxides produced during friction, lowering the severity of wear in polymer composites [42–45]. PTFE and graphite in polymer composites act as internal lubricants, resulting in less adhesion with the counterpart and a reduction in friction between the sliding surfaces [46, 47]. Other fillers like carbon fiber, aramids, and glass fibers increase polymer composite strength and stiffness, reducing wear loss [48–51]. High-temperature-stable polymers, such as epoxies, are commonly used for matrix parts to reduce heat generation caused by sliding surfaces. Furthermore, considering self-lubrication and other physical properties, polyamides, PTFE, high-molecular-weight polyethylene, polyether ether ketone (PEEK), polyphenylene, and polyoxymethylene are used as matrix phases to design sliding wear components. Apart from the individual properties of polymers and fillers, good compatibility and interfacial adhesion between the filler and matrix are needed to eliminate failures like fiber pullout and cracks during the wear process.

11.5 Sliding Wear in Biocomposites, General

Bio-based materials and their composites have been widely used by humans since ancient times. For nearly a century, the invention of synthetic polymers and subsequent industrial revolutions have outpaced the use of natural-based materials [52]. Rising environmental concerns have brought about a paradigm shift in the need for a change from the current scenario of energy usage. In this frame of reference, the use of biocomposites in place of traditional polymer composites has grown dramatically. Due to their wide availability of less harmful processing methods, recyclability, and economic benefits, much attention has been received to bio-based materials and related products around the globe. [8, 53] Polymeric biocomposites have been slowly replacing synthetic or traditional composites in vast applications such as automotive, sensors, biomedical, construction, packaging, etc. [54–57].

One of the major challenges associated with fabrication of biocomposites is their compatibility. In general, achieving higher compatibility between hydrophilic biofillers and hydrophobic polymers is quite difficult, especially when the interfacial adhesion and compatibility between filler and polymer matrix play a significant role in designing a component for wear applications. Nevertheless, modifications to the fiber or polymer matrix are primarily done to make the biocomposites suitable for the desired properties. Several attempts have been made to determine the wear behavior of biocomposites using a variety of biosourced fibers and materials. Also, the sliding wear behavior of natural fiber-based polymer biocomposites has received a lot of attention. However, naturally sourced reinforcing fillers from agricultural, waste, and animal sources in biocomposites are less studied for sliding wear properties. The below section comprehends the sliding wear studies based on biocomposites having fillers other than natural fibers.

11.5.1 Sliding Wear Property of Biofiller Incorporated Biopolymer Composite

The advancement of bio-based materials provides many opportunities to progress toward a less "carbon footprint" and more sustainable way of life. Biopolymers derived from renewable sources or synthesized chemically offer environmentally friendly degradation and serve as greener substitutes for conventional, nonrenewable-sourced polymers [14]. Starch, polylactic acid (PLA), polyglycolide

(PGA), polyhydroxybutyrate (PHB), and polycaprolactone (PCL) are some of the commonly studied biopolymers for biocomposites applications [58]. Among these biopolymers, PLA has piqued great attention from both scientific and industrial communities since its initial commercialization in 2002 [59]. Orozco and colleagues developed a starch-filled PLA-based biocomposite and investigated its mechanical and tribological properties [60]. Although starch as filler has better compatibility with PLA and is more processable, its moisture sensitivity and lower temperature stability may have an adverse effect on the composite properties. Keeping this in mind, the authors developed maleic anhydride-grafted PLA–starch composites with improved properties. When the sliding wear properties of the composite were measured using a scratch test, it was discovered that the composite containing 15 wt% starch and 85 wt% modified PLA showed less wear loss during testing. Furthermore, as the starch content in composites increased, the dynamic friction decreased, which was attributed to the self-lubricating property of starch [60]. PLA-based composites have been much studied for recent technological advancements like 3D printing techniques [61, 62]. Ertane and colleagues attempted to comprehend the sliding behavior of 3D-printed wheat stem-derived biocarbon filled PLA-based biocomposite [52]. Under dry sliding wear tests, the composite with 30 wt% biocarbon demonstrated a COF value of 0.5 with minimal variation and the lowest wear volume, owing to embrittlement and stiffness change caused by biocarbon reinforcement of PLA. Since the fillers were based on biogenic carbon, from an environmental perspective, downcycling of the prepared composite for agricultural usage is envisaged.

Several studies have endeavored to transform agricultural byproducts into bio-based polymers [63]. Bhuyan et al. developed a biocomposite with tung oil-derived thermoset polymer and maize grain-derived spent germ filler, as well as crosslinkers, to explore the impact of crosslinking agents on the sliding wear behavior of composites [64]. In the absence of a crosslinking agent, the spent germ filler was unable to improve the wear resistance of the composite, which was attributed to the plasticizing effect of the filler's unreacted crude corn oil content. On the other hand, the inclusion of crosslinking agent at 20 wt% improved the wear property and yielded better results.

Nanomaterials have been one of the most researched materials in the past few decades due to their enhanced physicomechanical properties. The incorporation of nanomaterials in polymer matrix exhibits better properties for a wide variety of applications [65–68]. For instance, nanocellulose derived from natural sources has significant importance for preparing biocomposites [69, 70]. Also, bio-derived thermoset resins like bio-epoxy have been widely studied. These resins provide excellent properties similar to synthetic epoxy resin, along with the advantage of being greener material [71, 72]. Barari et al. investigated the sliding wear behavior of biocomposites, which consisted of industrial waste-derived bio-epoxy reinforced with silylated cellulose nanofiber [73]. The researchers examined the sliding wear behavior under various applied loads and sliding speeds. When compared to the sliding wear behavior of composite at a higher applied load, composite at lower applied load showed improved wear resistance and a lower COF value, which was attributed to the formation of a transfer film on the mating surface. The reliance of wear behavior on the

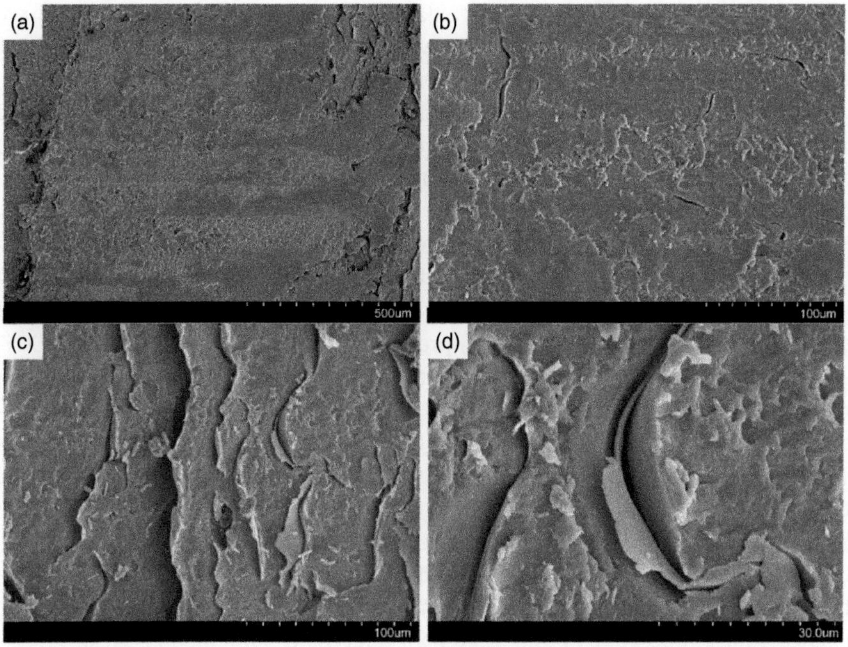

Figure 11.3 SEM analysis of worn surface of 1.4 wt% CNF (cellulose nano-fibers) reinforced with bio-epoxy nanocomposite at applied load of (a) 4 N and (b) 10 N, and worn surface analysis of bio-epoxy nanocomposite at applied load of 10 N at (c) 500× and (d) 1800× magnification. Source: Barari et al. [73]/with permission of Elsevier.

applied load, and the major wear mechanism, adhesion wear, were understood from the morphological analysis (Figure 11.3) of worn surface. Furthermore, no effect of sliding speed on the sliding behavior of biocomposite was observed.

11.5.2 Sliding Wear Property of Synthetic/Inorganic Filler Incorporated Biopolymer Composite

In an ecofriendly perspective, fully bio-based composites with bio-sourced or biodegradable fillers and polymers are extremely important for future sustainable steps. However, the fabrication and properties of these composites cannot be met for a wide range of applications. Biocomposites with bio-based fillers are limited due to their

(a) Hydrophilic nature and moisture sensitivity,
(b) Poor mechanical strength compared to synthetic fillers,
(c) Low load applications,
(d) Time consumption for extracting the fillers (e.g. silk extraction process),
(e) Biocompatibility, due to improper wetting,
(f) Irregular shape and size (impact the density and hardness),
(g) Easily prone to degradation [5].

As a consequence, biocomposites containing synthetic or inorganic fillers have been intensively investigated in order to overcome the limitations of biofillers. A few works are also reported on the sliding wear behavior of biocomposites with inorganic filler/synthetic fillers and biopolymers. Nanoclay, one of the most abundant materials on the planet with layered silicate structures, has been extensively studied for the preparation of polymer nanocomposite [74]. Bhuyan and colleagues conducted the early studies to investigate sliding wear behavior of biocomposites having inorganic fillers and bio polymer [75]. They prepared the biocomposites based on low-saturated soyabean oil and (4-vinylbenzyl)triethylammonium chloride, functionalizing montmorillonite as a reinforcing agent. Despite the fact that different probes (Si_3N_4 and diamond) produced different wear mechanisms, COF, and wear loss values during the wear analysis, the composite with an optimum montmorillonite (MMT) clay content of 1 wt% demonstrated superior wear properties. Similarly, taking into account the good miscibility of CAP and Chitosan, as well as the high mechanical properties of nanoclay, Iyer and colleagues investigated the sliding wear behavior of a silane-modified nonclay-reinforced CAP-Chitosan biocomposite [76]. At different applied loads of 5 and 10 N, the biocomposite with a filler proportion of 6 wt% displayed superior wear properties. However, beyond the filler content of 6 wt%, crack formation was observed during sliding due to possible chances of agglomeration of fillers and poor filler–matrix interactions sites.

Carbonaceous materials have long been used as reinforcing agents in PMCs owing to their high surface area, good physicomechanical properties, and high thermal and electrical conductivity [77, 78]. The self-lubricating properties of carbon materials, as well as their mechanism for reducing shear and mitigating plastic deformations during the wear process, exhibits them to potentially to be used as fillers for polymer composites in tribological applications [79]. However, the use of carbon materials in biocomposites for sliding wear applications has not been extensively explored. Deepthi et al. investigated the sliding wear behavior of functionalized multi-walled carbon nanotube (MWCNT) reinforced chitosan–starch blended biocomposite [80]. The wear loss of biocomposites was evaluated under three different sliding loads (6, 16, and 26 N) with varying MWCNT content (0.5, 1, 1.5, and 2 wt%) in polymer matrix. The studies show a lesser wear loss at a lower sliding load (6 N). Furthermore, it was observed that, at all loaded conditions, the composite with 1 wt% of MWCNT showed enhanced wear resistance. Similarly, Omrani and coworkers analyzed the wear properties of wood pulp-derived bio-epoxy/carbon fiber biocomposite [81]. The research focused on parameters such as sliding speed, normal load, and the fiber volume fraction of the prepared biocomposite to understand the sliding wear behavior of the prepared biocomposite. From the analysis, they observed that

(a) The wear properties (COF and wear loss) improved with the increase in fiber volume fraction.
(b) At higher loads, the detachment of fibers from polymer matrix led to improved lubrication between mating surfaces in order to reduce the COF; on the other hand, wear loss was increased.
(c) Inconsistent wear behavior at different sliding speeds.

With the emergence and development of 3D printing techniques, the design of complex polymer-based components and their properties have received considerable attention [82]. Przekop and coworkers developed a 3D-printed sliding element using PLA-graphite biocomposite [83]. Even though the influence of graphite improved wear resistance of the composite, some reductions in mechanical properties were observed, which were attributed to the rigid behavior of filler that impacted the property.

11.6 Conclusion

In this chapter, a short review of the classifications and techniques to improve sliding wear properties is reported. The wear properties of a polymeric material were classified based on the mechanism and type of wear as adhesive or sliding, abrasive, corrosive, erosive, fatigue, and fretting wear. Sliding wear occurs when metallic or nonmetallic surfaces of two bodies rub against each other in wet or dry conditions. The studies were reported based on the type of filler, such as biofillers or inorganic synthetic fillers.

Acknowledgment

HP would like to thank King Mongkut's University of Technology North Bangkok for the financial support under the foreign postdoctoral scholarship scheme, grant no. KMUTNB-Post-65-01.

References

1 Feldman, D. (2008). Polymer history. *Des. Monomers Polym.* 11: 1–15. https://doi.org/10.1163/156855508X292383.

2 Thakur, V.K., Thakur, M.K., Raghavan, P., and Kessler, M.R. (2014). Progress in green polymer composites from lignin for multifunctional applications: a review. *ACS Sustainable Chem. Eng.* 2: 1072–1092. https://doi.org/10.1021/sc500087z.

3 Pulikkalparambil, H., Rangappa, S.M., Siengchin, S., and Parameswaranpillai, J. (2021). Introduction to epoxy composites. In: *Epoxy Composites*, 1–21. Germany: Wiley.

4 Correa, J.P., Montalvo-Navarrete, J.M., and Hidalgo-Salazar, M.A. (2019). Carbon footprint considerations for biocomposite materials for sustainable products: a review. *J. Cleaner Prod.* 208: 785–794. https://doi.org/10.1016/j.jclepro.2018.10.099.

5 Vinod, A., Sanjay, M.R., Suchart, S., and Jyotishkumar, P. (2020). Renewable and sustainable biobased materials: an assessment on biofibers, biofilms, biopolymers and biocomposites. *J. Cleaner Prod.* 258: 120978. https://doi.org/10.1016/j.jclepro.2020.120978.

6 Vineeth, S.K., Gadhave, R.V., and Gadekar, P.T. (2019). Nanocellulose applications in wood adhesives—review. *Open J. Polym. Chem.* 09: 63–75. https://doi.org/10.4236/ojpchem.2019.94006.

7 Mohanty, A.K., Misra, M., and Hinrichsen, G. (2000). Biofibres, biodegradable polymers and biocomposites: an overview. *Macromol. Mater. Eng.* 276–277: 1–24. https://doi.org/10.1002/(SICI)1439-2054(20000301)276:1<1::AID-MAME1> 3.0.CO;2-W.

8 John, M. and Thomas, S. (2008). Biofibres and biocomposites. *Carbohydr. Polym.* 71: 343–364. https://doi.org/10.1016/j.carbpol.2007.05.040.

9 Sanjay, M.R., Madhu, P., Jawaid, M. et al. (2018). Characterization and properties of natural fiber polymer composites: a comprehensive review. *J. Cleaner Prod.* 172: 566–581. https://doi.org/10.1016/j.jclepro.2017.10.101.

10 Clavería, I., Gimeno, S., Miguel, I. et al. (2020). Tribological performance of nylon composites with nanoadditives for self-lubrication purposes. *Polymers (Basel)* 12: 2253. https://doi.org/10.3390/polym12102253.

11 Shi, Y., Feng, X., Wang, H., and Lu, X. (2008). The effect of surface modification on the friction and wear behavior of carbon nanofiber-filled PTFE composites. *Wear* 264: 934–939. https://doi.org/10.1016/j.wear.2007.06.014.

12 Johnson, B.B., Santare, M.H., Novotny, J.E., and Advani, S.G. (2009). Wear behavior of carbon nanotube/high density polyethylene composites. *Mech. Mater.* 41: 1108–1115. https://doi.org/10.1016/j.mechmat.2009.04.003.

13 Parameswaranpillai, J., Pulikkalparambil, H., Sanjay, M.R., and Siengchin, S. (2019). Polypropylene/high-density polyethylene based blends and nanocomposites with improved toughness. *Mater. Res. Express* 6: 075334. https://doi.org/10.1088/2053-1591/ab18cd.

14 George, A., Sanjay, M.R., Srisuk, R. et al. (2020). A comprehensive review on chemical properties and applications of biopolymers and their composites. *Int. J. Biol. Macromol.* 154: 329–338. https://doi.org/10.1016/j.ijbiomac.2020.03.120.

15 Chand, N. and Fahim, M. (2008). Introduction to tribology of polymer composites. In: *Tribology of Natural Fiber Polymer Composites*, 59–83. United Kingdom: Elsevier.

16 Zum Gahr, K.-H. (ed.) (1987). Sliding wear. In: *Microstructure and Wear of Materials*, 351–495. Netherlands: Elsevier.

17 Hutchings, I. and Shipway, P. (2017). Sliding wear. In: *Tribology: Friction and Wear of Engineering Materials*, 2e (ed. I. Hutchings and P. Shipway), 107–164. Netherlands: Elsevier.

18 Khruschov, M.M. (1974). Principles of abrasive wear. *Wear* 28: 69–88. https://doi.org/10.1016/0043-1648(74)90102-1.

19 El-Tayeb, N.S.M. and Yousif, B.F. (2007). Evaluation of glass fiber reinforced polyester composite for multi-pass abrasive wear applications. *Wear* 262: 1140–1151. https://doi.org/10.1016/j.wear.2006.11.015.

20 Arjula, S. and Harsha, A.P. (2006). Study of erosion efficiency of polymers and polymer composites. *Polym. Test.* 25: 188–196. https://doi.org/10.1016/j.polymertesting.2005.10.009.

21 Patnaik, A., Satapathy, A., Chand, N. et al. (2010). Solid particle erosion wear characteristics of fiber and particulate filled polymer composites: a review. *Wear* 268: 249–263. https://doi.org/10.1016/j.wear.2009.07.021.

22 Wang, J., Yan, F., and Xue, Q. (2009). Tribological behavior of PTFE sliding against steel in sea water. *Wear* 267: 1634–1641. https://doi.org/10.1016/j.wear.2009.06.015.

23 Gu, Y., Zheng, X., Liu, Q. et al. (2018). Investigating corrosion performance and corrosive wear behavior of sol–gel/MAO-coated Mg alloy. *Tribol. Lett.* 66: 101. https://doi.org/10.1007/s11249-018-1052-8.

24 Abdelbary, A. (2014). Methodology of testing in wear. In: *Wear of Polymers and Composites* (ed. A. Abdelbary), 159–183. Cambridge, UK: Elsevier.

25 Chandran, M. (2019). Synthesis, characterization, and applications of diamond films. In: *Carbon-Based Nanofillers and Their Rubber Nanocomposites* (ed. S.Y. Yaragalla, R. Mishra, S. Thomas, et al.), 183–224. Netherlands: Elsevier.

26 De Barros Bouchet, M.I. and Kano, M. (2007). Superlubricity of diamond/glycerol technology applied to automotive gasoline engines. In: *Superlubricity* (ed. A. Erdemir and J.-M. Martin), 471–492. Netherlands: Elsevier.

27 Abdelbary, A. (2014). Wear of internally lubricated polymers: silicone impregnated polyethylenes. In: *Wear of Polymers and Composites* (ed. A. Abdelbary), 113–132. Netherlands: Elsevier.

28 Hesse, D., Stanko, M., Hohenberg, P., and Stommel, M. (2020). Investigation of process control influence on tribological properties of FLM-manufactured components. *J. Manuf. Mater. Process* 4: 37. https://doi.org/10.3390/jmmp4020037.

29 Harrass, M., Friedrich, K., and Almajid, A.A. (2010). Tribological behavior of selected engineering polymers under rolling contact. *Tribol. Int.* 43: 635–646. https://doi.org/10.1016/j.triboint.2009.10.003.

30 Nahvi, S.M., Shipway, P.H., and McCartney, D.G. (2009). Particle motion and modes of wear in the dry sand–rubber wheel abrasion test. *Wear* 267: 2083–2091. https://doi.org/10.1016/j.wear.2009.08.013.

31 Molnar, W., Varga, M., Braun, P. et al. (2014). Correlation of rubber based conveyor belt properties and abrasive wear rates under 2- and 3-body conditions. *Wear* 320: 1–6. https://doi.org/10.1016/j.wear.2014.08.007.

32 More, S.R., Bhatt, D.V., and Menghani, J.V. (2017). Recent research status on erosion wear—an overview. *Mater. Today Proc.* 4: 257–266. https://doi.org/10.1016/j.matpr.2017.01.020.

33 Yashas Gowda, T.G., Sanjay, M.R., Parameswaranpillai, J. et al. (2021). Tribological applications of polymer composites. In: *Tribology of Polymer Composites*, 355–368. Netherlands: Elsevier.

34 Senthilkumar, K., Siva, I., Karthikeyan, S. et al. (2021). Mechanical, structural, thermal and tribological properties of nanoclay based phenolic composites. In: *Phenolic Polymers Based Composite Materials*, 123–138. Singapore: Springer Singapore.

35 Brostow, W., Kovačević, V., Vrsaljko, D., and Whitworth, J. (2010). Tribology of polymers and polymer-based composite. *J. Mater. Educ. V* 32: 273–290.

36 Ye, J., Burris, D., and Xie, T. (2016). A review of transfer films and their role in ultra-low-wear sliding of polymers. *Lubricants* 4: 4. https://doi.org/10.3390/lubricants4010004.

37 Bahadur, S. (2000). The development of transfer layers and their role in polymer tribology. *Wear* 245: 92–99. https://doi.org/10.1016/S0043-1648(00)00469-5.

38 Amenta, F., Bolelli, G., Pedrazzi, S. et al. (2021). Sliding wear behaviour of fibre-reinforced PTFE composites against coated and uncoated steel. *Wear* 486–487: 204097. https://doi.org/10.1016/j.wear.2021.204097.

39 Vadivel, H.S., Golchin, A., and Emami, N. (2018). Tribological behaviour of carbon filled hybrid UHMWPE composites in water. *Tribol. Int.* 124: 169–177. https://doi.org/10.1016/j.triboint.2018.04.001.

40 Wang, Y., Yin, Z., Li, H. et al. (2017). Friction and wear characteristics of ultrahigh molecular weight polyethylene (UHMWPE) composites containing glass fibers and carbon fibers under dry and water-lubricated conditions. *Wear* 380–381: 42–51. https://doi.org/10.1016/j.wear.2017.03.006.

41 Mazza, L., Trivella, A., Grassi, R., and Malucelli, G. (2015). A comparison of the relative friction and wear responses of PTFE and a PTFE-based composite when tested using three different types of sliding wear machines. *Tribol. Int.* 90: 15–21. https://doi.org/10.1016/j.triboint.2015.04.001.

42 Chauhan, S.R., Kumar, A., and Singh, I. (2010). Sliding friction and wear behaviour of vinylester and its composites under dry and water lubricated sliding conditions. *Mater. Des.* 31: 2745–2751. https://doi.org/10.1016/j.matdes.2010.01.020.

43 Basavarajappa, S., Ellangovan, S., and Arun, K.V. (2009). Studies on dry sliding wear behaviour of graphite filled glass–epoxy composites. *Mater. Des.* 30: 2670–2675. https://doi.org/10.1016/j.matdes.2008.10.013.

44 Upadhyay, R.K. and Kumar, A. (2019). Epoxy-graphene-MoS_2 composites with improved tribological behavior under dry sliding contact. *Tribol. Int.* 130: 106–118. https://doi.org/10.1016/j.triboint.2018.09.016.

45 Kumaresan, K., Chandramohan, G., Senthilkumar, M. et al. (2011). Dry sliding wear behaviour of carbon fabric-reinforced epoxy composite with and without silicon carbide. *Compos. Interfaces* 18: 509–526. https://doi.org/10.1163/156855411X610241.

46 Gürgen, S., Çelik, O.N., and Kuşhan, M.C. (2019). Tribological behavior of UHMWPE matrix composites reinforced with PTFE particles and aramid fibers. *Compos. Part B Eng.* 173: 106949. https://doi.org/10.1016/j.compositesb.2019.106949.

47 Annappa, A.R. and Basavarajappa, S. (2014). Studies on dry sliding wear behaviour of functionally graded graphite particle-filled glass-epoxy composites. *Compos. Interfaces* 21: 395–414. https://doi.org/10.1080/15685543.2014.870866.

48 Chand, N. and Sharma, M.K. (2008). Development and sliding wear behaviour of milled carbon fibre reinforced epoxy gradient composites. *Wear* 264: 69–74. https://doi.org/10.1016/j.wear.2007.01.043.

49 Huang, T., Lu, R., Ma, Y. et al. (2012). Study on the friction and sliding wear behavior of hybrid polytetrafluoroethylene/kevlar fabric composites filled with polyphenylene sulfide. *J. Macromol. Sci. Part B* 51: 109–124. https://doi.org/10.1080/00222348.2011.583195.

50 Wu, J. and Cheng, X.H. (2006). The tribological properties of Kevlar pulp reinforced epoxy composites under dry sliding and water lubricated condition. *Wear* 261: 1293–1297. https://doi.org/10.1016/j.wear.2006.03.014.

51 Siddhartha, S.A.K. (2015). Mechanical and dry sliding wear characterization of short glass fiber reinforced polyester-based homogeneous and their functionally graded composite materials. *Proc. Inst. Mech. Eng. Part L J. Mater. Des. Appl.* 229: 274–298. https://doi.org/10.1177/1464420713511429.

52 Ertane, E.G., Dorner-Reisel, A., Baran, O. et al. (2018). Processing and wear behavior of 3D printed PLA reinforced with biogenic carbon. *Adv. Tribol.* 2018: 1–11. https://doi.org/10.1155/2018/1763182.

53 Mahmud, S., Hasan, K.M.F., Jahid, M.A. et al. (2021). Comprehensive review on plant fiber-reinforced polymeric biocomposites. *J. Mater. Sci.* 56: 7231–7264. https://doi.org/10.1007/s10853-021-05774-9.

54 Zini, E. and Scandola, M. (2011). Green composites: an overview. *Polym. Compos.* 32: 1905–1915. https://doi.org/10.1002/pc.21224.

55 Qiu, Q., Chen, H., Wang, Y., and Ying, Y. (2019). Recent advances in the rational synthesis and sensing applications of metal-organic framework biocomposites. *Coord. Chem. Rev.* 387: 60–78. https://doi.org/10.1016/j.ccr.2019.02.009.

56 Pandit, A. and Kumar, R. (2021). A review on production, characterization and application of bacterial cellulose and its biocomposites. *J. Polym. Environ.* 29: 2738–2755. https://doi.org/10.1007/s10924-021-02079-5.

57 Zarna, C., Opedal, M.T., Echtermeyer, A.T., and Chinga-Carrasco, G. (2021). Reinforcement ability of lignocellulosic components in biocomposites and their 3D printed applications—a review. *Compos. Part C Open Access* 6: 100171. https://doi.org/10.1016/j.jcomc.2021.100171.

58 Van de Velde, K. and Kiekens, P. (2002). Biopolymers: overview of several properties and consequences on their applications. *Polym. Test.* 21: 433–442. https://doi.org/10.1016/S0142-9418(01)00107-6.

59 Murariu, M. and Dubois, P. (2016). PLA composites: from production to properties. *Adv. Drug Delivery Rev.* 107: 17–46. https://doi.org/10.1016/j.addr.2016.04.003.

60 Orozco, V.H., Brostow, W., Chonkaew, W., and López, B.L. (2009). Preparation and characterization of poly(lactic acid)-g-maleic anhydride + starch blends. *Macromol. Symp.* 277: 69–80. https://doi.org/10.1002/masy.200950309.

61 Bhagia, S., Bornani, K., Agrawal, R. et al. (2021). Critical review of FDM 3D printing of PLA biocomposites filled with biomass resources, characterization, biodegradability, upcycling and opportunities for biorefineries. *Appl. Mater. Today* 24: 101078. https://doi.org/10.1016/j.apmt.2021.101078.

62 Lee, C.H., Padzil, F.N.B.M., Lee, S.H. et al. (2021). Potential for natural fiber reinforcement in PLA polymer filaments for fused deposition modeling (FDM) additive manufacturing: a review. *Polymers (Basel)* 13: 1407. https://doi.org/10.3390/polym13091407.

63 Riedel, S.L. and Brigham, C. (2019). Polymers and adsorbents from agricultural waste. In: *Byproducts from Agriculture and Fisheries* (ed. B.K. Simpson, A.N.A. Aryee, and F. Toldrá), 523–544. India: Wiley.

64 Bhuyan, S., Sundararajan, S., Pfister, D., and Larock, R.C. (2010). Effect of filler composition and crosslinker concentration on the tribological behavior of spent germ particle-based polymeric composites. *Tribol. Int.* 43: 171–177. https://doi.org/10.1016/j.triboint.2009.05.012.

65 Pulikkalparambil, H., Siengchin, S., and Parameswaranpillai, J. (2018). Corrosion protective self-healing epoxy resin coatings based on inhibitor and polymeric healing agents encapsulated in organic and inorganic micro and nanocontainers. *Nano-Struct. Nano-Objects* 16: 381–395. https://doi.org/10.1016/j.nanoso.2018.09.010.

66 Parameswaranpillai, J.V., Salim, N., Pulikkalparambil, H. et al. (2022). *Micro- and Nano-Containers for Smart Applications*. Singapore: Springer Singapore.

67 Sreenath, P.R., Mandal, S., Panigrahi, H. et al. (2020). Carbon dots: fluorescence active, covalently conjugated and strong reinforcing nanofiller for polymer latex. *Nano-Struct. Nano-Objects* 23: 100477. https://doi.org/10.1016/j.nanoso.2020.100477.

68 Mazari, S.A., Ali, E., Abro, R. et al. (2021). Nanomaterials: applications, waste-handling, environmental toxicities, and future challenges—a review. *J. Environ. Chem. Eng.* 9: 105028. https://doi.org/10.1016/j.jece.2021.105028.

69 Mondal, S. (2018). Review on nanocellulose polymer nanocomposites. *Polym. Plast. Technol. Eng.* 57: 1377–1391. https://doi.org/10.1080/03602559.2017.1381253.

70 Kausar, A. (2020). A review of fundamental principles and applications of polymer nanocomposites filled with both nanoclay and nano-sized carbon allotropes—graphene and carbon nanotubes. *J. Plast. Film Sheeting* 36: 209–228. https://doi.org/10.1177/8756087919884607.

71 Pulikkalparambil, H., Varghese, S.A., Siengchin, S., and Parameswaranpillai, J. (2018). Thermally mendable and improved hydrophilic bioepoxy/PEG-PPG-PEG blends for coating application. *Mater. Res. Express* 6: 025307. https://doi.org/10.1088/2053-1591/aaeccf.

72 Rangappa, S.M., Parameswaranpillai, J., Yorseng, K. et al. (2021). Toughened bioepoxy blends and composites based on poly(ethylene glycol)-block-poly(propylene glycol)-block-poly(ethylene glycol) triblock copolymer and sisal fiber fabrics: a new approach. *Constr. Build. Mater.* 271: 121843. https://doi.org/10.1016/j.conbuildmat.2020.121843.

73 Barari, B., Omrani, E., Dorri Moghadam, A. et al. (2016). Mechanical, physical and tribological characterization of nano-cellulose fibers reinforced bio-epoxy composites: an attempt to fabricate and scale the 'Green' composite. *Carbohydr. Polym.* 147: 282–293. https://doi.org/10.1016/j.carbpol.2016.03.097.

74 Guo, F., Aryana, S., Han, Y., and Jiao, Y. (2018). A review of the synthesis and applications of polymer–nanoclay composites. *Appl. Sci.* 8: 1696. https://doi.org/10.3390/app8091696.

75 Bhuyan, S., Sundararajan, S., Lu, Y., and Larock, R.C. (2010). A study of the physical and tribological properties of biobased polymer–clay nanocomposites at different clay concentrations. *Wear* 268: 797–802. https://doi.org/10.1016/j.wear.2009.12.021.

76 Iyer, S.B., Sunitha, G., Dube, N.M., and Sailaja, R.R.N. (2011). Sliding wear characteristics of biodegradable nanocomposites of cellulose acetate phthalate and chitosan reinforced with functionalized nanoclay. *Trans. Indian Ceram. Soc.* 70: 153–158. https://doi.org/10.1080/0371750X.2011.10600163.

77 Díez-Pascual, A.M. (2020). Carbon-based polymer nanocomposites for high-performance applications. *Polymers (Basel)* 12: 872. https://doi.org/10.3390/polym12040872.

78 Pulikkalparambil, H., Parameswaranpillai, J., Jacob George, J. et al. (2017). Physical and thermo-mechanical properties of bionano reinforced poly(butylene adipate-co-terephthalate), hemp/CNF/Ag-NPs composites. *AIMS Mater. Sci.* 4: 814–831. https://doi.org/10.3934/matersci.2017.3.814.

79 Tahir, N.A.M., Bin, A.M.F., Tamaldin, N. et al. (2018). A brief review on the wear mechanisms and interfaces of carbon based materials. *Compos. Interfaces* 25: 491–513. https://doi.org/10.1080/09276440.2018.1380472.

80 Deepthi, M.V., Ananthapadmanabha, G.S., Sampathkumaran, P. et al. (2012). Preparation of bio-nanocomposites of chitosan/thermoplastic starch reinforced with multiwalled carbon nanotubes. In: *2012 IEEE 10th International Conference on the Properties and Applications of Dielectric Materials*, 1–4. Bangalore, India: IEEE.

81 Omrani, E., Barari, B., Dorri Moghadam, A. et al. (2015). Mechanical and tribological properties of self-lubricating bio-based carbon-fabric epoxy composites made using liquid composite molding. *Tribol. Int.* 92: 222–232. https://doi.org/10.1016/j.triboint.2015.06.007.

82 Saroia, J., Wang, Y., Wei, Q. et al. (2020). A review on 3D printed matrix polymer composites: its potential and future challenges. *Int. J. Adv. Manuf. Technol.* 106: 1695–1721. https://doi.org/10.1007/s00170-019-04534-z.

83 Przekop, R.E., Kujawa, M., Pawlak, W. et al. (2020). Graphite modified polylactide (PLA) for 3D printed (FDM/FFF) sliding elements. *Polymers (Basel)* 12: 1250. https://doi.org/10.3390/polym12061250.

12

Friction and Wear Properties of Biocomposites for Dental, Orthopedic, and Biomedical Applications

Piyush Gaur[1], Chandrasekar Muthukumar[2], and V. Parthasarathy[3]

[1] *University of Petroleum and Energy Studies, Mechanical Engineering Cluster, School of Engineering, Bidholi Campus, Via Premnagar, Dehradun, Uttarakhand 248007, India*
[2] *Hindustan Institute of Technology and Science, Department of Aeronautical Engineering, Rajiv Gandhi Salai, OMR, Padur, Kelambakkam, Chennai 603103, Tamil Nadu, India*
[3] *Rajalakshmi Institute of Technology, Department of Physics, Poonamallee, Chennai 600124, Tamil Nadu, India*

12.1 Introduction

Composite materials are created by combining two or more constituents with frequently different compositions and properties that do not dissolve or combine. Natural composites are generally found in both animals and plants. Composites are categorized as ceramic matrix composites (CMC), metal matrix composites (MMC), and polymer matrix composites (PMC), and they are extensively employed in structural, aerospace, automotive, and biomedical applications [1, 2]. MMCs find their application in structural applications due to their advantageous properties, such as high specific strength, low density, and easier fabrication [3]. At the same time, polymer composites with enhanced properties can replace conventional materials in automotive and structural applications [4]. Composite materials can be tailored to meet specific needs by varying their mechanical, physical, and chemical properties. This significant advantage allows composite materials to be applied in the human body, particularly as implants and prosthetic devices, i.e. biomaterials. Human organs have distinct properties and functions. As a result, replacing the infected/damaged/injured parts presents numerous challenges. Therefore, the body parts are replaced using biomaterials.

Understanding the interaction between materials and tissues has advanced over the last decade with the rapid advancement in materials technology. Biomaterials are natural or synthetic, living or non-living materials that interact with biological systems. The use of biomaterials dates back to ancient civilizations. Some examples include using artificial eyes, teeth, glues, ear waxes, and tissues in missing/defective body parts by Chinese, Indian, and Egyptian civilizations. Typically, biomaterials are composed of multiple components and are currently extensively utilized in medical devices, biomaterials engineering, drug delivery, biomedicine, orthodontics,

Tribological Properties, Performance, and Applications of Biocomposites, First Edition.
Edited by Chandrasekar Muthukumar, Senthilkumar Krishnasamy,
Senthil Muthu Kumar Thiagamani, and Ganesan Chinnachamy.

dentistry, and orthopedics, among other fields, owing to their desirable properties such as biocompatibility, nontoxicity, and corrosion resistance [5, 6]. The proper functioning of any biomaterial is associated with many factors like biocompatibility, particularly surface and structural compatibilities [7, 8], mechanical properties of implants, such as elastic modulus and stiffness [9], deformation features, chemical, biological, and physical suitability, health condition, surgical techniques, and patient's activities. Table 12.1 enlists important/critical factors to consider when selecting a material for use in the human body [10]. Biomaterials are utilized in the form of implants such as bone plates, sutures, grafts, joint sockets, artificial heart valves, dental implants, and medical devices such as electronic biosensors, artificial pacemakers, and artificial organs like heart, liver, and blood vessels to replace or restore the functions of damaged or diseased tissues or body parts. This results in an improvement in the overall quality of life for patients. In the biomedical industry, it is estimated that the global market for biomaterials is somewhere in the neighborhood of $12 billion per year, with an average global growth rate of 7 to 12% per year [10]. Biomaterials are expected to function properly within the hostile conditions of the human body's internal environment. The internal environment of the human body depends on the patient's activities and conditions.

Table 12.1 Important material properties for biomaterial selection in biomedical applications.

	Description		
Factors	**Mechanical properties**	**Physical properties**	**Chemical/biological properties**
Material properties level 1	Elastic modulus, Poisson's ratio, shear modulus, yield strength, compressive strength and tensile strength	Density	Chemical composition (bulk and surface)
Material properties level 2	Hardness, flexural strength and flexural modulus	Surface topology and texture roughness	Adhesion
Specific functional requirements (based on application)	Stiffness, rigidity, fatigue strength, fracture toughness, friction and wear, creep resistance, abrasion resistance and proof stress	Thermal expansion, refractive index, electrical conductivity, form and geometry, Opacity	Bioinert, biostability, bioactive, bio functionality and biodegradation
Processing and fabrication	Reproducibility, quality, sterilizability, packaging, secondary processability		
Characteristics of host: tissue, organ, species, sex, age, health condition, race, activity, and systemic response			
Medical/surgical procedure, period of application/usage, cost			

Source: Adapted from Ramakrishna et al. [10].

Clinical research experience clearly shows that not all commonly used engineering materials are appropriate for biomaterial use. However, until recently, most biomedical implants and medical devices were still designed and manufactured using single-phase, homogenous, and isotropic elastic materials [10], which are classified as (a) metal, (b) ceramics, (c) polymers, and (d) composites made from the combinations of a, b, and c. All these materials can also be bioactive, bioinert, and biodegradable. Metals, as implants, are characterized by a high level of mechanical strength, ductility, and wear resistance. The most common metal alloys used as bioimplants are stainless steel, titanium-based alloys, and cobalt–chromium [10]. Metal alloys suffer from the significant drawback of producing severe imaging artifacts in three-dimensional (3D) imaging procedures, such as nuclear magnetic resonance imaging and X-ray computer tomography (CT) [11, 12]. The stiffness mismatch between soft and hard tissues is another severe drawback of using metal implants in the human body. Few metal alloys, like cobalt–chromium and stainless steel alloys, are more prone to corrosion, releasing metal ions that trigger allergic reactions [13, 14]. However, titanium alloys are used in various applications due to their excellent biocompatibility inside the human body. Furthermore, biomaterials like ceramics are well known for their high compression resistance, good corrosion resistance, and biocompatibility. Instead of all these properties, they suffer from low fracture strength, brittleness, absence of resilience, and difficulty in fabrication, all negative aspects of ceramics [10]. Because of the limitations of traditional biomaterials, researchers and engineers have been motivated to develop polymer-based biocomposite materials as an alternative option for use in human body applications. Thus, the term "*biocomposites*" is specifically used to refer to those composite materials that have the potential to be used in bioengineering applications. *Biocomposites* are typically created by incorporating various reinforcing phases to obtain diversified biological and mechanical properties to improve the performance of implants and biomedical devices [15–17]. Biocomposites have significant advantages over other biomaterials, such as the absence of corrosion and the release of metal ions such as chromium or nickel [17, 18]. They also offer high resistance against fatigue loading and higher fracture toughness. Advanced polymer biocomposites also help reduce adverse effects by matching the stiffness of the implant material to that of the host tissues as they are low modulus materials, thus promoting bone tissue remodeling [19, 20]. Numerous orthopedic applications have been suggested for fiber-reinforced polymers due to their low elastic modulus and high strength [20]. Therefore, it is hypothesized that composite materials have a more significant structural biocompatibility potential compared to conventional biomaterials. For particular applications, such as dental implants, polymer composites can provide superior aesthetic properties.

All biomaterials, including biocomposites, must satisfy several requirements, including adequate strength, high corrosion resistance, adhesion, functionality, low friction, high wear resistance, and biocompatibility [21]. In contrast, the numerous developed biomaterials fail to meet all of the abovementioned criteria. According to reports, wear and corrosion are the leading contributors to the failure of biomaterials. Some biomedical applications of tribology include the heart valves [22] wear of dentures [23, 24], and screws and plates for bone fracture repair

[25], lubrication of artificial heart pumps [26], friction between garments and friction between skin and garments [27, 28], and pericardium and pleural surfaces. Wear is material loss due to mechanical (*surface-to-surface friction*) or chemical stimuli. On the surface of a material, mechanical wear causes the formation of microcracks, micro-cuts, and micro fatigue [29–32]. Friction and wear performance are essential for dental and orthopedics implant materials, which always have to deal with friction between opposing and surrounding tissues [29]. It was found that different architectural types and orientations of the composites resulted in different wear mechanisms exhibited by the materials, which in turn caused the opposing tissues to experience varying degrees of wear [33]. Wear is critical in controlling and determining a biomaterial's long-term clinical performance [34]. This chapter, thus, discusses the tribological aspects of biocomposites used in dental, orthopedic, and biomedical applications. The tribo-evaluation of materials used for advanced human body applications like artificial implants, filler materials, hip arthroplasty, bioprosthesis, and other phenomena is discussed and presented in subsequent sections.

12.2 Desired Properties and Classification of Biomaterials

12.2.1 Desired Properties of Biomaterials

Any implant or biomaterial is developed and designed to perform specific functions in the human body for an extended period of time. Hence, any implant inside the human body must satisfy various material requirements, as shown in Figure 12.1 and explained below for more clarity.

- *Mechanical properties*: Mechanical properties of hard and soft tissues must be considered when designing and developing any biomaterial (Implants). Hence, matching the modulus of biomaterial with the target tissue is crucial to prevent stress shielding of the implant [10]. For example, the matching of modulus of elasticity of biomaterials with bone helps to prevent the stress shielding of the

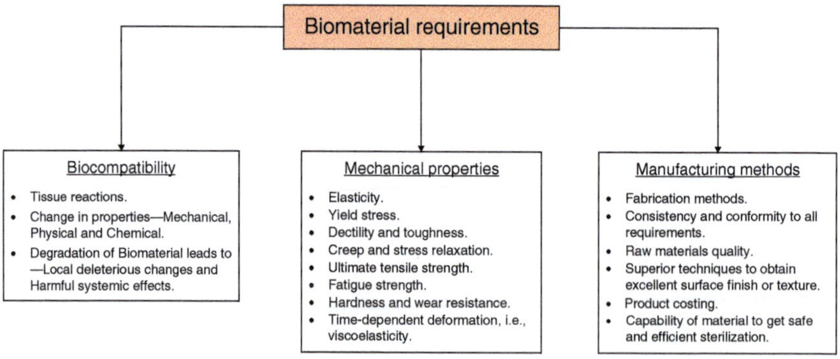

Figure 12.1 Implant material requirements in the human body. Source: Adapted from Long and Rack [35].

bones, which is in the range of 4–30 GPa [36–38]. Moreover, the material should have a high strength and low modulus to lengthen the implant's service period and prevent it from becoming loose, eliminating the requirement for subsequent surgical intervention [10].

- *Biocompatibility*: The newly developed substance must be harmless to living systems and should not inflict any bodily harm; this requirement encompasses all of the potentially harmful effects that a substance can have on the various parts of a biological system (ionic composition of plasma, bone, extra- and intracellular tissues) [28, 36–38]. Hence, an optimal interaction is possible between the host tissue and biomaterial only when both surface and structural compatibility are met.
- *High corrosion resistance*: A material for a perfect implant should have high corrosive resistance because the release of metal ions into the body occurs with any biomaterial with low corrosion resistance, which leads to allergic reactions. Example—Using nickel-based biomaterials in orthopedic applications could cause an allergy to the body.
- *s*: The biomaterial must also possess a low friction coefficient and high wear resistance when it slides against the surrounding tissues. The implant can become dislodged due to increased friction coefficient or wear resistance [36, 39]. Furthermore, the generated material wear debris can cause inflammation and destroy the implant's supporting tissues.
- *Nontoxic and long fatigue life*: The biomaterials should not alter the DNA of the genome and cause damage to the surrounding cells, i.e. they should neither be genotoxic nor cytotoxic. The biomaterial is also expected to have high fatigue resistance to prevent failure from fluctuating loads. This type of failure has been reported for cardiovascular and hip prosthesis applications [40–42].

12.2.2 Classification of Biomaterials

Biomaterial-based implants and medical devices (*orthopedic, dental, bone cement, etc.*) are categorized as (1) metallic materials, (2) ceramics, (3) polymers, and (4) composites, each of which is described below.

12.2.2.1 Metallic Biomaterials

Metallic-based biomaterials and implants offer superior mechanical properties and higher mechanical reliability. They are mainly used in the fabrication of biomedical devices and implants for tissue replacement for hard and soft tissues, such as artificial hip joints, bone plates, cardiovascular stents, dental implants, load-bearing applications, and internal fixation devices [43, 44]. The primary considerations in selecting metal and its alloys as biomaterials are their excellent electrical and thermal properties and mechanical properties, including higher tensile and yield strengths, higher fatigue life, higher creep, corrosion, and wear resistance [25, 40, 45, 46]. The following tables provide a summary of the many applications as well as the chemical compositions of various metallic alloys used in biomedical applications. Common metallic materials typically used for manufacturing bio-implants are stainless steel, titanium alloys, and cobalt-based alloys as shown in Table 12.2 [48].

Table 12.2 Various metallic bioimplants used in the human body.

		Comparison of metallic biomaterials used in the human body		
Metals and alloys	**Selected examples**	**Advantages**	**Disadvantages**	**Principal applications [47]**
Stainless steels	316L stainless steel	High wear resistance	Allergic issues with Ni & Cr; Modulus of Co much higher than bone	Fracture fixation, surgical instruments and stents
Cobalt and Cr alloys	Co–Cr–Mo Cr–Ni–Cr–Mo	High wear resistance	Allergic issues with Ni & Cr; Modulus of Co much higher than bone	Bone and joint replacement, dental restorations, heat valves and dental implants
Titanium-based alloys	Cp–Ti Ti–Al–V Ti–Al–Nb Ti-13Nb-13Zr Ti–Mo–Zr–Fe	High biocompatibility. Low Young's modulus, low density and excellent corrosion resistance	Toxic effect on Al and V on long term and poor tribological properties	Bone and joint replacement, pacemaker encapsulation, dental implants and fracture fixation
	Hg–Ag–Sn amalgam	Easy *in situ* formability to a desired shape susceptible to corrosion in the oral environment	Concerns related to Hg toxicity	Dental restorations
	Platinum and Pt–Ir	High corrosion resistance under extreme voltage and charge transfer conditions		Electrodes
	Ni–Ti	Low Young's Modulus	Ni cause allergy	Bone plates, orthodontic wires and stents

Source: Adapted from Niinomi [40].

Stainless Steel Stainless alloys are Fe-based alloys that were the first used in orthopedic applications. Because of their well-known resistance to corrosion against various agents, various steels fall under this category, generally known as stainless steels. It finds various applications owing to its ease of availability, excellent fabrication properties, low cost, great strength, and biocompatibility. The austenitic stainless steel SUS 316L is the most commonly used material in the biomedical field. The "L" in the designation denotes the low carbon content, which is crucial for its high corrosive resistance; and also has 10.5% Cr as the primary alloying element, thereby preventing the formation of rust [49, 50]. However, the presence of Ni in this alloy creates allergic reactions. The latest research has been directed toward developing high-nitrogen-content austenitic stainless steel to prevent these allergic reactions [40]. Few stainless-steel applications, along with their classification, have been tabulated as illustrated in Table 12.3.

Titanium-Based Alloys Titanium and its alloys have emerged as a potential choice for implants, particularly for orthopedic and dental applications [51–54]. Ti alloys offer good mechanical properties, excellent corrosive behavior due to the presence of titanium oxide layer, enhanced biocompatibility, and lightweight, nontoxic, and nonmagnetic behavior. All these abovementioned properties make titanium a perfect choice for manufacturing human body implants. The titanium alloys and their applications in the human body are tabulated below (Table 12.4). Despite the above advantages, these alloys suffer from poor friction and wear properties, primarily due to "low resistance to plastic shearing and poor work hardening [55].

Cobalt-Based Alloys 60 years ago, the surgical implant was based on cast cobalt-based alloys. The better understanding of the cast Co–Cr–Mo system provided the technical justification for using this alloy in biomedical applications. Due to high wear and corrosion resistance and biocompatibility, Co–Cr alloys are used to make artificial knee and hip joints. Co–Cr–Mo alloys, which are strong and ductile, are used to make hip joints and increase Co alloys' wear resistance; Carbide dispersion is

Table 12.3 Application of stainless steel used in biomedical applications.

Types of stainless steel in use for medical applications		
Types of stainless steel	**Cr content (%)**	**Medical applications**
Austenitic stainless steel	16–26	Canulae, guide pins, dental impression trays, hypodermic needles, holloware, storage cabinets, steam sterilizers, knee and hip implants
Ferritic stainless steel	11–30	Solid handles for instruments, guide pins and fasteners
Martensitic stainless steel	10.5–18	Bone curettes, dental burs, chisels and gouges, dental chisels, explorers, forceps, root elevators and scalers, orthodontic pliers and scalpels. Hemostats and reactors

Source: Adapted from Niinomi [40].

Table 12.4 Titanium commonly used alloys and their applications in different human body areas.

Titanium alloys for biomedical applications

S.no	Alloy	Microstructure
1	Pure Ti	(ASTM F67-89)
2	Ti-6Al-7Nb (ASTM F1295-92, ISO5832-11)	α + β type (Swiss)
3	Ti-6Al-4V (ASTM F1108-88)	α + β type
4	Ti-6Al-4V ELI (ASTM F136-84, F620-87)	α + β type
5	Ti-5Al-3Mo-4Zr	α + β type (Japan)
6	Ti-5Al-2.5Fe (ISO5832-10)	α + β type (Germany)
7	Ti-15Zr-4Nb-2Ta-0.2Pd	α + β type (Japan)
8	Ti-15Sn-4Nb-2Ta-0.2Pd	α + β type (Japan)
9	Ti-12Mo-6Zr-2Fe (ASTM F1813-97)	β type (USA), Low modulus
10	Ti-13Nb-13Zr (ASTM F1713-96)	Near β type (USA). Low modulus

Mechanical properties and clinical applications of Ti-based metallic materials

S.no	Alloy	Elastic modulus (Gpa)	0.2% offset yield strength (MPa)	Ultimate tensile strength (MPa)	Elongation (%)	Clinical applications
1	Ti-6Al-4V	110	860	930	10–15	Total joint replacement arthroplasty primarily for knees and hips
2	Pure Ti	102–110	170–480	240–550	15–24	Pacemaker cases, housings for ventricular-assist devices, dental implants, implantable infusion drug pumps, maxillofacial and craniofacial implants, spinal surgery, screws and staples
3	Ti-13Nb-13Zr	79–84	836–908	973–1037	10–16	Orthopedic implants
4	Ti-6Al-7Nb	105	795	860	10	Femoral hip stems, spinal components, fracture fixation plates, fasteners, rods, nails, screws and wire

Source: Adapted from Hussein et al. [34].

used [40]. Deformation-induced transformation of metastable phase to martensitic phase improves Co alloy wear resistance [40]. Wrought Co–Cr alloys can be used for implant devices that require high strength Co–Cr alloy is used to make stents and other surgical implants because it is biocompatible with blood and soft tissues. Ni presence in wrought Co–Cr alloys has been observed to cause allergic reactions.

12.2.2.2 Ceramic Biomaterials

Ceramics are considered promising biomaterials owing to their low thermal and electrical conductivity, high resistance to corrosion and compression, and biocompatibility. Most importantly, the bone is the constitution of ceramics like hydroxyapatite (HA) or tricalcium phosphate (TCP). Ceramics were first discovered in powdered form and were typically employed as filler materials; however, it was later found that that promoted bone generation. Table 12.5 shows the mechanical properties of a few commonly used ceramics in the human body. More details about ceramic-based biomaterials can be found in [56, 58, 59].

12.2.2.3 Composite Biomaterials

As stated earlier, a composite material may result from combining two or more materials to obtain a novel material with properties more significant than its constituents. The sole purpose of this combination is to improve mechanical properties like fatigue strength, hardness, tensile strength, and modulus. The elastic modulus of composites is significantly different from that of their constituents. Moreover, they are anisotropic [10]. The performance of the implants and their interaction with the surrounding tissues can be improved by reinforcing them with ceramic fillers. For example, in bio-implants made up of metal-based composites, ceramic coatings are provided for improved biocompatibility and stabilization of the bone growth on the implant's surface [56]. The implants manufactured using polymer matrix-reinforced ceramic fillers such as HA and TCP exhibit excellent biocompatibility and mechanical properties closely matching those of the bone, thereby reducing the effect of stress shielding [60, 61]. A schematic Figure 12.2 shows below the potential use of composites to repair, reconstruct, and replace human tissues.

Table 12.5 Mechanical properties of commonly used ceramics biomaterials [56, 57].

Material	Young's modulus (GPa)	Ultimate tensile strength (GPa)
Zirconia	220	2000
Alumina	380	4000
Bioinert glass	75	1000
Graphite	25	138
Vitreous (carbon)	31	172
Pyrolytic (carbon)	28	900
Hydroxyapatite	117	600
Vitro-ceramic	118	1080

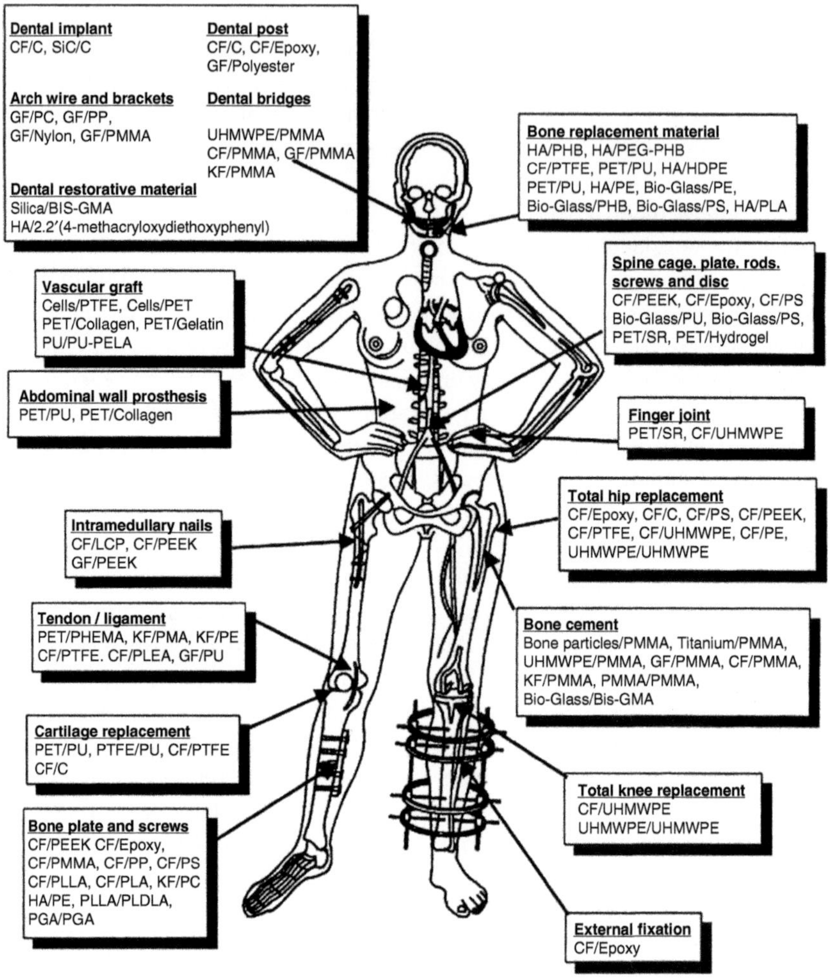

CF: carbon fibers, **C**: carbon, **GF**: glass fibers, **KF**: kevlar fibers, **PMMA**: Polymethylmethacrylate, **PS**: polysulfone, **PP**: Polypropylene, **UHMWPE**: ultra-high-molecular weight polyethylene, **PLDLA**: poly(L-DL-lactide), **PLLA**: poly (L-lactic acid), **PGA**: polglycolic acid, **PC**: polycarbonate, **PEEK**: polyetheretherketone, **HA**: hydroxyapatite, **PMA**: polymethylacrylate, **BIS-GMA**: bis-phenol A glycidyl methacrylate, **PU**: polyurethane, **PTFE**: polytetrafluoroethylene, **PET**: polyethyleneterephthalate, **PEA**: poltethylacrylate, **SR**: silicone rubber, **PELA**: Block co-polymer of lactic acid and polyethylene glycol, **LCP**: liquid crystalline polymer, **PHB**: polyhydroxybutyrate, **PEG**: polyethyleneglycol, **PHEMA**: poly(20hydroxyethyl methacrylate)

Figure 12.2 Various applications of different polymers and composite biomaterials used in the human body. Source: Ramakrishna et al. [10]/with permission of Elsevier.

12.3 Wear of Biomaterials

The discussion above has clearly illustrated that implants' friction and wear resistance, whether metallic, ceramic, or polymer composites, play a crucial role in maintaining proper functioning inside the human body. Hence, the friction and wear resistance of the implant become critically important from a clinical point of view. Several studies have been performed on friction and wear of metallic, ceramic, and polymer-based composite implants. However, this chapter will only discuss polymeric-based biocomposites' friction and wear properties in the human body. Before that, we first summarize the various wear test characterization configurations used by multiple researchers, followed by the experimental techniques used to characterize implants' friction and wear properties and a review of numerous results for the wear and friction of various biocomposites used in dental and orthopedics.

12.3.1 Wear Testing Methods

Characterizing biomaterials' friction and wear properties is critical to using the desired standardized test procedure. Commonly used techniques proposed in the literature to describe the wear properties of implants are block on a disc [62, 63], ball on a disc [64–67] and pin-on-disc [68]. The configurations are shown in the Figure 12.3. All the above tests were performed under a normal body temperature of 37 °C or similar conditions. These tests are operated in simulated body fluid (Ringer solution) to simulate real human body conditions. However, a few trials have been reported in the dry state as well.

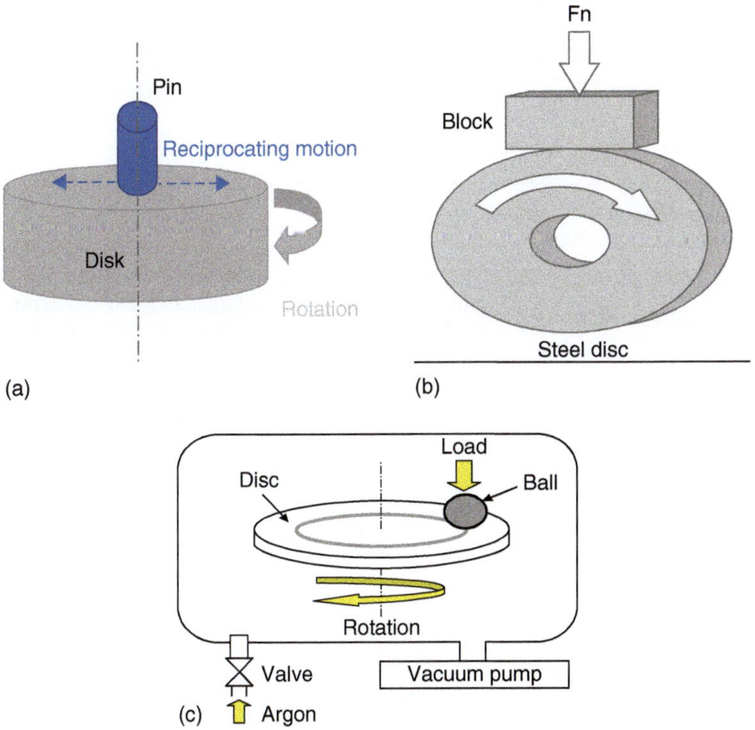

Figure 12.3 Schematic arrangement of (a) pin-on-disc. Source: Shen et al. [69]/with permission of Elsevier, (b) block-on- disc. Source: Adapted from Đurić et al. [70], (c) ball of disc wear test [71] configurations. Source: Edalati et al. [71]/with permission of Elsevier.

In the pin-on-disc method, the material to be tested is shaped like a pin and pressed against a grinding wheel that is turning. The process is good because the surface pressure stays the same. The set-up looks like a linear friction bearing, meaning the material is lost. It is so hard to line up the pin if the pin does not stand vertically in the corner of the pin contact on the spinning disc. The side of the pin takes off the lubricant that it touches. In ball-on-disc, the specimen is made in the shape of a ball, and high pressure may be used. During the test, less coolant is lost from the ball specimen than from the pin specimen. Set up in a way that is similar to a radial friction bearing. When compared to the spinning disc, the contact area of the ball is smaller than that of the pin. Wear makes the area of contact bigger. The problem with this method is that it is hard to measure the amount of wear on the ball specimen. In the block-on-disc method, the specimen is a block. This method is good for simulating a wide range of harsh field conditions, such as high heat, high speed, and high stacking pressure. Table 12.6 summarizes some common merits and demerits of different testing configurations.

12.3.2 Friction and Wear Characterization Techniques for Biomaterials

A material's friction and wear behavior have been evaluated using various characterization techniques. The energy-dispersive X-ray spectroscopy (EDXS) is used to estimate the chemical composition of certain regions. To determine the wear mechanism or carry out a post-test examination, a scanning electron microscope/light microscope (SEM/LM) was utilized [72]. The 3D surface roughness

Table 12.6 Merits and demerits of different wear test configurations.

S. no	Test configuration	Merits	Demerits
1	Ball on disc	High pressure on surfaces, Lubricant skin off due to ball, Setup is similar to the linear and radial friction bearing	Sliding track on the disc is larger as compared to the contact surface of the ball; Small contact ratio, wear results in a large contact area; strenuous to determine the wear volume of the ball
2	Block-on-disc	Capable of simulating multiple harsh field in service conditions, e.g. loading pressure. high speed and temperature	N/A
3	Pin-on-disc	Surface pressure remains intact after running; wear volume and rate can be determined easily and closely simulates a linear friction bearing	Pin alignment is difficult; misalignment in a pin can result in edge contacts leading to high loads, long run time

Source: Adapted from Hussein et al. [34].

of biomaterial was evaluated with the help of a 3D Talysurf instrument [64]. The chemical composition of the surface was investigated using X-ray photoelectron spectroscopy (XPS). An optical and 3D profilometer is employed to determine the volume loss that occurred in the wear track. In recent times, an atomic force microscope (AFM) has been applied in studying various tribological phenomena [73, 74], including friction, surface roughness, scratching, and adhesion, to name a few. A depth-sensing indentation system can be used with an AFM to measure various interesting mechanical properties, including the modulus of elasticity and the hardness of a material [75–79].

12.4 Friction and Wear Properties of Biocomposites Used in Different Biomedical Applications

12.4.1 Dental Applications

Human teeth are one of the most sensitive regions from a wear and tear point of view. Human teeth are divided into crown and root, separated by gum. The teeth help chew the food, which aids in digestion. The food particles get compressed during the chewing process. Due to the variety of foods available, the hardness of food particles varies. The teeth should be strong enough to chew various food particles. Nowadays, dental treatments are necessary for the normal function of teeth without causing toothache in humans. Cavity filling, shattered tooth replacement, cavity lining, and other dental procedures are available. Metallic and polymeric biomaterials were employed to treat dental cavities and damaged tooth replacements. The material chosen from the available biomaterials is determined by its physical and mechanical qualities and toxicity. Biomaterials such as stainless steel, zirconia, amalgam, alumina, acrylic resins, and others were commonly employed to reconstruct fractured teeth [80]. Because the presence of mercury in amalgams makes them harmful to the human body, their use is restricted. When utilized as implants, the metals are not only unsightly but also poisonous. Such issues do not exist with polymers. Because of the esthetical appearance of polymers, amalgam is recommended for restoring the backside of teeth, whereas resins are preferred for the front part of the teeth. Polyether-ether ketone (PEEK) is the most specialized dental polymeric substance [3, 81, 82].

Restorative dentistry includes the preservation of decayed teeth, the restoration of defective, missing, or traumatized teeth, and the fabrication of partial or complete dentures. Dental materials used in restorative applications must meet biocompatibility, mechanical properties, and aesthetic considerations. Table 12.7 [83] lists the common nonmetallic dental materials used for restorative dental work. The data on dental materials' wear and frictional behavior is used to predict the amount of restoration wear during mastication or normal tooth contacts and the level of frictional forces transmitted to the supporting oral tissues. Two-body and three-body abrasions are the predominant wear mechanisms in the mouth. The first occurs during the physiological rubbing of teeth in saliva, whereas the second involves the presence of abrasive particles from food and/or toothpaste [84]. The abrasion resistance

Table 12.7 Application and merits of nonmetallic implants in dental applications.

Material type	Applications	Advantages
Resin composites (macro filled, micro filled and hybrid)	Anterior and posterior restorations	Aesthetics, absence of mercury, high corrosion resistance
Acrylic resin	Dentures and artificial teeth	Aesthetics
Ceramics (alumina/glass/feldspathic)	Partial and full dentures	Aesthetics, high hardness

of restorative materials and denture teeth is among the most important tribological properties. Abrasion, pin-on-disc (plate), and scratch tests are typical laboratory procedures for evaluating the abrasion resistance of candidate dental materials [82, 83]. Regarding the tribological properties of restorative materials and denture teeth, their resistance to wear is one of the most crucial aspects to look for. Abrasion, pin-on-disc (plate), and scratch tests are laboratory evaluations typically carried out to determine the wear resistance of potential dental material. During the abrasion tests, the materials of interest are rubbed against abrasive papers while water or slurry is used as a lubricant. The amount of volume lost is then measured [64]. Less strenuous tests, using abrasives such as calcium carbonate [85–87] or a mixture of alumina and glass [88], are sometimes used because they appear to provide a more realistic simulation. Additionally, various specialized tests have been developed to estimate the resistance of restorations to the abrasiveness of dentifrices [89, 90].

The pin-on-disc type test setup is adapted to evaluate the wear between two sliding bodies, one of which is usually a loaded pin made of human enamel, and the other is a disc or plate made up of dental material tested. The wear depth and track on the discs lost by the pin are measured, and the wear rates or friction coefficients are determined. On the other hand, the scratch test, i.e. the hardness test method, determines the material's resistance to penetration. Under specified load and in a specified environment, a diamond hemisphere [91, 92] or a cone [93] is slid across the surface of the specimen, and the scratch width is measured. The extent of surface failure is depicted by the scratch width, which can then be correlated with the wear resistance. Following microscopic examination of the wear scars, the mode of failure, i.e. elastic, ductile, or brittle, and the transition load between ductile deformation and brittle fracture are determined. The major limitation of these experimental tests is the difficulty of extrapolating the results to material behavior *in vivo*. Thus, a clinical evaluation of the material performance in real dental applications is always conducted after the laboratory tests. Clinical evaluation involves visual observation, exploration, and photographic assessment of restorations and SEM examination of the dental specimens. In the early 1970s, the US Public Health Service [92, 93] came up with the first guidelines for how restorations should be evaluated clinically and how to measure how much they have broken down. After that, special ways have been made to make it easier to measure the wear of restorations *in vivo* [94–97].

12.4.1.1 Friction and Wear of Dental Resins

Denture teeth, veneers, and restorative materials made of acrylic resins and, later, composite resins with macro-, micro-, colloidal, and hybrid fillers have gained widespread acceptance as restorative materials, primarily for anterior teeth. Their benefits over the traditionally used amalgams in restoration include improving aesthetics and eliminating mercury, which is present in amalgams but not in theirs. Amalgams can contain potentially harmful mercury. On the other hand, the wear resistance of dental composites is often inadequate for their use in posterior tooth fillings, which are subject to higher mastication stresses than front tooth fillings. This is a significant limitation of dental composites. The friction and wear behavior of dental resins made of composites can be studied *in vivo* and *in vitro*.

***Wear of Dental Resins Composites* In Vitro** Using abrasion [91, 93, 98] and scratch tests [91, 92, 95], researchers have compared the *in vitro* tribological characteristics of early filled and unfilled resins. Abrasion tests were performed on silicon carbide paper, and the results showed that composite resins containing a silanated filler had the highest wear resistance. Additionally, the acrylic resin had a higher wear resistance than diacrylate resin [91]. In terms of its abrasion resistance, the acrylic resin also performed better than the diacrylate resin [91, 92]. The wear resistance of dimethyl acrylate resins depended on the type of abrasive paper used. The dimethyl acrylate resins offered a lower resistance on silicon carbide paper and a higher resistance on alumina and quartz paper. The resistance to penetration, frictional torque, and wear modes in the water of various filled and unfilled restorative resins were compared through a scratch test [88, 89]. The results of the tests demonstrated that the strength of a resin's resistance to being broken could be improved by incorporating the appropriate silanated fillers. The matrix material that determined the mode of deformation damage was significantly more severe for cross-linked diacrylate resin than linear [98]. The deformation mode also changed, ranging from brittle fracture to plastic deformation, depending on the dimethacrylate resin used [91].

Over the years, the composite resins have undergone various modifications to improve their friction and wear resistance [99]. Bis-GMA, UDMA, and TEGDMA are the types of dental restorative matrix materials that are most commonly used. The incorporation of the reinforcing agents into the matrix increases both the modulus of elasticity and strength of the matrix as well as the wear resistance. Quartz, silica, and silica glass are reinforcement materials that can be used [100, 101]. The dental polymer matrix is also reinforced with silica (20–50 nm) using bis-GMA (weight percent 70) and TEGDMA (weight percent 30). The percentage of silica used as reinforcement ranged from 20% to 50% by mass. Compared to conventional composites containing silica with particle sizes ranging from 10 to 40 mm, the addition of reinforcement improved the material's mechanical properties [102–104]. Also, these composite hybrid resins, reinforced by the mixture of submicron and macrofillers particles, have resulted in better *in vitro* wear resistance compared to conventional resins [105, 106]. The friction and wear properties of small particles, micro-filled resins, and hybrid resins were also compared using abrasion tests with SiC and calcium carbonate as abrasives. Out of these, the hybrid resins displayed the best wear resistance and were found to be unaffected by the storage water, contrary to the results obtained for the other two resins [106]. All the resin composites tend to absorb water, which negatively affects the matrix filler bonds [101, 106, 107].

Hence, materials have been researched and are used in the current scenario for effective dental restoration. Bis-GMA, epoxy, and polyamide are the most common types of resin used for dental posts. Glass fiber and carbon fiber are used to reinforce the resin. Post materials like epoxy-carbon fiber and Bis-GMA-glass fiber composites were used a lot [82]. The carbon fiber/epoxy composite showed superior properties as compared to Bis-GMA glass fiber composite, but it could cause skin irritation, allergies, and asthma, so researchers are looking for new materials for dental posts. Research is now being done on polyamide-carbon fiber as a material for dental posts owing to its high corrosion and wear resistance.

***Wear of Dental Restorations* In Vivo** It has been determined that amalgams outperform dental composites in terms of *in vivo* wear performance in posterior restorations when compared to amalgams [108–111]. It was discovered that conventional composite resins had a lower resistance to wear than amalgams, and the wear degradation began not long after the restoration was completed [108–110, 112]. Even though the recent development of better composites has led to their more confident and frequent use for posterior tooth restorations, evaluation tests still show that these materials have higher wear than amalgams. This is the case even though the use of composites has increased significantly in recent years [102, 111–114]. A two-stage process occurs when composite resins wear in the body [91, 92, 110, 115]. Abrasion is performed on the polymeric matrix in the first stage, which reveals the filler particles. During the process's second stage, the forces acting on the exposed particles cause the particles to rotate, chip, and become dislodged. The difficulty in accurately simulating the wear behavior of dental materials in laboratory tests was confirmed by comparing the patterns of wear seen *in vivo* and *in vitro* [92, 108].

12.4.2 Orthopedic Applications

The primary function of bones in the human body is to provide essential support within the body. A typical human bone is composed of crystals of HA coagulated with collagen fibers. Due to the lower shear and tensile strengths of the bone, it is subject to fracture under the action of impact load. Fixations are usually provided to treat fractured bones. In the internal fixation technique, the fracture of the bone is repaired from the interior. The interior site must be exposed for treatment, whereas the wound must be opened with external fixation. The alignment of the bones in external fixation is done with mechanical devices like casts, braces, and other fixation devices. The fracture site was not required to be opened. Plaster components were employed to fabricate the external fixation system with poor durability and wear resistance. They were polymeric composites used to replace these plaster components [2]. Historically, orthopedic implants were initially manufactured and derived from Co-based alloys and Co-based stainless steel. These resources possessed numerous benefits, including high corrosion resistance, strength, and economical. The utilization of Ti-based alloys has increased over time. Because titanium's modulus of elasticity was comparable to human bone, the metal implants had a limited fatigue life and were subjected to wear debris generation [21, 116, 117]. Various bone treatments incorporate fiber reinforcement, namely, spinal rods and plates, bone plates, and total hip replacement, because it offers adequate strength and hardness [21]. Carbon nanotubes were also added to the CF-epoxy composite as hybrid reinforcement in the last 10 years. The carbon nanotubes as reinforcement enhance the mechanical properties with a decrease in the coefficient of thermal expansion [118].

Natural load-bearing articulating joints are considered very efficient, like bearings, allowing the relative motion of bones under an applied load with minimal friction and wear of contacting surfaces. The articulating surfaces are covered with a thick layer of cartilage tissue lubricated with synovial fluid to generate a coefficient of friction between 0.005 and 0.025. The anatomical function of the joint can be affected by numerous factors, such as arthritis, injury to the joint, and any other degenerative diseases. Joint replacement surgery is often conducted in cases of excruciating pain and to restore the function of the diseased joints. Although the anatomical and physiological functions of the natural joint are impossible to reproduce artificially, some compromise has to be reached to design and build an optimally functioning artificial implant. The long-term performance of the total joint replacement prosthesis depends on the mechanical properties used for the loaded components, the quality of the implant fixation to the surrounding bone structure, and the rate of wear and creep of the articulating components. In the current scenario, nonmetallic biocomposites such as fiber-reinforced plastics (FRP) and carbon fiber-reinforced carbon (CFRC) are seen as the possible alternative to replacing metal-based implants. The metallic stems have a specific drawback since they are not biomechanically compatible with the host bone due to their high elastic moduli. The metallic stem geometry combined with its high stiffness results in stresses mainly carried by the implants, which can cause stress shielding at the bone. This can lead to bone resorption, aseptic loosening, and premature failure of the implant materials.

The loosening of the bone/implant surfaces is one of the primary causes of implant failure due to various sources. One of the primary reasons for this type of failure is the stiffness mismatch between bone and implant, which has a detrimental effect on the bone and may result in prosthesis fracture. The loosening of the implant can also result from the disruption of the cement mantle due to the improper properties of the acrylic cement or necrosis of the surrounding tissues caused by elevated temperatures. Another critical factor associated with aseptic loosening is the presence of wear particles. Other factors contributing to loosening effects are poor stabilization and relative motion at the interface between implant and tissues under load and the formation of interfacial connective tissues. Also, the articulating component's good friction and wear characteristics are critical to the proper functioning of the implant. The implant parts more susceptible to wear damage are the acetabular cups and femoral heads in total hip replacements and tibial, patellar, or femoral components of knee prostheses. A schematic of bearing components of joint prostheses is shown in Figure 12.4.

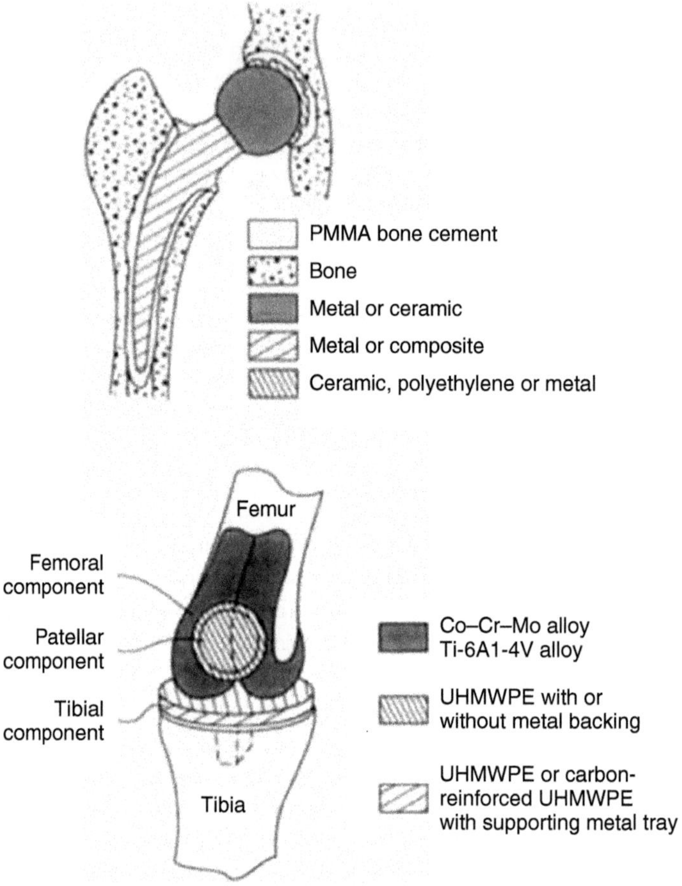

Figure 12.4 Schematic representation of total hip and knee joint replacements and bearing components. Source: Adapted from Stachowiak [119].

When evaluating bioimplants' friction and wear properties, the whole tribosystem must be considered, including materials, environment, contact geometries, the magnitude of load, surface speed, and temperature. Numerous experimental methods have been developed to evaluate bioimplants' friction and wear properties *in vitro*. These are classified as standard laboratory test rigs and joint simulators. The most common laboratory test rigs used to estimate wear and friction are listed in Table 12.6. In all of the mentioned tests, the dependence of the wear and frictional coefficients on operating parameters mentioned above is determined for a particular combination of materials. The results obtained from simple wear test rigs only provide the initial valuable information for the preselection of materials and different combinations of materials. None of the wear tests so far yield information about friction and wear of the implant because of the complexities encountered in reproducing the actual contact conditions. However, in many cases, these tests can provide accurate information about the probable mechanisms involved *in vivo* and aid in appropriate material selection [120, 121]. On the other hand, newly designed prosthetic components and whole units are usually tested using joint simulators, in which the prostheses or the assembly are subjected to the load and the motion cycles to reproduce the natural working cycle of the joint. The tribological performance of the prosthesis can be assessed along with the joint simulator's design and manufacturing aspects. The tribological evaluation of the materials using joint simulators involves measuring the wear rates, frictional torque, and changes in the physical properties of the articulating components. The worn surfaces are examined, and the morphology of the wear particles is studied. Simulator tests to study the tribological characteristics, if correctly applied, can furnish reliable information about the friction and wear of the bioimplant [120, 122–126].

The operating conditions in simulated tests are only a close approximation of the conditions operating in the body. Realistically, the loading acting on the joints is complex and multiaxial, and the activation of supporting muscles is challenging to reproduce. Also, prostheses developed for humans cannot be tested on animals. Hence, the final trials of the developed prostheses are conducted as clinical trials on humans. The tribological performance of the prostheses is also obtained from the *ex vivo* examination. The ASTM F 561 standard describes examining and analyzing orthopedic implants. The dismantled components are examined using microscopic technology, and wear is measured using surface profilometry technology. In addition, a histological examination is performed to determine the reaction of the surrounding body tissue to the implant material.

12.4.2.1 Friction and Wear of Biocomposites in Orthopedics Applications

Currently, prosthetic joints' most common bearing structure is a polymer/metal or polymer/ceramic combination such as a Pari joint (low-friction arthroplasty). These material combinations have friction coefficients up to three times lower than their metal/metal counterparts [122, 127]. The polymer component of the bearing is usually stationary, and the metal or ceramic components move against it. The life of such articulating joint is limited by the rate of penetration of metal or ceramic into the polymer, which in turn depends upon the combined effects of polymer creep and wear properties. Of the many polymeric materials currently available for biomedical applications, ultra-high molecular weight polyethylene (UHMWPE) seems to offer the best combination of low friction, wear resistance, and biocompatibility [120, 128, 129]. Over 30 years of clinical research have confirmed the suitability of UHMWPE in total replacement joints. Besides simple UHMWPE, the composite with short carbon fibers has also found application in artificial implants, especially knee prostheses. Among other polymers, polyoxymethylene (Delrin) has been used with some success as part of the hip joint, but attempts to use polytetrafluoroethylene (PTFE) and polyester were unsuccessful, resulting in early failure of the prosthesis. Although UHMWPE is widely accepted, research efforts are ongoing to improve the material's wear properties and creep resistance. Along with UHMWPE, research efforts have also been redirected to using other polymers like polyoxymethylene, also known as Delrin, which has been used successfully as an articulating hip component. Few attempts to use PTFE and polyester were unsuccessful and resulted

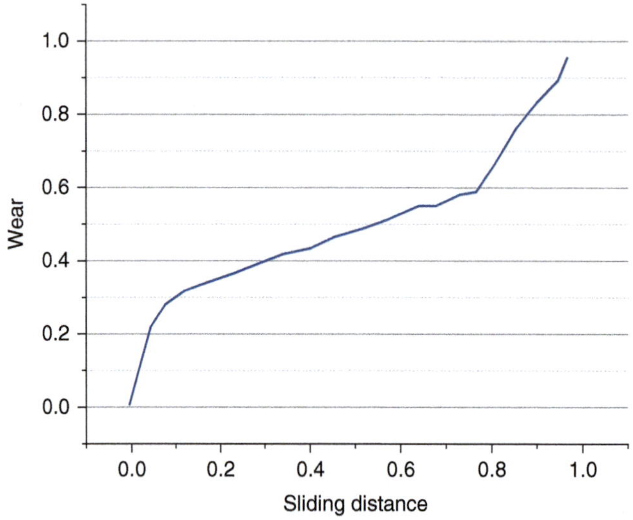

Figure 12.5 Wear rate vs. sliding distance curve for UHMWPE (No shown in the curves are superficial numbers).

in early prosthesis failure. Although the use of UHMWPE is now widely accepted, research efforts are still ongoing to improve its wear and creep properties.

UHMWPE's wear and friction properties are analyzed with respect to several factors, such as operating conditions, environment, type of counterface material, roughness, and the mechanical properties of polyethylene. The operating conditions affecting the wear properties of UHMWPE include sliding distance, contact load, and speed. In wear tests, the rates are usually higher at the start and then decrease with the increase in the sliding distance (Figure 12.5) [124, 130]. After covering a considerable sliding distance, the wear curve again changes its slope, and at this stage, the wear induces the fatigue wear mechanism in UHMWPE [120]. The contact stresses affect the wear behavior of polyethylene differently [122, 130–134]. The variation in wear rate with contact pressure can be given by this exponential relationship [122, 133, 134]:

$$W = W_0\, e^{(AP)}$$

where W is the wear rate defined as the depth of the wear scar per unit sliding distance, P is the contact pressure (MPa), and W_0 and A_0 are material constants.

Under light loads, polyethylene has low wear rates; however, these rates increase when subjected to heavier loads. The constants W_0 and A_0 change depending on the experimental conditions as well as the properties of the polyethylene material [130, 135]. The pressure–velocity (PV) limit, often used in engineering applications, is also used as a design criterion in artificial implants with polymer components [136–138]. Permissible PK values for artificial joints are higher than those for dry sliding because artificial joints are lubricated. The problem, however, is that walking leads to short-term insufficient lubrication; for example, in the hip joint, the critical value of the PK can be exceeded [131, 139, 140]. This is a problem because walking can cause hip joint degeneration. The wear of polyethylene increases sharply when subjected to high contact loads. This is critical, especially in knee prostheses, which can experience contact stresses in addition to yield. This may be one of the reasons why the wearing behavior of polyethylene is considered satisfactory for hip prostheses but is generally considered insufficient for knee prostheses.

The effect of the environment has a considerable effect on the wear and friction properties of the UHMWPE. The wear resistance of UHMWPE is usually lower under dry sliding than in the presence of aqueous lubrication [131]. When sliding on stainless steel in the air, the wear coefficient of UHMWPE is almost three times higher than when sliding with water or plasma lubricant [131]. While one study found that the type of liquid medium does not affect the wear behavior of UHMWPE [131], other studies found the opposite [122, 132, 141]. When sliding on stainless steel, UHMWPE's wear and friction values in water and saline solutions were significantly higher than in serum. These experiments also included the detection of the presence of corrosion products [122, 132]. Bovine serum wear damage was minimal on steel and UHMWPE sliding surfaces [122]. Serum proteins appear to form a barrier layer on the surface of the steel, which not only reduces adhesion but also prevents the transfer of the polymer film. Moreover, the type of counterface material like stainless steel AISI 316J, and 316LM used in the femoral components was highly susceptible to corrosion when it came into contact with body fluids and the internal environment. To counter the excessive corrosion, these femoral materials have been removed by more advanced and developed alloys like cobalt and titanium alloys, for example, Ti-6Al-4V alloys and cast Co–Cr–Mo alloys which offer a superior combination of corrosion resistance, mechanical strength and biocompatibility, compared to that of the stainless-steel alloys. The titanium alloys also do not contain the known Allergic components such as nickel, cobalt, or chromium. Hence, any adverse tissue reaction to the leaching metallic ions would reduce automatically. The wear properties of UHMWPE are comparable to those of stainless steel, cobalt–chromium and titanium alloys when tested under otherwise identical conditions [142–145]. The tribological behavior of UHMWPE has been studied [146, 147] using the zero-wear model [147], which is defined as the amount of wear that does not cause a significant change in the initial surface roughness. The three alloys investigated were stainless steel, Co–Cr–Mo alloys, and Ti alloys. The performance of all three sets was comparable, and each set's maximum shear stress value was found to be the same even after accounting for wear. Blood serum in contact with highly stressed UHMWPE showed no significant difference in wear compared to contact with cobalt–chromium and titanium alloys [141]. However, there is some uncertainty regarding the wear properties of the combination of UHMWPE and titanium alloy, as the wear performance of this system has been described as both unsatisfactory [130, 148] and satisfactory [135]. In laboratory tests in dry and wet environments [130] and saline solution [132], the formation of a black deposit and its sinking into the polyethylene surface and damage to the surface of the titanium alloy were observed. Wear tests [54] confirmed that UHMWPE had poor wear properties compared to titanium alloys after exposure to a counter-shaped knee prosthesis contact space. The tested combination of

UHMWPE and Ti-6A1-4V had high friction, and both parts showed signs of wear. Increased wear of titanium alloys is particularly noticeable in the presence of abrasive materials such as acrylic cement particles [132]. This is because the acrylic cement particles have a very rough surface. The slippage appears to have caused the problem associated with the collapse of the passivating titanium oxide thin film. For this reason, numerous attempts have been made to achieve a wear-resistant surface layer using different methods. These methods include passivation [130] nitriding [141], TiN ion plating [149], nitrogen ion implantation [75, 150–152] and oxygen diffusion hardening [149]. Although these surface treatments typically reduced wear damage to the titanium alloy samples, the resulting wear-resistant layers had very poor durability. Since the early 1970s, high-density alumina ceramics have been used in a wide variety of prosthetic designs to replace metal components in femoral joints [153, 154]. The tribological advantages of polyethylene/alumina configurations over hinged polyethylene/metal systems have been demonstrated by extensive laboratory testing in both dry and wet environments [121, 127, 132, 149, 155, 156]. Compared to sliding on stainless steel [127, 132], Co–Cr–Mo alloy [132] and Ti-6A1-4V alloy, friction and wear of UHMWPE sliding on alumina were significantly reduced. Wear is reduced by a factor of 2–20. The typical tribological characteristics of polyethylene rubbing against various metallic and ceramic counterfaces are tabulated in Table 12.8.

The other factor that significantly influences the wear behavior of polyethylene is the surface roughness of the hard surface. The impact of this factor depends on whether the conditions are dry or wet. When sliding dry, polyethylene has the lowest coefficient of wear [121, 155, 158], and the surface roughness of the mating surface is about $R_a = 0.1\ \mu m$ [121, 155, 158, 159]. In a dry environment, the weak effect of

Table 12.8 Wear results test summary of conventional UHMWPE and other polymers.

Typical counterface	Friction coefficient	Wear coefficient ($mm^3\ Nm^{-1}$)	Test conditions	Ref.
EN58J	N/A	$2.5–3.4 \times 10^{-7}$	4 MPa, 0.24 m s^{-1}, dry	[120]
EN58J	N/A	$0.9–2 \times 10^{-7}$	45 N, 0.02 m s^{-1}, dry	[120]
316L	N/A	1.7×10^{-7}	3.4 MPa, 0.04 m s^{-1}, water	[131]
316L	N/A	1.7×10^{-7}	3.4 MPa, 0.04 m s^{-1}, plasma	[131]
316L	N/A	5.0×10^{-7}	3.4 MPa, 0.04 m s^{-1}, dry	[131]
316L	0.04–0.16	9.1×10^{-9}	3.4 MPa, 0.04 m s^{-1}, serum	[141]
316L	0.03–0.09	8.9×10^{-9}	6.9 MPa, 0.05 m s^{-1}, serum	[141]
Co–Cr	0.05–0.11	7.5×10^{-9}	6.9 MPa, 0.05 m s^{-1}, serum	[141]
Alumina	0.06–0.11	6.2×10^{-9}	3.4 MPa, 0.05 m s^{-1}, serum	[157]
Zirconia	0.04	5.6×10^{-9}	3.4 MPa, 0.05 m s^{-1}, serum	[132]
Alumina	0.05	1.0×10^{-9}	3.4 MPa, 0.05 m s^{-1}, serum	[132]
316L	0.07	1.8×10^{-9}	3.4 MPa, 0.05 m s^{-1}, serum	[132]

surface roughness on wear has been attributed to the formation of a thin polymer transport film on the metal surface. This film is believed to be the cause of the poor result. The film will cover any underlying markings or surface irregularities, giving the surface a more even appearance. The formation of transfer films is inhibited in humid environments, and the wear of polyethylene increases with increasing mating surface roughness for both metals [28, 30, 58, 73] and ceramics [68, 74]. The formation of transport membranes is disrupted in humid environments. After contact with UHMWPE using bovine serum as a lubricant, the initial abrasion marks on the metal plates eventually smoothed out, and wear rates eventually decreased over long sliding distances [28]. It appears that the formation of partial films, particularly in water and Ringer's solution, is responsible for the similarities in the long-term wear rates of surfaces with different degrees of roughness [58]. When it comes to polyethylene wear, there is hardly any difference between dry and wet conditions on very smooth surfaces.

12.5 Conclusion

This chapter outlines the importance of wear and friction properties in biocomposites for dental and orthopedic applications. Biomaterials with high wear resistance and low friction coefficients are more suitable for orthopedic and dental implants. The matching of elastic modulus of implants with the target tissue helps in preventing the stress shielding of implants. The polymer-based implants reinforced with ceramic fillers like HA and calcium phosphates possesses good biocompatibility and also exhibits mechanical properties similar to that of bone However, the proper functioning of the dental and orthopedic implants depends on friction and wear resistance according to the clinical trials. The various analytical techniques, such as pin on a disc, ball on a disc, and block on a disc are recommended to evaluate the wear properties of the biocomposite-based implants in a stimulated body fluid at 37 °C. The dental resin composites-reinforced with macro, micro and hybrid fillers are used as dental restorative materials owing to their improved friction and wear resistance. The dental resin composites are subjected to abrasion and scratch tests to evaluate their friction and wear behavior. The UHMWPE is suggested as promising material for developing orthopedic implants owing to its low friction, wear resistance, and biocompatibility as compared to conventional metal-based implants. The UHMWPE reinforced with short carbon fiber finds application in developing knee prostheses. However, active research is going on to improve the wear and creep properties of UHMWPE.

References

1 Parikh, H.H. and Gohil, P.P. (2015). Tribology of fiber reinforced polymer matrix composites—a review. *J. Reinf. Plast. Compos.* https://journals.sagepub

.com/doi/full/10.1177/0731684415591199?casa_token=JXjz1RBhXIEAAAAA%3AOWwkVdy3p3d0YY8;pDbZng8Ff0rtybGZCDN3jzKwFJiK6UC4QL1M37NHe576W6xi3SfqVTgGKQ9VHbUQ.

2 R. Deshmukh and S. S. Kulkarni, *A Review on Biomaterials in Orthopedic Bone Plate Application*, undefined, , 2022. [Online]. Available: https://www.semanticscholar.org/paper/A-Review-on-Biomaterials-in-Orthopedic-Bone-Plate-Deshmukh-Kulkarni/823ff27ce7f95b768c8add4da2349b69b22433c5.

3 Jena, P.C., Parhi, D.R., Pohit, G., and Samal, B.P. (2015). Crack assessment by FEM of AMMC beam produced by modified stir casting method. *Mater. Today Proc.* 2 (4): 2267–2276. https://doi.org/10.1016/j.matpr.2015.07.263.

4 Whitlow, T., Jones, E., and Przybyla, C. (2016). In-situ damage monitoring of a SiC/SiC ceramic matrix composite using acoustic emission and digital image correlation. *Compos. Struct.* 158: 245–251. https://doi.org/10.1016/j.compstruct.2016.09.040.

5 Hofmann, G.O. (1992). Biodegradable implants in orthopaedic surgery—a review on the state-of-the-art. *Clin. Mater.* 10 (1): 75–80. https://doi.org/10.1016/0267-6605(92)90088-B.

6 Lee, J.-H., Kim, H.-W., and Seo, S.-J. (2016). Polymer-ceramic bionanocomposites for dental application. *J. Nanomater.* 2016: e3795976. https://doi.org/10.1155/2016/3795976.

7 Moszner, N. and Salz, U. (2001). New developments of polymeric dental composites. *Prog. Polym. Sci.* 26 (4): 535–576. https://doi.org/10.1016/S0079-6700(01)00005-3.

8 Evans, S.L. and Gregson, P.J. (1998). Composite technology in load-bearing orthopaedic implants. *Biomaterials* 19 (15): 1329–1342. https://doi.org/10.1016/s0142-9612(97)00217-2.

9 Huang, Z.M., Ramakrishna, S., and Tay, A.A.O. (2000). Unified micromechanical model for estimating elastic, elasto-plastic and strength behaviors of knitted fabric reinforced composites. *J. Reinf. Plast. Compos.* 19 (8): 642–656. https://doi.org/10.1177/073168440001900803.

10 Ramakrishna, S., Mayer, J., Wintermantel, E., and Leong, K.W. (2001). Biomedical applications of polymer-composite materials: a review. *Compos. Sci. Technol.* 61 (9): 1189–1224. https://doi.org/10.1016/S0266-3538(00)00241-4.

11 Roberson, J.R. (1987). Excessive metal release due to loosening and fretting of sintered particles on porous-coated hip prostheses. Report of two cases. *J. Bone Joint Surg. Am.* 69 (2): 310.

12 Blumenthal, N.C., Cosma, V., Jaffe, W., and Stuchin, S. (1994). A new technique for quantitation of metal particulates and metal reaction products in tissues near implants. *J. Appl. Biomater. Off. J. Soc. Biomater.* 5 (3): 191–193. https://doi.org/10.1002/jab.770050303.

13 Claes, L.E. (1992). Mechanical characterization of biodegradable implants. *Clin. Mater.* 10 (1–2): 41–46. https://doi.org/10.1016/0267-6605(92)90083-6.

14 Halstead, A., Jones, C.W., and Rawlings, R.D. (1979). A study of the reaction of human tissue to proplast. *J. Biomed. Mater. Res.* 13 (1): 121–134. https://doi.org/10.1002/jbm.820130113.

15 Turner, C.H. (1992). On Wolff's law of trabecular architecture. *J. Biomech.* 25 (1): 1–9. https://doi.org/10.1016/0021-9290(92)90240-2.

16 Yang, K., Ren, Y., and Wan, P. (2012). High nitrogen nickel-free austenitic stainless steel: a promising coronary stent material. *Sci. China Technol. Sci.* 55 (2): 329–340. https://doi.org/10.1007/s11431-011-4679-3.

17 Yang, K. and Ren, Y. (2010). Nickel-free austenitic stainless steels for medical applications. *Sci. Technol. Adv. Mater.* 11 (1): 014105. https://doi.org/10.1088/1468-6996/11/1/014105.

18 Nicholson, J.W. (1998). Adhesive dental materials—a review. *Int. J. Adhes. Adhes.* 18 (4): 229–236. https://doi.org/10.1016/S0143-7496(98)00027-X.

19 Törmälä, P., Vainionpää, S., Kilpikari, J., and Rokkanen, P. (1987). The effects of fibre reinforcement and gold plating on the flexural and tensile strength of PGA/PLA copolymer materials in vitro. *Biomaterials* 8 (1): 42–45. https://doi.org/10.1016/0142-9612(87)90027-5.

20 Sheikh, Z., Najeeb, S., Khurshid, Z. et al. (2015). Biodegradable materials for bone repair and tissue engineering applications. *Materials* 8 (9): 5744–5794. https://doi.org/10.3390/ma8095273.

21 Patel, N.R. and Gohil, P.P. *A Review on Biomaterials: Scope, Applications & Human Anatomy Significance*. IJETAE Publication House.

22 Reul, H., Schmitz, C., Pfaff, E.M. et al. (2002). In-vitro assessment of the wear development mechanism and stabilization of wear in the Edwards MIRA/Sorin bicarbon mechanical heart valve orifice ring. *J. Heart Valve Dis.* 11 (3): 409–418; discussion 418.

23 Litonjua, L.A., Andreana, S., Bush, P.J., and Cohen, R.E. (1985). Tooth wear: attrition, erosion, and abrasion. *Quintessence Int. Berl. Ger.* 34 (6): 435–446. June 2003.

24 Tsujimoto, A., Barkmeier, W.W., Fischer, N.G. et al. (2018). Wear of resin composites: current insights into underlying mechanisms, evaluation methods and influential factors. *Jpn. Dent. Sci. Rev.* 54 (2): 76–87. https://doi.org/10.1016/j.jdsr.2017.11.002.

25 Shahgaldi, B.F. and Compson, J. (2000). Wear and corrosion of sliding counterparts of stainless-steel hip screw-plates. *Injury* 31 (2): 85–92. https://doi.org/10.1016/s0020-1383(99)00241-7.

26 Walowit, J.A., Malanoski, S.B., Horvath, D. et al. (1992). The analysis, design, and testing of a blood lubricated hydrodynamic journal bearing. *ASAIO J. Am. Soc. Artif. Intern. Organs* 43 (5): M556–M559. October 1997.

27 Sivamani, R.K., Goodman, J., Gitis, N.V., and Maibach, H.I. (2003). Coefficient of friction: tribological studies in man – an overview. *Skin Res. Technol.* 9 (3): 227–234. https://doi.org/10.1034/j.1600-0846.2003.02366.x.

28 Sivamani, R.K., Goodman, J., Gitis, N.V., and Maibach, H.I. (2003). Friction coefficient of skin in real-time. *Skin Res. Technol.* 9 (3): 235–239. https://doi.org/10.1034/j.1600-0846.2003.20361.x.

29 Saravanan, I. and Devaraju, A. (2019). Wear mechanism of UHMWPE polymer composites for bio medical applications. *Mater. Res. Express* 6 (10): 105355. https://doi.org/10.1088/2053-1591/ab3ed9.

30 Zafar, M.S. (2019). Wear behavior of various dental restorative materials. *Mater. Technol.* 34 (1): 25–31. https://doi.org/10.1080/10667857.2018.1462978.

31 Zhou, Z.R. and Zheng, J. (2008). Tribology of dental materials: a review. *J. Phys. Appl. Phys.* 41 (11): 113001. https://doi.org/10.1088/0022-3727/41/11/113001.

32 Zheng, J., Weng, L.Q., Shi, M.Y. et al. (2013). Effect of water content on the nanomechanical properties and microtribological behaviour of human tooth enamel. *Wear* 301 (1): 316–323. https://doi.org/10.1016/j.wear.2012.12.043.

33 Gao, K., Jianjuna, G., Hui, S. et al. (2022). Friction and wear behavior of bioinspired composites with nacre-like lamellar and brick-and-mortar architectures against human enamel. *J. Mater. Sci. Technol.* 128: 133–141. https://doi.org/10.1016/j.jmst.2022.04.027.

34 Hussein, M.A., Mohammed, A.S., and Al-Aqeeli, N. (2015). Wear characteristics of metallic biomaterials: a review. *Materials* 8 (5): 2749–2768. https://doi.org/10.3390/ma8052749.

35 Long, M. and Rack, H.J. (1998). Titanium alloys in total joint replacement—a materials science perspective. *Biomaterials* 19 (18): 1621–1639. https://doi.org/10.1016/s0142-9612(97)00146-4.

36 Galmiche, R., Migaud, H., and Beaulé, P.-E. (2020). Hip anatomy and biomechanics relevant to hip replacement. In: *Personalized Hip and Knee Joint Replacement* (ed. C. Rivière and P.-A. Vendittoli). Cham (CH): Springer Accessed: Aug. 04, 2022. [Online]. Available: http://www.ncbi.nlm.nih.gov/books/NBK565771/.

37 Kennedy, F.E. (2013). 3—Biomechanics of the hip and knee: implant wear. In: *Wear of Orthopaedic Implants and Artificial Joints* (ed. S. Affatato), 56–92. Woodhead Publishing https://doi.org/10.1533/9780857096128.1.56.

38 Katz, J.L. (1980). Anisotropy of Young's modulus of bone. *Nature* 283 (5742): 5742. https://doi.org/10.1038/283106a0.

39 Ramsden, J.J., Allen, D.M., Stephenson, D.J. et al. (2007). The design and manufacture of biomedical surfaces. *CIRP Ann.* 56 (2): 687–711. https://doi.org/10.1016/j.cirp.2007.10.001.

40 Niinomi, M. (2002). Recent metallic materials for biomedical applications. *Metall. Mater. Trans. A* 33 (3): 477–486. https://doi.org/10.1007/s11661-002-0109-2.

41 Shemtov-Yona, K. and Rittel, D. (2016). Fatigue of dental implants: facts and fallacies. *Dent. J.* 4 (2): 16. https://doi.org/10.3390/dj4020016.

42 Teoh, S.H. (2000). Fatigue of biomaterials: a review. *Int. J. Fatigue* 22 (10): 825–837. https://doi.org/10.1016/S0142-1123(00)00052-9.

43 Park, J. and Lakes, R.S. (2007). *Biomaterials: An Introduction*. Springer Science & Business Media.

44 Karanjai, M., Sundaresan, R., Rao, G.V.N. et al. (2007). Development of titanium based biocomposite by powder metallurgy processing with in situ forming of Ca–P phases. *Mater. Sci. Eng. A* 447 (1): 19–26. https://doi.org/10.1016/j.msea.2006.10.154.

45 Geetha, M., Singh, A.K., Asokamani, R., and Gogia, A.K. (2009). Ti based biomaterials, the ultimate choice for orthopaedic implants—a review. *Prog. Mater. Sci.* 54 (3): 397–425. https://doi.org/10.1016/j.pmatsci.2008.06.004.

46 Barfeie, A., Wilson, J., and Rees, J. (2015). Implant surface characteristics and their effect on osseointegration. *Br. Dent. J.* 218 (5): 5. https://doi.org/10.1038/sj.bdj.2015.171.

47 Williams, D., Cahn, R.W., and Bever, M.B. (1990). *Concise Encyclopedia of Medical and Dental Materials*. Elsevier Science.

48 Paital, S.R. and Dahotre, N.B. (2009). Calcium phosphate coatings for bio-implant applications: materials, performance factors, and methodologies. *Mater. Sci. Eng. R Rep.* 66 (1): 1–70. https://doi.org/10.1016/j.mser.2009.05.001.

49 Prasad, K., Bazaka, O., Chua, M. et al. (2017). Metallic biomaterials: current challenges and opportunities. *Materials* 10 (8): 884. https://doi.org/10.3390/ma10080884.

50 Fazel-Rezai, R. (2011). *Biomed. Eng. Theory Appl.* https://doi.org/10.5772/2629.

51 Gegner, J. (2013). *Tribol. Fund. Adv.* https://doi.org/10.5772/3427.

52 Henriques, V.A.R., Galvani, E.T., Petroni, S.L.G. et al. (2010). Production of Ti–13Nb–13Zr alloy for surgical implants by powder metallurgy. *J. Mater. Sci.* 45 (21): 5844–5850. https://doi.org/10.1007/s10853-010-4660-8.

53 Ribeiro, A.L.R., Junior, R.C., Cardoso, F.F. et al. (2009). Mechanical, physical, and chemical characterization of Ti–35Nb–5Zr and Ti–35Nb–10Zr casting alloys. *J. Mater. Sci. Mater. Med.* 20 (8): 1629–1636. https://doi.org/10.1007/s10856-009-3737-x.

54 Kaur, M. and Singh, K. (2019). Review on titanium and titanium based alloys as biomaterials for orthopaedic applications. *Mater. Sci. Eng. C* 102: 844–862. https://doi.org/10.1016/j.msec.2019.04.064.

55 Alam, M.O. and Haseeb, A.S.M.A. (2002). Response of Ti–6Al–4V and Ti–24Al–11Nb alloys to dry sliding wear against hardened steel. *Tribol. Int.* 35 (6): 357–362. https://doi.org/10.1016/S0301-679X(02)00015-4.

56 Festas, A., Ramos, A., and Davim, J. (2020). Medical devices biomaterials—a review. *Proc. Inst. Mech. Eng. Part J. Mater. Des. Appl.* 234 (1): 218–228. https://doi.org/10.1177/1464420719882458.

57 Katti, K.S. (2004). Biomaterials in total joint replacement. *Colloids Surf. B Biointerfaces* 39 (3): 133–142. https://doi.org/10.1016/j.colsurfb.2003.12.002.

58 Eliaz, N. (2019). Corrosion of metallic biomaterials: a review. *Materials* 12 (3): 407. https://doi.org/10.3390/ma12030407.

59 Davis, R., Singh, A., Jackson, M.J. et al. (2022). A comprehensive review on metallic implant biomaterials and their subtractive manufacturing. *Int. J. Adv. Manuf. Technol.* 120 (3): 1473–1530. https://doi.org/10.1007/s00170-022-08770-8.

60 Su, Y., Luo, C., Zhang, Z. et al. (2018). Bioinspired surface functionalization of metallic biomaterials. *J. Mech. Behav. Biomed. Mater.* 77: 90–105. https://doi.org/10.1016/j.jmbbm.2017.08.035.

61 Pires, A.L.R., Bierhalz, A.C.K., and Moraes, Â.M. (2015). Biomaterials: types, applications, and market. *Quím. Nova* https://doi.org/10.5935/0100-4042.20150094.

62 Gialanella, S., Ischia, G., and Straffelini, G. (2008). Phase composition and wear behavior of NiTi alloys. *J. Mater. Sci.* 43 (5): 1701–1710. https://doi.org/10.1007/s10853-007-2358-3.

63 Chen, Q. and Thouas, G.A. (2015). Metallic implant biomaterials. *Mater. Sci. Eng. R Rep.* 87: 1–57. https://doi.org/10.1016/j.mser.2014.10.001.

64 Suresh, K.S., Geetha, M., Richard, C. et al. (2012). Effect of equal channel angular extrusion on wear and corrosion behavior of the orthopedic Ti–13Nb–13Zr alloy in simulated body fluid. *Mater. Sci. Eng. C* 32 (4): 763–771. https://doi.org/10.1016/j.msec.2012.01.022.

65 Xu, L., Xiao, S., Tian, J., and Chen, Y. (2013). Microstructure, mechanical properties and dry wear resistance of β-type Ti–15Mo–xNb alloys for biomedical applications. *Trans. Nonferrous Met. Soc. China* 23 (3): 692–698. https://doi.org/10.1016/S1003-6326(13)62518-2.

66 Muñoz, A.I. and Mischler, S. (2011). Effect of the environment on wear ranking and corrosion of biomedical CoCrMo alloys. *J. Mater. Sci. Mater. Med.* 22 (3): 437–450. https://doi.org/10.1007/s10856-010-4224-0.

67 Mamoun, F., Omar, A., Mohamed, L. et al. (2016). Comparative study on tribological behavior of Ti-6Al-7Nb and SS AISI 316L alloys, for total hip prosthesis. In: *TMS 2014: 143rd Annual Meeting & Exhibition, Cham*, 237–246. https://doi.org/10.1007/978-3-319-48237-8_32.

68 Li, S.J., Yang, R., Li, S. et al. (2004). Wear characteristics of Ti-Nb-Ta-Zr and Ti-6Al-4V alloys for biomedical applications. *Wear* 257 (9–10): 869–876. https://doi.org/10.1016/j.wear.2004.04.001.

69 Shen, G., Fang, F., and Kang, C. (2018). *Tribological Performance of Bioimplants: A Comprehensive Review*, 17.

70 S. Đurić, D. Adamovic, B. Nedic, B. Trifunović, and D. Jovanović, *Society of Production Engineering SPMS 2018 37. Savetovanje Proizvodnog mašinstva Srbije ICPE-S 2018 37 th International Conference on Production Engineering of Serbia MATERIAL SELECTION OF PROTECTIVE COATINGS*. 2019.

71 Edalati, K., Ashida, M., Horita, Z. et al. (2014). Wear resistance and tribological features of pure aluminum and Al–Al_2O_3 composites consolidated by high-pressure torsion. *Wear* 310 (1): 83–89. https://doi.org/10.1016/j.wear.2013.12.022.

72 Okazaki, Y. and Gotoh, E. (2005). Comparison of metal release from various metallic biomaterials in vitro. *Biomaterials* 26 (1): 11–21. https://doi.org/10.1016/j.biomaterials.2004.02.005.

73 Bhushan, B. (2008). Nanotribology, nanomechanics and nanomaterials characterization. *Philos. Trans. R. Soc. Math. Phys. Eng. Sci.* 366 (1869): 1351–1381. https://doi.org/10.1098/rsta.2007.2163.

74 Choubey, A., Basu, B., and Balasubramaniam, R. (2004). Tribological behaviour of Ti-based alloys in simulated body fluid solution at fretting contacts. *Mater. Sci. Eng. A* 379 (1): 234–239. https://doi.org/10.1016/j.msea.2004.02.027.

75 Iwabuchi, A., Lee, J.W., and Uchidate, M. (2007). Synergistic effect of fretting wear and sliding wear of Co-alloy and Ti-alloy in Hanks' solution. *Wear* 263 (1): 492–500. https://doi.org/10.1016/j.wear.2007.01.102.

76 Last, J.A., Russell, P., Nealey, P.F., and Murphy, C.J. (2010). The applications of atomic force microscopy to vision science. *Invest. Ophthalmol. Vis. Sci.* 51 (12): 6083–6094. https://doi.org/10.1167/iovs.10-5470.

77 Variola, F. (2015). Atomic force microscopy in biomaterials surface science. *Phys. Chem. Chem. Phys.* 17 (5): 2950–2959. https://doi.org/10.1039/C4CP04427D.

78 Fessel, G., Frey, K., Schweizer, A. et al. (2011). Suitability of Thiel embalmed tendons for biomechanical investigation. *Ann. Anat. Anat. Anz.* 193 (3): 237–241. https://doi.org/10.1016/j.aanat.2011.03.007.

79 T. Nguyen, V. Humblot, V. Migonney, R. Levy, *Atomic Force Microscopy Characterization of Biomaterials Modified with Poly(styrene Sulfonate)*, 2021, https://doi.org/10.26434/chemrxiv.14687556.v1.

80 Zhang, M. and Matinlinna, J.P. (2012). E-glass fiber reinforced composites in dental applications. *Silicon* 4 (1): 73–78. https://doi.org/10.1007/s12633-011-9075-x.

81 Jena, P.C. (2018). Free vibration analysis of short bamboo fiber based polymer composite beam structure. *Mater. Today Proc.* 5 (2, Part 1): 5870–5875. https://doi.org/10.1016/j.matpr.2017.12.185.

82 Krishnakumar, S. and Senthilvelan, T. (2021). Polymer composites in dentistry and orthopedic applications—a review. *Mater. Today Proc.* 46: 9707–9713. https://doi.org/10.1016/j.matpr.2020.08.463.

83 Cranin, A.N. (1987). Handbook of biomaterials evaluation. Scientific, technical and clinical testing of Implant Materials. Andreas F. von Recum, Editor, New York, 1986, MacMillan. With a forward by Solomon Pollack and contributions by 61 authors and 10 section editors. Index of 14 pp. $120.00. *J. Biomed. Mater. Res.* 21 (9): 1167–1168. https://doi.org/10.1002/jbm.820210909.

84 Sarikaya, R., Song, L., Ye, Q. et al. (2020). Evolution of network structure and mechanical properties in autonomous-strengthening dental adhesive. *Polymers* 12 (9): E2076. https://doi.org/10.3390/polym12092076.

85 Michelotti, G., Niedzwiecki, M., Bidjan, D. et al. (2020). Silane effect of universal adhesive on the composite-composite repair bond strength after different surface pretreatments. *Polymers* 12 (4): E950. https://doi.org/10.3390/polym12040950.

86 Okamoto, M., Matsumoto, S., Sugiyama, A. et al. (2020). Performance of a biodegradable composite with hydroxyapatite as a scaffold in pulp tissue repair. *Polymers* 12 (4): E937. https://doi.org/10.3390/polym12040937.

87 Arbildo-Vega, H.I., Lapinska, B., Panda, S. et al. (2020). Clinical effectiveness of bulk-fill and conventional resin composite restorations: systematic review and meta-analysis. *Polymers* 12 (8): 8. https://doi.org/10.3390/polym12081786.

88 Ehrnford, L., Derand, T., Larsson, L.-A., and Svensson, A. (1980). An abrasion test for composite resins. *J. Dent. Res.* 59 (4): 716–720. https://doi.org/10.1177/00220345800590040801.

89 Snyder, R.G., Foust, D.R., and Bowman, B.M. (1977). Study of impact tolerance through free-fall investigations. *Rep. Insur. Inst. Highw. Saf. Wash. DC* 324.

90 Roulet, J.F. and Roulet-Mehrens, T.K. (1982). The surface roughness of restorative materials and dental tissues after polishing with prophylaxis and polishing pastes. *J. Periodontol.* 53 (4): 257–266. https://doi.org/10.1902/jop.1982.53.4.257.

91 Powers, J.M., Roberts, J.C., and Craig, R.G. (1976). Wear of filled and unfilled dental restorative resins. *Wear* 39 (1): 117–122. https://doi.org/10.1016/0043-1648(76)90227-1.

92 Bloem, T.J., McDowell, G.C., Lang, B.R., and Powers, J.M. (1988). In vivo wear. Part II: Wear and abrasion of composite restorative materials. *J. Prosthet. Dent.* 60 (2): 242–249. https://doi.org/10.1016/0022-3913(88)90324-1.

93 Harrison, A., Huggett, R., and Hunt, H.H. (1977). Prosthetic technology in commercial dental laboratories i n the United States. *J. Dent.* 5 (2): 131–140. https://doi.org/10.1016/0300-5712(77)90074-4.

94 Fitzgerald, R.H., Brindley, G.W., and Kavanagh, B.F. (1988). The uncemented total hip arthroplasty. Intraoperative femoral fractures. *Clin. Orthop.* 235: 61–66.

95 Wilder, A.D., May, K.N., and Leinfelder, K.F. (1983). Three-year clinical study of UV-cured composite resins in posterior teeth. *J. Prosthet. Dent.* 50 (1): 26–30. https://doi.org/10.1016/0022-3913(83)90160-9.

96 Atkinson, T.S., Ewers, B.J., and Haut, R.C. (1999). The tensile and stress relaxation responses of human patellar tendon varies with specimen cross-sectional area. *J. Biomech.* 32 (9): 907–914.

97 Lambrechts, P., Vanherle, G., Vuylsteke, M., and Davidson, C.L. (1984). Quantitative evaluation of the wear resistance of posterior dental restorations: a new three-dimensional measuring technique. *J. Dent.* 12 (3): 252–267. https://doi.org/10.1016/0300-5712(84)90071-X.

98 Kleczewska, J. and Bieliński, D.M. (2007). Friction and wear of resin-based dental materials. *Arch. Civ. Mech. Eng.* 7 (4): 87–96. https://doi.org/10.1016/S1644-9665(12)60227-1.

99 Leinfelder, K.F. (1985). Composite resins. *Dent. Clin. North Am.* 29 (2): 359–371. https://doi.org/10.1016/S0011-8532(22)02168-1.

100 Atai, M. and Motevasselian, F. (2009). Temperature rise and degree of photopolymerization conversion of nanocomposites and conventional dental composites. *Clin. Oral Investig.* 13 (3): 309–316. https://doi.org/10.1007/s00784-008-0236-2.

101 Curtis, A.R., Palin, W.M., Fleming, G.J.P. et al. (2009). The mechanical properties of nanofilled resin-based composites: the impact of dry and wet cyclic pre-loading on bi-axial flexure strength. *Dent. Mater.* 25 (2): 188–197. https://doi.org/10.1016/j.dental.2008.06.003.

102 Chng, C.K., Foong, K., Gandedkar, N.H. et al. (2014). A new esthetic fiber-reinforced polymer composite resin archwire: a comparative atomic force microscope (AFM) and field-emission scanning electron microscope (FESEM) study. *Prog. Orthod.* 15 (1): 39. https://doi.org/10.1186/s40510-014-0039-8.

103 Moharamzadeh, K., Brook, I.M., and Van Noort, R. (2009). Biocompatibility of resin-based dental materials. *Materials* 2 (2): 514–548. https://doi.org/10.3390/ma2020514.

104 Tanimoto, Y., Inami, T., Yamaguchi, M. et al. (2015). Preparation, mechanical, and *in vitro* properties of glass fiber-reinforced polycarbonate composites for orthodontic application: preparation of GFRPS for orthodontic applications. *J. Biomed. Mater. Res. B Appl. Biomater.* 103 (4): 743–750. https://doi.org/10.1002/jbm.b.33245.

105 Feilzer, A.J., De Gee, A.J., and Davidson, C.L. (1987). Setting stress in composite resin in relation to configuration of the restoration. *J. Dent. Res.* 66 (11): 1636–1639. https://doi.org/10.1177/00220345870660110601.

106 Sarrett, D.C., Söderholm, K.-J.M., and Batich, C.D. (1991). Water and abrasive effects on three-body wear of composites. *J. Dent. Res.* 70 (7): 1074–1081. https://doi.org/10.1177/00220345910700071201.

107 Söderholm, K.J., Zigan, M., Ragan, M. et al. (1984). Hydrolytic degradation of dental composites. *J. Dent. Res.* 63 (10): 1248–1254. https://doi.org/10.1177/00220345840630101701.

108 Phillips, R.W., Avery, D.R., Mehra, R. et al. (1973). Observations on a composite resin for class II restorations: three-year report. *J. Prosthet. Dent.* 30 (6): 891–897. https://doi.org/10.1016/0022-3913(73)90283-7.

109 Leinfelder, K.F., Sluder, T.B., Sockwell, C.L. et al. (1975). Clinical evaluation of composite resins as anterior and posterior restorative materials. *J. Prosthet. Dent.* 33 (4): 407–416. https://doi.org/10.1016/s0022-3913(75)80037-0.

110 Kusy, R.P. and Leinfelder, K.F. (1977). Pattern of wear in posterior composite restorations. *J. Dent. Res.* 56 (5): 544. https://doi.org/10.1177/00220345770560052101.

111 Eames, W.B., Strain, J.D., Weitman, R.T., and Williams, A.K. (1939). Clinical comparison of composite, amalgam, and silicate restorations. *J. Am. Dent. Assoc.* 89 (5): 1111–1117. Nov. 1974, https://doi.org/10.14219/jada.archive.1974.0556.

112 Sturdevant, J.R., Lundeen, T.F., Sluder, T.B. et al. (1988). Five-year study of two light cured posterior composite resins. *Dent. Mater.* 4 (3): 105–110. https://doi.org/10.1016/S0109-5641(88)80001-0.

113 Norman, R.D., Wright, J.S., Rydberg, R.J., and Felkner, L.L. (1990). A 5-year study comparing a posterior composite resin and an amalgam. *J. Prosthet. Dent.* 64 (5): 523–529. https://doi.org/10.1016/0022-3913(90)90121-r.

114 Swift, E.J. (1987). Wear of composite resins in permanent posterior teeth. *J. Am. Dent. Assoc.* 115 (4): 584–588. https://doi.org/10.1016/S0002-8177(87)54013-2.

115 Koran, A., Craig, R.G., and Tillitson, E.W. (1972). Coefficient of friction of prosthetic tooth materials. *J. Prosthet. Dent.* 27 (3): 269–274. https://doi.org/10.1016/0022-3913(72)90034-0.

116 Suzuki, T., Smith, W.R., Stahel, P.F. et al. (2010). Technical problems and complications in the removal of the less invasive stabilization system. *J. Orthop. Trauma* 24 (6): 369–373. https://doi.org/10.1097/BOT.0b013e3181c29bf5.

117 Cooper, G.J., Townend, D.J., Cater, S.R., and Pearce, B.P. (1991). The role of stress waves in thoracic visceral injury from blast loading: modification of stress transmission by foams and high-density materials. *J. Biomech.* 24 (5): 273–285.

118 Kamae, T. and Drzal, L.T. (2012). Carbon fiber/epoxy composite property enhancement through incorporation of carbon nanotubes at the fiber–matrix interphase—Part I: The development of carbon nanotube coated carbon fibers and the evaluation of their adhesion. *Compos. Part Appl. Sci. Manuf.* 43 (9): 1569–1577. https://doi.org/10.1016/j.compositesa.2012.02.016.

119 Stachowiak, G.W. (1993). Chapter 14—Friction and wear of polymers, ceramics and composites in biomedical applications. In: *Composite Materials Series*, vol. 8 (ed. K. Friedrich), 509–557. Elsevier https://doi.org/10.1016/B978-0-444-89079-5.50018-0.

120 Brown, K.J., Atkinson, J.R., Dowson, D., and Wright, V. (1976). The wear of ultrahigh molecular weight polyethylene and a preliminary study of its relation to the in vivo behaviour of replacement hip joints. *Wear* 40 (2): 255–264. https://doi.org/10.1016/0043-1648(76)90102-2.

121 Dowson, D. and Wallbridge, N.C. (1985). Laboratory wear tests and clinical observations of the penetration of femoral heads into acetabular cups in total replacement hip joints: I: Charnley prostheses with polytetrafluoroethylene acetabular cups. *Wear* 104 (3): 203–215. https://doi.org/10.1016/0043-1648(85)90048-1.

122 Weightman, B.O., Paul, I.L., Rose, R.M. et al. (1973). A comparative study of total hip replacement prostheses. *J. Biomech.* 6 (3): 299–311. https://doi.org/10.1016/0021-9290(73)90052-3.

123 Walker, L., Sohn, M.S., and Hu, X.Z. (2002). Improving impact resistance of carbon-fibre composites through interlaminar reinforcement. *Compos. Part Appl. Sci. Manuf.* 33 (6): 893–902. https://doi.org/10.1016/S1359-835X(02)00010-6.

124 Dumbleton, J.H., Miller, D.A., and Miller, E.H. (1972). A simulator for load bearing joints. *Wear* 20 (2): 165–174. https://doi.org/10.1016/0043-1648(72)90379-1.

125 Buechel, F.F. and Pappas, M.J. (2015). *Principles of Human Joint Replacement*. Cham: Springer International Publishing https://doi.org/10.1007/978-3-319-15311-7.

126 Riener, R., Hoogen, J., Schmidt, G. et al. (2000). Knee joint simulator based on haptic, visual and acoustic feedback. *IFAC Proc.* 33 (26): 543–547. https://doi.org/10.1016/S1474-6670(17)39201-7.

127 Beutler, H., Lehmann, M., and Stähli, G. (1975). Wear behaviour of medical engineering materials. *Wear* 33 (2): 337–350. https://doi.org/10.1016/0043-1648(75)90287-2.

128 Atkinson, J.R., Dowling, J.M., and Cicek, R.Z. (1980). Materials for internal prostheses: the present position and possible future developments. *Biomaterials* 1 (2): 89–96. https://doi.org/10.1016/0142-9612(80)90005-8.

129 Hussain, M., Naqvi, R.A., Abbas, N. et al. (2020). Ultra-high-molecular-weight-polyethylene (UHMWPE) as a promising polymer material for biomedical applications: a concise review. *Polymers* 12 (2): 2. https://doi.org/10.3390/polym12020323.

130 Rostoker, W. and Galante, J.O. (1981). The influence of titanium surface treatments on the wear of medical grade polyethylene. *Biomaterials* 2 (4): 221–224. https://doi.org/10.1016/0142-9612(81)90061-2.

131 Dumbleton, J.H. and Shen, C. (1976). The wear behavior of ultrahigh molecular weight polyethylene. *Wear* 37 (2): 279–289. https://doi.org/10.1016/0043-1648(76)90034-X.

132 Kumar, P., Oka, M., Ikeuchi, K. et al. (1991). Low wear rate of UHMWPE against zirconia ceramic (Y-PSZ) in comparison to alumina ceramic and SUS 316L alloy. *J. Biomed. Mater. Res.* 25 (7): https://doi.org/10.1002/jbm.820250703.

133 Sawae, Y., Murakami, T., and Chen, J. (1998). Effect of synovia constituents on friction and wear of ultra-high molecular weight polyethylene sliding against prosthetic joint materials. *Wear* 216 (2): 213–219. https://doi.org/10.1016/S0043-1648(97)00196-8.

134 Willmann, G., Früh, H.J., and Pfaff, H.G. (1996). Wear characteristics of sliding pairs of zirconia (Y-TZP) for hip endoprostheses. *Biomaterials* 17 (22): 2157–2162. https://doi.org/10.1016/0142-9612(96)00042-7.

135 Rose, R.M., Goldfarb, H.V., Ellis, E., and Crugnola, A.M. (1983). On the pressure dependence of the wear of ultrahigh molecular weight polyethylene. *Wear* 92 (1): 99–111. https://doi.org/10.1016/0043-1648(83)90010-8.

136 Barbour, P.S.M., Barton, D.C., and Fisher, J. (1995). The influence of contact stress on the wear of UHMWPE for total replacement hip prostheses. *Wear* 181–183: 250–257. https://doi.org/10.1016/0043-1648(95)90031-4.

137 Saikko, V. (2006). Effect of contact pressure on wear and friction of ultra-high molecular weight polyethylene in multidirectional sliding. *Proc. Inst. Mech. Eng. [H]* 220 (7): 723–731. https://doi.org/10.1243/09544119JEIM146.

138 Landy, M.M. and Walker, P.S. (1988). Wear of ultra-high-molecular-weight polyethylene components of 90 retrieved knee prostheses. *J. Arthroplasty* 3: S73–S85. https://doi.org/10.1016/S0883-5403(88)80013-5.

139 Merola, M. and Affatato, S. (2019). Materials for hip prostheses: a review of wear and loading considerations. *Materials* 12 (3): 495. https://doi.org/10.3390/ma12030495.

140 Baykal, D., Siskey, R.S., Haider, H. et al. (2014). Advances in tribological testing of artificial joint biomaterials using multidirectional pin-on-disk testers. *J. Mech. Behav. Biomed. Mater.* 0: 117–134. https://doi.org/10.1016/j.jmbbm.2013.05.020.

141 McKellop, H., Clarke, I.C., Markolf, K.L., and Amstutz, H.C. (1978). Wear characteristics of UHMW polyethylene: a method for accurately measuring extremely low wear rates. *J. Biomed. Mater. Res.* 12 (6): 895–927. https://doi.org/10.1002/jbm.820120611.

142 Liu, H., Ji, H., and Wang, X. (2013). Tribological properties of ultra-high molecular weight polyethylene at ultra-low temperature. *Cryogenics* 58: 1–4. https://doi.org/10.1016/j.cryogenics.2013.05.001.

143 Panin, S.V., Kornienko, L.A., Alexenko, V.O. et al. (2020). Increasing wear resistance of UHMWPE by loading enforcing carbon fibers: effect of irreversible and elastic deformation, friction heating, and filler size. *Materials* 13 (2): 2. https://doi.org/10.3390/ma13020338.

144 Aldousiri, B., Shalwan, A., and Chin, C.W. (2013). A review on tribological behaviour of polymeric composites and future reinforcements. *Adv. Mater. Sci. Eng.* 2013: e645923. https://doi.org/10.1155/2013/645923.

145 Luebke, R.J. and Chan, K.C. (1985). Effect of microwave oven drying on surface hardness of dental gypsum products. *J. Prosthet. Dent.* 54 (3): 431–435. https://doi.org/10.1016/0022-3913(85)90566-9.

146 Rhee, S.H. and Dumbleton, J.H. (1976). The application of the zero wear model to joint prostheses. *Wear* 36 (2): 207–224. https://doi.org/10.1016/0043-1648(76)90006-5.

147 Bayer, R.G., Clinton, W.C., Nelson, C.W., and Schumacher, R.A. (1962). Engineering model for wear. *Wear* 5 (5): 378–391. https://doi.org/10.1016/0043-1648(62)90075-3.

148 Navarro, M., Michiardi, A., Castaño, O., and Planell, J.A. (2008). Biomaterials in orthopaedics. *J. R. Soc. Interface* 5 (27): 1137.

149 Streicher, R.M., Weber, H., Schön, R., and Semlitsch, M. (1991). New surface modification for Ti-6Al-7Nb alloy: oxygen diffusion hardening (0DH). *Biomaterials* 12 (2): 125–129. https://doi.org/10.1016/0142-9612(91)90190-L.

150 Ito, A., Tateishi, T., Niwa, S., and Tange, S. (1993). In-vitro evaluation of the cytocompatibility of the wear particles generated by UHMWPE/zirconia friction. *Clin. Mater.* 12 (4): 203–209. https://doi.org/10.1016/0267-6605(93)90074-H.

151 Shetty, R.H. (2000). Surface hardening of orthopedic implants. In: *Biomaterials Engineering and Devices: Human Applications: Volume 2. Orthopedic, Dental, and Bone Graft Applications* (ed. D.L. Wise, D.J. Trantolo, K.-U. Lewandrowski, et al.), 191–202. Totowa, NJ: Humana Press https://doi.org/10.1007/978-1-59259-197-8_11.

152 McKellop, H.A. and Röstlund, T.V. (1990). The wear behavior of ion-implanted Ti-6Al-4V against UHMW polyethylene. *J. Biomed. Mater. Res.* 24 (11): 1413–1425. https://doi.org/10.1002/jbm.820241102.

153 Salzer, M., Knahr, K., Locke, H. et al. (1979). A bioceramic endoprosthesis for the replacement of the proximal humerus. *Arch. Orthop. Trauma. Surg.* 93 (3): 169–184. https://doi.org/10.1007/BF00523668.

154 Friedrich, K. (2012). *Advances in Composite Tribology*. Elsevier.

155 Dowson, D. (1995). A comparative study of the performance of metallic and ceramic femoral head components in total replacement hip joints. *Wear* 190 (2): 171–183. https://doi.org/10.1016/0043-1648(96)80015-9.

156 Sharma, A. and Sharma, G. (2018). Biomaterials and their applications. *AIP Conf. Proc.* 1953 (1): 080041. https://doi.org/10.1063/1.5032847.

157 McKellop, H., Clarke, I., Markolf, K., and Amstutz, H. (1981). Friction and wear properties of polymer, metal, and ceramic prosthetic joint materials evaluated on a multichannel screening device. *J. Biomed. Mater. Res.* 15 (5): 619–653. https://doi.org/10.1002/jbm.820150503.

158 Anderson, J.C. (1986). Chapter 10—The wear and friction of commercial polymers and composites. In: *Composite Materials Series*, vol. 1 (ed. K. Friedrich), 329–362. Elsevier https://doi.org/10.1016/B978-0-444-42524-9.50014-4.

159 Dowson, D., Taheri, S., and Wallbridge, N.C. (1987). The role of counterface imperfections in the wear of polyethylene. *Wear* 119 (3): 277–293. https://doi.org/10.1016/0043-1648(87)90036-6.

13

Wear and Friction Behavior of Biocomposites Fabricated Through Additive Manufacturing

Manickam Ramesh[1], Kanagaraj Niranjana[2], and Manickam Tamil Selvan[1]

[1] *KIT-Kalaignarkarunanidhi Institute of Technology, Department of Mechanical Engineering, Coimbatore 641402, Tamil Nadu, India*
[2] *KIT-Kalaignarkarunanidhi Institute of Technology, Department of Aeronautical Engineering, Coimbatore 641402, Tamil Nadu, India*

13.1 Introduction

Conventional composites comprise a reinforcing phase with a binding agent; materials selection, manufacturing processes, and microstructural parameters can tailor their physical and chemical properties. Microstructural features may classify additives as microparticles, continuous fibers; two-dimensional (2D) weaved textiles, or even 3D shapes [1, 2]. It is common practice for 3D-integrated preforms to have fibers or threads aligned in more than one orientation [3]. Since they are cost-effective, have good structural stability, can withstand breakage, and are easily designed. In light of recent developments in additive manufacturing (AM) methods, we decided to look into the feasibility of directly fabricating multidirectional molds in layers. AM, more often known as 3D printing, has piqued the interest of business researchers, academic and research sectors, and hobbyists. It has recently been possible to generate high-quality prints using faster and less expensive AM processes. Additionally, polymer materials have enabled 3D printing with a broader range of features [4, 5].

When two solid surfaces come into contact with one another, there is a risk of deterioration to both contact layers. Wear is the process by which solid material is removed from a surface, either one or both of which were previously in solid-state contact. The word "wear rate," which refers to the mass, volume, or height loss of material removed per unit time or sliding distance, is used to quantify wear. It has both light and heavy wear [6, 7]. The result of low-intensity wear is a coating that is smooth and free of large granules of wear debris (usually between 0.01 and 1 μm in size). Extreme wear, on the other hand, creates bigger wear debris (20–200 m) and a roughened worn surface, both of which are visible to the human eye. The wear mechanism may be adhesion, abrasion, spalling, erosive, pitting, fatigue, or corrosive [8].

Tribological Properties, Performance, and Applications of Biocomposites, First Edition.
Edited by Chandrasekar Muthukumar, Senthilkumar Krishnasamy,
Senthil Muthu Kumar Thiagamani, and Ganesan Chinnachamy.

The removal of material from solid surfaces can be accomplished through mechanisms known as friction and wear. These mechanisms involve contact and sliding. Polymers are widely employed in dry sliding circumstances because of their low frictional resistance; these conditions are easier to comprehend than wet ones. The inefficiency of elevated process equipment, leading to material wastage, can be traced back in large part to the neglect of tribological interactions and a general lack of knowledge. As a result of its widespread application, researchers have been inspired to explore the underlying principles of friction and wear in polymers [9].

13.2 Additive Manufacturing of Biocomposites

The AM process has some distinctive features. This is a revolutionary class of methods that is also followed by the terms rapid prototyping and freeform fabrication. Each of these terms describes an identical entity. AM is defined as a process of joining materials to make objects from 3D model data, usually layer upon layer, according to ASTM-I F2792: Conventional Jargon for Described AM Techniques [10]. Subtractive manufacturing refers to the traditional method of making 3D objects by cutting away layers of material one by one [11]. Direct fabrication, in which design data are directly translated into a real product without the need for tooling or machining, and additive, layer-wise processing are the two primary characteristics that define AM (Figure 13.1). Because of these qualities, AM is provided with a number of one-of-a-kind capabilities that are incomparable to traditional manufacturing processes [13].

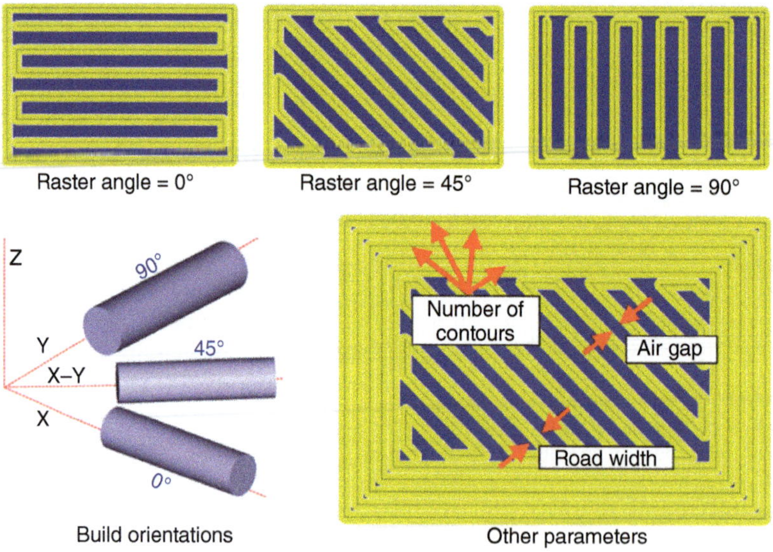

Figure 13.1 Process parameters of AM. Source: Mohamed et al. [12]/with permission of Elsevier.

The process of AM makes it possible to fabricate objects with intricate shapes. When using traditional techniques, which are limited in their ability to explore crevices and other inside surfaces, a complicated item is frequently constructed by combining separate primary sections. This approach frequently results in early structural breakdown at material junctions [14, 15]. When using AM, however, objects are created using the same process regardless of the degree of complexity they contain. This procedure includes splitting the intended structure into a given amount of layers with a defined thickness and printing the split portions layer, upon layer progressively from the bottom up to the top. The thickness of each layer is also present [16]. Because of this capability, considerable degrees of freedom may be used in the creation of geometric structures. Functional characteristics are often accompanied by increasing structural complexity; however, additive printing makes it simpler to include these elements by automating the production process and eliminating some of the phases [17].

It is possible to develop multi-functionality and gradient functionality characteristics through multi-material AM. During the processing stage, this is achieved by altering the elemental composition and placement inside a component [13]. Polylactic acid (PLA) is a bio-based polymer produced entirely without the use of any synthetic chemicals in the fermentation of plant materials like corn, potatoes, sugar, etc. High brittleness and low impact strength necessitate modifying PLAs with excellent mechanical characteristics. In order to get the desired results of the modification, the addition of fiber filler is required [18]. The automotive industry, industrial construction, and many other commercial markets are just a few of the many that benefit from reinforced PLA or filler-added PLA. When synthetic materials acrylonitrile butadiene styrene (ABS) and biomaterials PLA were subjected to the same conditions for wear testing using a pin-on-disc apparatus, it was discovered that ABS exhibited a higher wear rate and frictional force than PLA did for the same set of printing parameters and part orientations [19].

In another study, PLA was supplemented with bioactive carbon, and wear was measured using the translator oscillation apparatus SRV-3. A PLA specimen with 30% carbon volume had the minimum wear volume after a dry sliding test against an Al_2O_3 ball. Furthermore, a variety of wear processes, including fatigue and erosion, were identified, each of which was affected by a particular attribute of the material that biogenic carbon had. The mean amount of the coefficient of friction did not alter when natural materials were added to PLA; however, the highest value of the coefficient of friction did decrease, and the wear behavior of PLA improved [20]. Utilizing the FDM 3D printing technique would allow for the processing of a wide variety of thermoplastic polymers. The biodegradable polymer (PLA) is highly regarded as a viable replacement for traditional polymers made from petroleum [21]. The processing of ABS results in an offensive odor, despite the fact that ABS possesses outstanding mechanical qualities [22]. Thermoplastic material known as polyethylene terephthalate-glycol (PETG) polyester is created by polycondensing terephthalic acid and ethylene glycol. It is a saturated form of polyester [23]. The benefits that are offered by the PLA and ABS materials are brought together in the

PETG substance. It is both more flexible and capable of producing items with a high degree of strength [24].

The use of polymer components in engineering materials is becoming more popular as a result of the material's beneficial properties, which include self-lubrication, chemical stability, and vibration dampening. However, tribological investigations of 3D-printed structures are still in their infancy at this point [25]. The majority of the research that has been conducted on PLA has concentrated on the mechanical characteristics of PLA as a polymer material in and of itself [26]. There are additional instances of it in research on the mechanical structure of additive polymers. Tribological features of printed structures are rarely discussed in published studies. This demonstrates that the subject being discussed is relevant at the present time. Accordingly, it is worthwhile to initiate deeper study into the influence of 3D printing on the abrasive wear characteristics of polymers [3].

13.3 Fabrication of Biocomposites Using AM

AM allows for greater flexibility in the fabrication process. The lead time may be considerably cut down by using additive printing since it immediately creates an item from the design file. This also allows rapid design demos and makes it very cost-effective to tailor small lot sizes that have difficult shapes [27, 28]. In the last stage of 3D printing, the substance is positioned exactly where it is needed, and any excess material may often be recovered and reused, minimizing the quantity of material that is lost. Compared to traditional methods, AM of multidirectional preforms has a number of distinct advantages [29]. The schematic representation of different printers is depicted in Figure 13.2.

13.4 Types of Wear Behavior Based on Its Processes, Effects, and Environment

According to the findings of a few researchers, the friction and wear characteristics of polymer components formed of fused filament fabrication (FFF) have been investigated. In particular, the effect of the following three printing factors has been taken into consideration: wear and friction coefficient of FFF components are affected by load, speed, and orientation. The wear changes more with sliding speed than with orientation for a given load [31]. Components made with FFF contain asperities, but their frictional properties improve with use: as the surface roughness diminishes, the coefficient of friction (COF) lowers and stabilizes, and the remaining material is worn down to powder. Wear patterns like abrasion, fatigue, plowing, corrugation, erosion, and cavitation commonly cause damage to contact surfaces [32]. Changes in the spacing of the solids in contact can cause abrasive wear to alter the body's shape in an irreversible manner. In most cases, the rate of wear will first decrease rapidly and then increase suddenly to a maximum, where it will stabilize and follow a constant linear rate [33]. This was determined by conducting a sliding wear test in

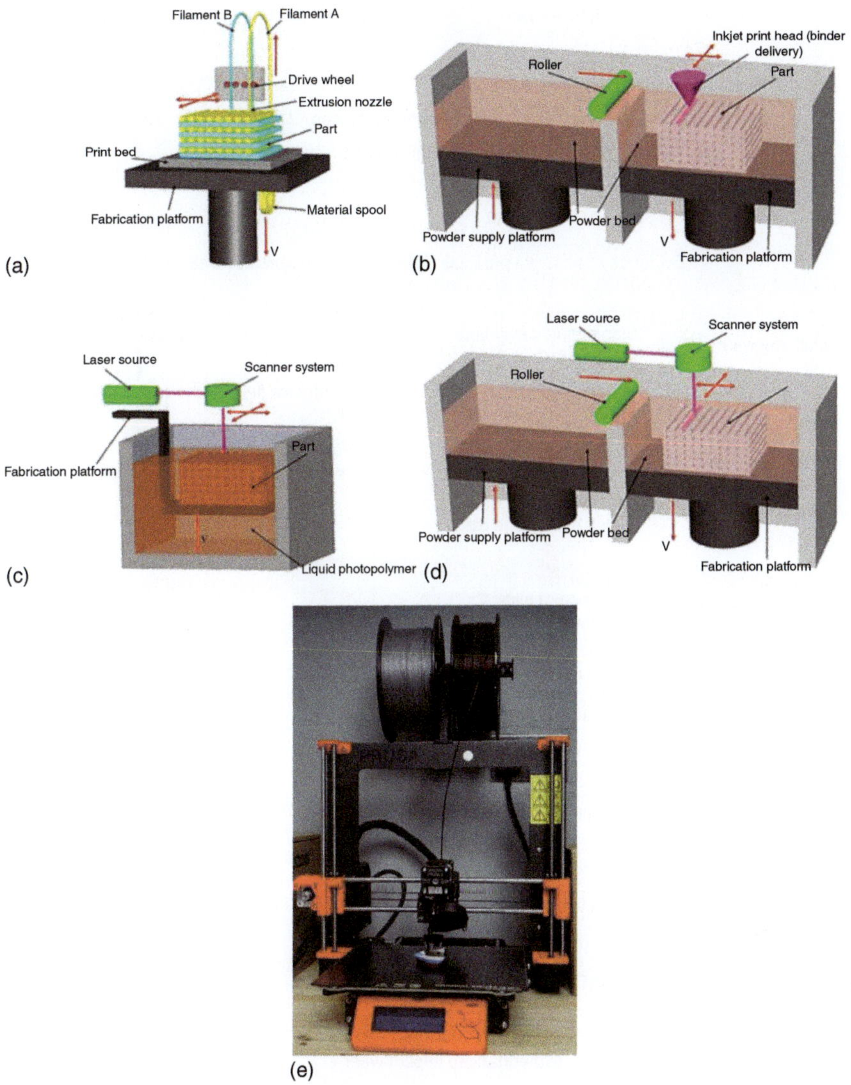

Figure 13.2 Schematic representation of (a) FDM [18]; (b) Inkjet printing [18]; (c) Vat polymerization [18]; (d) Powder bed fusion. Source: Ngo et al. [18]/ with permission of Elsevier; and (e) FDM 3D printer. Source: Bankupalli et al. [30]/with permission of Elsevier.

accordance with ASTM G95 [4]. Different types of wear and their characteristics are explained in Figure 13.3.

13.4.1 Adhesion Wear

The contact area in the adhesion wear mechanism has high adhesion but a weak cohesive bond. In most metal materials, binding might be altered by adding a little amount of alloy to the bulk-processed material [35]. Although adhesion will be

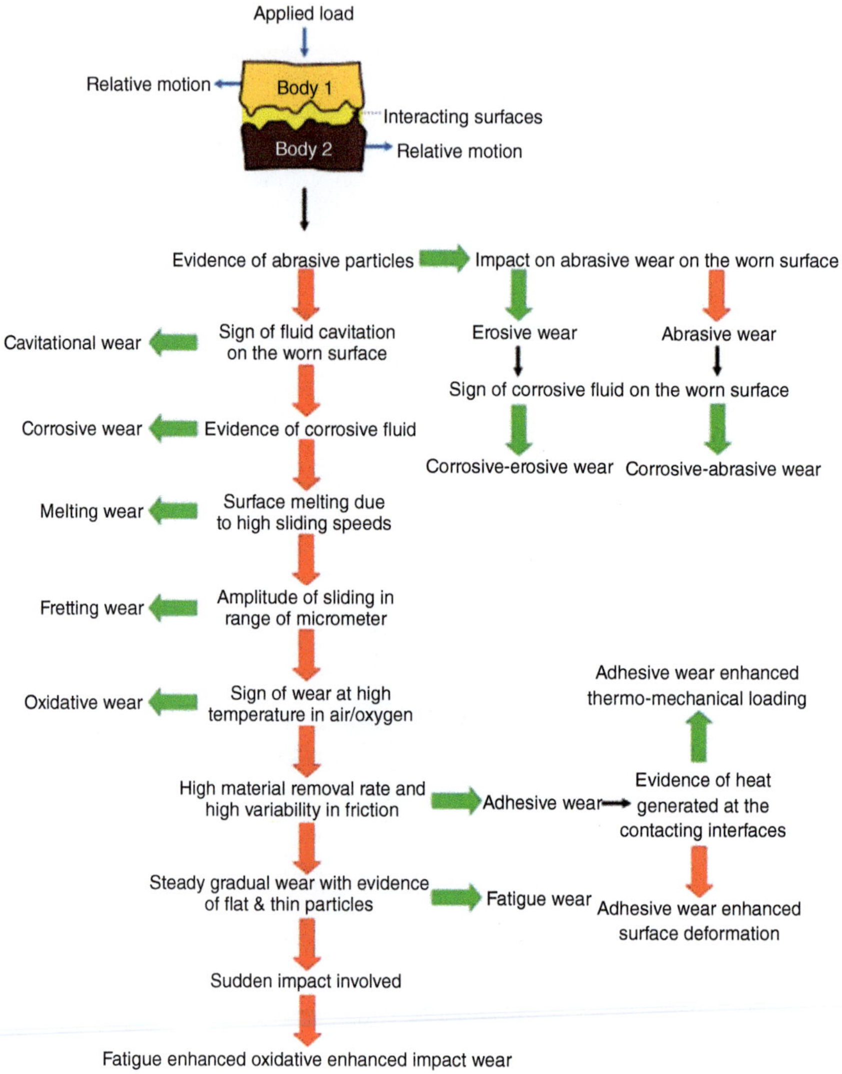

Figure 13.3 Different types of wear and its characteristics. Source: Nirmal et al. [34]/with permission of Elsevier.

at its best when using similar materials, the two components should not have the same chemical structure. The outside of both should be free of oxide and lubricant [26]. When the surface of one component of a working machine transfers to another component, this is visible. It has the potential to cause significant wear issues on the pair's external surface, such as hollowness, gaps, and valleys [21].

13.4.2 Abrasive Wear

It is possible to observe the movement of substance from the interfaces as a consequence of the abrasion wear mechanism, which finally leads to the building

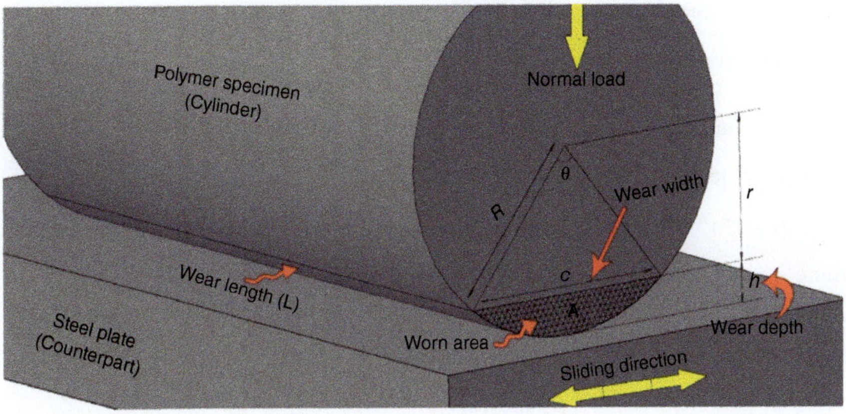

Figure 13.4 Abrasive wear dimension.

of a trench on both surfaces. This groove is ultimately caused by the abrasion-wear mechanism. This process is sensitive to several factors, including the abrasive's firmness, size, and shape [33]. The abrasive wear dimension is given in Figure 13.4.

It is possible to define abrasive wear as:

- In the first case, known as two-body abrasive wear, a harder interface rubs away material from a softer one, leaving behind a smooth abrasive surface.
- The other case, known as three-body abrasive wear on one or both surfaces, is caused by a stiff particle that becomes trapped between two smooth ones.

Hardness is a crucial property that has to be considered for abrasive wear. The rate of abrasive wear can be regulated, at least in part, by the material's hardness. When the hardness of a material is increased, abrasive wear rates are shown to decrease, according to experimental data [36]. When it comes to abrasive wear, particle size may be a significant factor for many different types of materials. When a hard surface with sharp points slides across a smooth surface, the elimination of matter from the rubbing surfaces can be observed. This results in wear-related consequences such as plowing, cutting, and trimming [37].

13.4.3 Erosive Wear

Erosion by wear occurs when the outer surface of substances is subjected to the imposition of hard fluid granules or possibly a combination of both. The speed and momentum of these particles are rather impressive. The most noticeable effect that wear and erosion have on a surface is material removal. The creation of river marks like structural change is a visible example of the erosive wear that occurs during this process. The process by which water molecules move at high velocities causes the loss of rigid particles bonded together in the material and, as a result, forms a pathway for the flow. This is known as erosive wear. If the cohesive link between the surfaces of a material is more vital, then the erosion wear in the sample will be less [38].

There are three primary factors that affect erosive wear are:

- Relative solidity
- Velocity
- Angle of impact

It is possible to quantify it using the equation:

$$E = c^* v^{n*} f(a) \tag{13.1}$$

where

v velocity of the water
a angle of impact
n coefficient, which varies depending on the materials

13.4.4 Fatigue Wear

In most cases, it takes place as a consequence of the cyclical repetition of stresses and finally results in the creation of a vacuum on the surface where the two materials are in contact with one another. Deformations of contact surfaces are generally brought on by cyclic loading [39]. Mechanical failure due to fatigue arises whenever a material is repetitively subjected to extremely high stresses; this causes the material's surface to develop a series of microscopic voids, which ultimately combine to form a larger crack. Increasing the relative humidity range is the most favorable condition in the various studies. Tires and other bearings on cars exhibit evidence of fatigue wear on roads (sliding). The cracking process typically consists of three stages: crack initiation, growth, and fracture [40].

13.4.5 Corrosive or Oxidative Wear

It happens in an oxidative atmosphere during dry sliding between machine parts. Oxygen and other additions or chemical agents in the atmosphere produce a reaction between the solid surfaces known as oxidation. The wear produced by oxidation or corrosion is one sort of wear that needs to be minimized. Inconel oxidation is another name for this phenomenon. The rough surface of the material being moved about generates an oxide layer during this process, further decreasing its toughness and rendering it mechanically incompatible [41].

13.4.6 Fretting Wear

Fretting wear can be caused when there is a low-amplitude oscillatory motion between the surfaces of two samples. The mechanism for fretting wear can be observed in the ceiling fan's trembling. Fretting wear depends on several parameters, including material contact, slip range, and pressure between materials.

13.5 Determining the Level of Specimen Deterioration

The amount of weight lost during and after a test is one of the most straightforward methods for calculating wear. If the materials being compared are similar and precise measurements are collected, this is an easy and straight-forward process. In this way, the wear rate can be determined as a function of time by converting the mass loss to volume loss. When a coating is applied, any loss in mass due to penetration is a result of both the coating and the underlying material(s). When plastic flows or deforms, wear may occur, but there may be no loss of mass as a result of the process. This may be the case when the material deforms or when plastic flow occurs [39].

It is possible to employ equations that are derived from the geometry of the wear scar in order to perform the task of computing the volume of wear that has occurred. If the contour of the scar is regular and symmetrical, then it will be simple to get the appropriate degree of precision using this approach. It will be more complicated if the scar's shape is irregular and asymmetrical. The equipment that was created for measuring the depth and width of the wear scar or the impact zone may be used to make estimates of the quantity of material removed at regular intervals. These estimates can be made using the equipment. This can be accomplished by using the equipment. Suppose the wear scar or crater can be produced accurately in 3D or solid modeling software. In that case, the volume wear may be estimated straight from the program in the case that we are discussing. It is feasible to measure the dimensions of the abrasive instrument as an additional step toward minimizing the volumetric loss that arises from the procedure. This may be done in order to reduce the amount of volume lost. In multi-layered coated systems, the amount of wear that has occurred may indicate the system's overall total wear loss. This can be determined by measuring the amount of wear that has taken place. This method is well known and well accepted; it involves determining the quantity of material that has been lost by determining the depth of the wear scar that has been left behind. These values should be exemplary for a surface profilometer to measure and record [39]. Various tribo-machines for wear testing are given in Figure 13.5 [34].

A technique that is more accurate has been developed, and it makes use of both a profilometer and computerized system equipment. The computer then automatically calculates the volume of the wear scar and displays it as an isometric plot. It is common practice to employ a wear coefficient when classifying a material's resistance to contact wear. Calculating a wear coefficient, denoted by the letter K, is the approach that is most frequently utilized. This coefficient is calculated on the premise that the amount of wear increases linearly and proportionately with the interface load and the sliding distance. It is proposed to use this wear coefficient as a criterion for wear testing, and it is now widely utilized in the field.

$$K = \frac{\text{Volume}}{\text{Load} \times \text{Sliding Distance}} \ \text{mm}^3\ \text{Nm}^{-3} \tag{13.2}$$

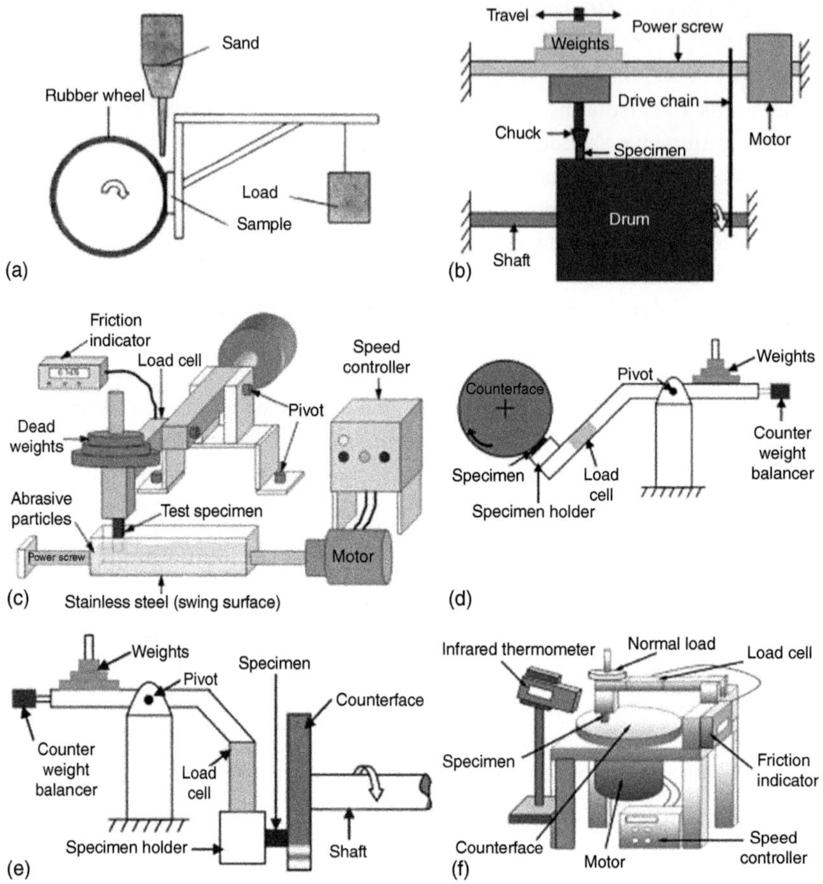

Figure 13.5 Various tribo-machines for wear testing. Source: Nirmal et al. [34]/with permission of Elsevier.

13.6 Wear and Frictional Characteristics of AM Products

The compiled wear and frictional characteristics of AM composite products are depicted in Table 13.1.

13.7 Method of Testing the Wear and Friction in the AM Parts

In order to choose the correct wear test, there are a set of claims that need to be evaluated in detail, including the following:

- Whether or not the qualities of interest can be measured by the wear testing method
- Whether or not the testing can withstand the load

Table 13.1 Wear and frictional characteristics of AM composite products.

Materials used		AM techniques and name of printer used	Process parameters	Types of wear testing	Tribometer used	Tribological properties findings			Observation	Ref.
matrix	Filler					Wear rate	COF	Frictional force		
PLA	Graphite	FFF		Dry sliding wear tests	• Time of the test: 2 h 27 min • Load: 5.49 N • Pin pressure on the disc 0.11 MPa • Roughness of the disc Ra (0.35–0.45) • Velocity of the disc 0.34 $m\,s^{-1}$ • Length of the pin path 3 km • Temperature of environment 23.8 °C • Humidity 26%	15.2 $\mu m\,km^{-1}$ (PLA) 15.2 $\mu m\,km^{-1}$ (50% PLA–graphite).	0.492 (PLA) 0.288 (50% PLA–graphite).		• It has been observed that PLA has a higher wear rate. • PLA–graphite was much more brittle and caused difficulties upon dosing the filament.	[42]

(Continued)

Table 13.1 (Continued)

Materials used		AM techniques and name of printer used	Process parameters	Types of wear testing	Tribometer used	Tribological properties findings			Observation	Ref.
matrix	Filler					Wear rate	COF	Frictional force		
PLA	Graphene	FDM	—	Dry sliding wear tests	—		0.1–0.35 (PLA blend) 0.45 (PLA).		• The melting and cooling of the printed filament following extrusion causes the printed PLA to have a lower crystallinity than the raw ingredients. • PLA and graphene (146 MPa) has more excellent toughness than PLA (123 MPa). • Improved wear resistance of 14% when compared to pure PLA is a result of the combination of PLA-increased graphene's hardness and the presence of graphene in the planar direction. • It has 65% of friction coefficient behavior. • The promising results in the feasibility of 3D printed PLA graphene for potential use as scaffolds for orthopedics applications.	[25]

ABS and PLA	—	FDM	Color of the filament: 1. White and gray (PLA) 2. White (ABS) Printing temperature ranges Low 190–195 °C Opt 200–205 °C Max 215–220 °C.	• Tribotester • Surface roughness of steel counterpart 3.2 Rz [μm], Ra [μm] • Test duration 130 s • Load 180 N • Alternating motion frequency, 10 Hz • Average speed 0.05 m s^{-1} • Stroke length 15 mm • Relative humidity, Rh 50%	ABS: 0.3 PLA (gray): 0.1.	• The average wear for the gray sample across all temperature gradients is almost 3 mm.	[3]

(Continued)

Table 13.1 (Continued)

Materials used		AM techniques and name of printer used	Process parameters	Types of wear testing	Tribometer used	Tribological properties findings			Observation	Ref.
matrix	Filler					Wear rate	COF	Frictional force		
ABS and PLA color (white, gray, and black) Build orientation (0°, 45°, and 90°)		FDM Nozzle diameter 0.4 mm Printing temperature 195 °C Build temperature 60 °C			• Dynamic and static friction Linear alternating reciprocating cylinder-on-plate friction and wear tester • Experimental temperature: ambient, • Relative humidity: 48 ± 2%, • Reciprocating sliding wear conditions: • Load 150 and 200 N, • Duration: 60 min, Frequency 10 Hz, stroke of the reciprocating wear test 6 mm				• Wear depth maximum in the black filament. • Varying the build orientation and filament color with two different applied loads, it's been observed the minimum coefficient of friction is 0.34 ± 0.08.	[16]

PLA		FDM	Part orientation (0 °, 45°, 60°,90 °)	Dry sliding wear	• Pin-on-disc testing • Load: 50 N • Speed: 1000 rpm • Time: 10 min • Diameter: 50 mm			• Wear rate of PLA is lesser than ABS the maximum wear rate of 1300 mm was found at 90° orientations for PLA, • Tensile strength—63 ± 3 MPa, • Flexural strength—79 ± 31 MPa.	
PLA	Biocarbon from wheat stems	FDM		Abrasive and fatigue wear (dry sliding wear)	• Sliding in the linear reciprocating mode, at about 20 °C (±0.2 °C) and 40% (±2%) relative humidity, sliding distance of 1000 μm, load of 1 N, testing time	Of 150 min.	PLA	• There were signs of different wear mechanism like fatigue wear or abrasion. • COF is smooth and most homogeny for PLA with 30 vol% biocarbon. 30 vol% reinforced PLA is more friction-resistant.	[20]

(Continued)

Table 1.1 (Continued)

Materials used		AM techniques and name of printer used	Process parameters	Types of wear testing	Tribometer used	Tribological properties findings			Observation	Ref.
matrix	Filler					Wear rate	COF	Frictional force		
ABS		FDM	Layer height: 0.18 mm Print speed: $60\ mm\ s^{-1}$ Fill density: 100% Fill pattern: line Extruder temperature: 220 °C	Ball-on-disc tribometer	All tests were performed at different loads: 20–55 N Sliding distance: 1000 m • Sliding speed of 100 rpm	Wear rate of the ABS that is 3D-printed decreases with a decrease in the applied load.	COF increases with an increase in the applied load.		• The outcomes demonstrate that the wear rate of the 3D-printed ABS decreases with the load. • Be that as it may, a high coefficient of friction is observed with the applied load. • The intrinsic surface finish and the resultant high value of COF for the surfaces that are 3D-printed is yet a challenge that is to be overcome.	[43]

Poly-ether-ether-ketone (PEEK)	Carbon fiber	FDM	Nozzle temperature: 415 °C Heated bed temperature:120 °C Printing speed:60 mm s^{-1} Extrusion width: 0.48 mm Layer height: 0.1 mm Infill percentage: 100%	Pin-on-flat linear reciprocating tribometer	Normal load: 250 N, contact pressure: 6.4 MPa Reciprocating length: 25 Mm Sliding speed: 50 Mm S^{-1} • Sliding cycles: 1, 2, 5, 10, and 20 k	—	—	—	• There is a direct correlation between the reduction in friction coefficient and wear rate during the steady wear phase (11.24% and 74.68%, respectively) and the increased engineering significance of 3D printed components.	[44]
Porous polyimide	—	FDM	Nozzle diameter:500 μm Motor speed: 108 R min^{-1} Layer height: 500 μm Printing speed: 1.5 Mm S^{-1} Extrusion pressure:0.36 MPa	Friction test machine (CEMTHT 07–135)	Load: 5–20 N Sliding frequency: 1 Hz • Sliding distance of 36 m wear time of 1 h at room temperature					[45]

(Continued)

Table 13.1 (Continued)

Materials used		AM techniques and name of printer used	Process parameters	Types of wear testing	Tribometer used	Tribological properties findings			Observation	Ref.
matrix	Filler					Wear rate	COF	Frictional force		
ABS and verogray poly-mers		Polyjet technology	Build ori-entation Part spacing Surface quality on product function-ality	Pin-on-disc	• Loads of 1, 5, and 10 N were applied under dry sliding contact with a 52 100 steel counter face at room temperature	Lower in the direction per-pendicular to sliding direction Higher wear in parallel sliding.	The development of a low shear tribo-film reduces the coefficient of friction under light loads.		• The findings show that for 1 N load, the alignment of the interface to the sliding direction has a significant impact on the friction coefficient. This reliance, however, is shown to be a consequence of the bulk material characteristics rather than the surface roughness for larger stresses of 5 and 10 N.	[46]

Poly(methyl methacrylate)	Nanodiamonds	Digital light processing		Reciprocating tribometer	The adversaries were needles with a 5 mm diameter constructed of either grade V titanium alloy (TI; Ti-6Al-4V) or stainless steel (SS; SUS 304). Perpendicular load 5 N frequency 0.5 Hz PBS was used as the lubricant 104 cycles were run • Total sliding distance of 40 m	Based on the results of the wear resistance test, both the pure non-functionalized nanodiamonds and the amine-functionalized nanodiamonds groups showed significantly better resilience to interface loss.	The nanodiamond amine-functionalized and nanodiamonds pure non-functionalized nanomaterials both had significantly lower COF; 0.01) than stainless steel work surfaces.		• It is found that qualities including COF, wear performance, and resistance to *Streptococcus mutans* were greatly enhanced by the addition of 0.1 wt. % nanodiamonds to the poly(methyl methacrylate)-based resin for 3D printing, with no discernible effect related with the functionalization of the nanodiamonds. • When compared to the controls for both the stainless steel and the titanium countersurfaces (p 0.001). The resistance to *S. mutans* biofilm formation after 48 h was also considerably increased by nanodiamonds and amine-functionalized nanodiamonds (p 0.01) compared to the control group.	[6]

(Continued)

Table 13.1 (Continued)

Materials used						Tribological properties findings				
matrix	Filler	AM techniques and name of printer used	Process parameters	Types of wear testing	Tribometer used	Wear rate	COF	Frictional force	Observation	Ref.
PEEK	Carbon fiber	FDM	Diameter of nozzle 0.3 mm Nozzle temperature 420 °C Build temperature 150 °C Layer height: 0.2 mm Print speed: $10\ mm\ s^{-1}$ Infill density: 100%	Extended plate-on-ring	• Normal force 50 N • Sliding velocity $1\ m\ s^{-1}$	With an outer circumference of 60 mm and a thickness of 5 mm, the counter body ring was crafted from 100Cr6 iron and had a median surface roughness of Ra = 0.2 m.	It's clear from the data that the side orientation has a major impact on the friction coefficient and wear rate.		• The wear rate is the sliding perpendicular direction shows a lower wear rate. Comparing the worn surfaces, efficient distribution of shear stress on carbon fibers is responsible for the much reduced friction and wear. • In addition, the friction performance of the coating material is greatly enhanced by absorbing nanosilica into the interface. These effects may be traced back to two mechanisms: nanosilica rolling at the interface and decreased adhesion between tribo-pairs. It is hoped that printing with optimum processing would allow the external nanoparticle donator to be incorporated into the synthetic structures without causing the production of flaws.	[35]

- Whether or not the effects of temperature and humidity can affect the results
- How long does the test take to complete?
- Whether or not the testing environment simulates the actual operating environment
- Whether or not there is rolling, sliding, etc., between the materials.

Wear resistance, hardness, rigidity, etc., are all calculated after a battery of tests is run to determine the material's wear rate. These tests are necessary in order to avoid overlooking the mechanisms of wear and tear as well as to improve the outcomes. Standardized wear testing procedures include.

13.7.1 Pin-on-Disk or Tribometer

Researchers argued that their technique was one of the most common ones used to determine wear rates. It is widely considered to be the most reliable and preferred method of wear testing for a wide range of materials, according to numerous studies. While running the tests, there are bound to be variations in a variety of process parameters. As a result, it has advanced to the forefront of wear testing procedures [47]. To elaborate, consider where a pressurized pin carrying the necessary material is depicted traveling in a circular motion on a rough surface, the rotational speed of which can be measured in revolutions per second (Figure 13.6) [48].

13.7.2 Pin-on-Drum

Pin-on-drum, upon performing its many tasks, elaborately increases the rate of wear by preventing crossing over an abrasion area when the appropriate force is given (imagine a turntable needle collecting polishing pads from a slightly curved sample moving in a spiral route). This is the one major improvement that Khrushchev made over the pin-on-disc approach. This sample replicates the rate of wear that happens when polishing pads are pulverized and crushed by projecting their movement in such a manner that the pin face may abrade from all sides [39]. This kind of test is typically alluded to as an extremely stressful abrasion evaluation

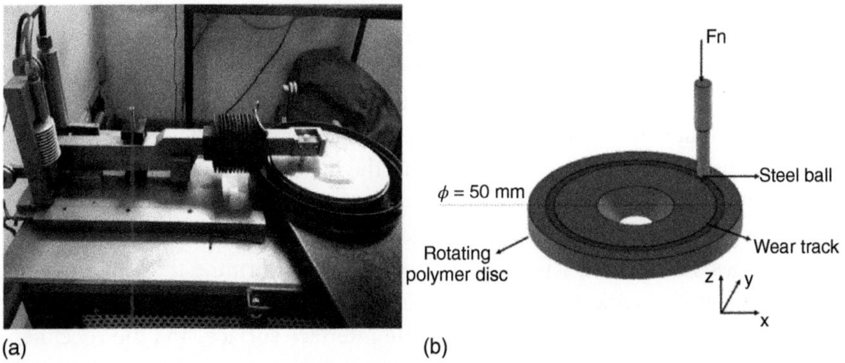

Figure 13.6 (a) Pin-on-disc or tribometer and (b) ball-on-disc.

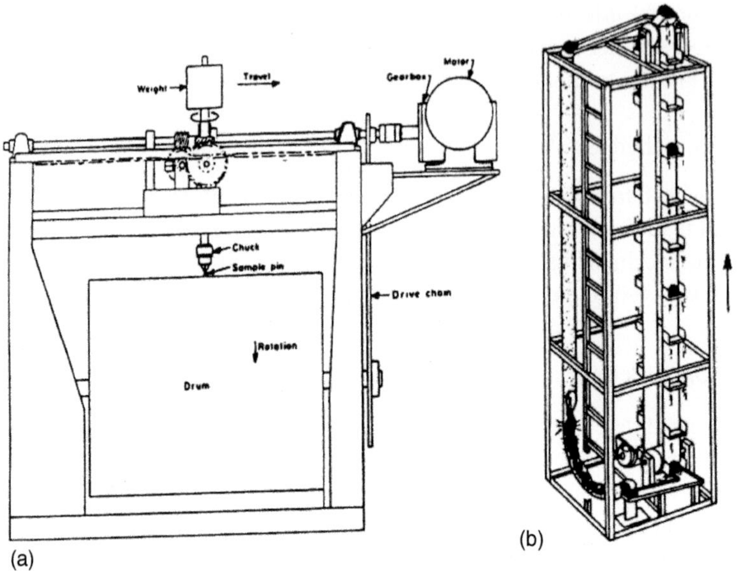

Figure 13.7 (a) Pin-on-drum and (b) impact tester. Source: Kennedy and Hashmi [39]/with permission of Elsevier.

due to the extreme force needed to shatter the abrasive particles. Under high-stress, two-body wear circumstances, it effectively covered a broad variety of substances, including composites, alloys, metals, and resins, and it helped determine the impacts of different production factors such as abrasive size and substance, load, and velocity [4]. Using a modified milling machine to maintain a consistent surface speed, climax research services created a pin-on table that researchers in Melbourne could use to calculate wear rates with more ease. Pin-on-drum relies on a head that rotates a specimen as it moves along a rotating drum covered in abrasive material (Figure 13.7a).

13.7.3 Repeated Impact Wear Test

The Izod method, which employs pendulum-like heavy equipment to determine a material's resilience, was once the standard approach to performing an impact test (Figure 13.7b). However, its utility is restricted in that it is mostly employed in determining how much energy was absorbed by the specimen during the collision. As a result, it is useless when experimenting with the material's response at the point of impact or the force exerted. However, the impact test has been improved recently by reducing the velocities of impact on specimens that are being subjected to regulated angles. As a result, scientists are now able to determine the toughness of a material when it is subjected to dynamic fracture. When the specimen's mass, the impact velocity, the specimen's shape on the interface surface, and the specimen's mechanical behavior were all within our control, it became clear that impact tests were a useful tool [49].

13.7.4 Acoustic Emission Monitoring Test

The aim of this tEst was to investigate the effect of the impact, which did cause tool wear, on the deformation of the specimen's surface roughness. Kubo and Hashimoto performed an experiment using modified scratch tests with a load that was gradually increased over time. Emission refers to the unique kind of acoustic energy released by a solid when subjected to pressure or force. In addition, it was computed with the aid of miniature sensors attached to the test article, which transformed the stress waves into electric impulses. With increasing component degradation, more energy is lost. Stress wave formation as a result of impact can be studied using this nondestructive method. It finds extensive application in composites reinforced with natural fibers. Adhesion-like thin-film properties can be studied with this method. Scratch films are examined using a camera, diamond indenter tip, and scanning electron microscopy [39].

13.7.5 Rubbing Test

A device comprising a pair of crossover hoops, spinning at a rate of 100 rev minute^{-1} is utilized in this test, which is carried out in accordance with the standards established by the ASTM organization on materials that may or may not be comparable to one another. It is best to do this test on a rough, non-lubricating surface so that the amount of material lost when a spinning cylinder is placed against a stationary cylinder at an acute angle can be accurately measured. In addition, the mass loss, the volume loss, and the radioactivity loss (for radioactive prototype) are the three methods that are utilized the most frequently in order to calculate the wear rate.

13.8 Conclusion

- Technology advancements in AM are in high demand due to the fourth industrial revolution because of the efficiency with which it can boost output while simultaneously producing high-quality prints and complicated components at a lower cost. By using other long-lasting materials in their stead, the building industry hopes to cut costs, reduce carbon dioxide emissions, and conserve scarce raw materials.
- Therefore, this work observed previous research that investigated the link between the various process printing factors, composite materials, mechanical behavior, tribo-testers, and tribological properties of polymer test specimens.
- Even though the bulk of the studies that were reviewed relied on scientific results, there were also several that compared the frictional characteristics of a number of different biopolymers.
- Polymers have great strength but poor adherence, plastic deformation, bending, and interlayering, all of which contribute to heat bonding within the material. This is because polymers have weaker interlayering.

- Tribological properties of 3D-printed polymers and composites were investigated, as were 3D-printing procedures, and different 3D-printed materials were compared. In addition to 3D printing polymers, researchers have also generated 3D printing composites; the properties of these composites are currently being examined for prospective applications in particular industries, such as the automobile and aerospace industries.
  - ➢ 3D printing can help achieve these aims because it improves geometrical freedom and building safety.
  - ➢ The gap between idea and finished product is narrowing because of the rise in popularity of 3D printing.
  - ➢ There are still issues to be worked out before this 3D printing technique can be used in buildings, including inconsistencies in rheological requirements, insufficient interlayer bonding, poor structural performance, and poor material qualities.
  - ➢ 3D-printed polymers exhibit anisotropic porosity but stronger elasticity than absorption. The expanded interfacial bonding, plastic deformation, and asperity fractures that occur when 3D-printed polymers are subjected to sliding environments contribute to a reduction in friction and wear.
- These results are equivalent to those that were obtained via the mechanisms of wear and degradation, which consisted of abrasion, asperity break, ploughing, delamination, and cracking, respectively. Wear, friction, temperature, surface roughness, contact conditions, and test conditions are some of the factors that are taken into consideration, as well as the adhesive wear mechanisms that each type of composite material experiences.
- Thus, the stated work's gaps and restrictions have been investigated and fixed. Though extensive, the study done falls short of providing a complete picture. In order to assess whether or not 3D-printed 3D polymer specimens and components can achieve their intended mechanical criteria, it is essential to collect and have an understanding of the facts pertaining to the association between process variables and structural components and their mechanical and tribological performance.
- The tribological and mechanical characteristics of 3D-printed materials may now match the highly expected expectations of both end-users and manufacturers. This has been a long-awaited development in the field. Despite the extensive study of 3D printed polymers, comparatively little has been done to characterize their tribological properties.

References

1 Hofmann, M. (2014). 3D printing gets a boost and opportunities with polymer materials. *ACS Macro Lett.* 3 (4): 382–386. https://doi.org/10.1021/mz4006556.

2 Ramesh, M., Maniraj, J., and Ramesh, S. Thermal properties of the grass/cane fiber-based hybrid composites. In: *Natural Fiber-Reinforced Composites: Thermal Properties and Applications*. Germany: Wiley https://doi.org/10.1002/9783527831562.ch8.

3 Hanon, M.M., Kovács, M., and Zsidai, L. (2019). Tribology behaviour investigation of 3D printed polymers. *Int. Rev. Appl. Sci. Eng.* 10 (2): 173–181. https://doi.org/10.1556/1848.2019.0021.

4 Kumar, A., Choudhary, A., Tiwari, A. et al. (2021). An investigation on wear characteristics of additive manufacturing materials. *Mater. Today Proc.* 47: 3654–3660. https://doi.org/10.1016/j.matpr.2021.01.263.

5 Ramesh, M. and Niranjana, K. (2021). Effect of process parameters on fused filament fabrication printed composite materials. In: *High-Performance Composite Structures*, 155–178. Singapore: Springer.

6 Mangal, U., Min, Y.J., Seo, J.-Y. et al. (2020). Changes in tribological and antibacterial properties of poly(methyl methacrylate)-based 3D-printed intra-oral appliances by incorporating nanodiamonds. *J. Mech. Behav. Biomed. Mater.* 110: 103992. https://doi.org/10.1016/j.jmbbm.2020.103992.

7 Ramesh, M., Ramnath, R.A., and Deepa, C. (2020). Friction and wear properties of carbon nanotube-reinforced polymer composites. In: *Tribology of Polymer Composites*. Netherlands: Elsevier https://doi.org/10.1016/B978-0-12-819767.

8 Gharde, S., Surendren, A., Korde, J.M. et al. Recent advances in additive manufacturing of bio-inspired materials. In: *Biomanufacturing*. Cham: Springer https://doi.org/10.1007/978-3-030-13951-3_2.

9 Zhiani Hervan, S., Altınkaynak, A., and Parlar, Z. (2021). Hardness, friction and wear characteristics of 3D-printed PLA polymer. *Proc. Inst. Mech. Eng. Part J J. Eng. Tribol.* 235 (8): 1590–1598. https://doi.org/10.1177/1350650120966407.

10 ISO/ASTM (2013). Additive manufacturing—general principles terminology (ASTM52900). *Rapid Manuf. Assoc.* 10–12. https://doi.org/10.1520/F2792-12A.2.

11 Guo, N. and Leu, M.C. (2013). Additive manufacturing: technology, applications and research needs. *Front. Mech. Eng.* 8 (3): 215–243. https://doi.org/10.1007/s11465-013-0248-8.

12 Mohamed, O.A., Masood, S.H., Bhowmik, J.L., and Somers, A.E. (2017). Investigation on the tribological behavior and wear mechanism of parts processed by fused deposition additive manufacturing process. *J. Manuf. Processes* 29: 149–159. https://doi.org/10.1016/j.jmapro.2017.07.019.

13 Quan, Z., Wu, A., Keefe, M. et al. (2015). Additive manufacturing of multi-directional preforms for composites: opportunities and challenges. *Mater. Today* 18 (9): 503–512. https://doi.org/10.1016/j.mattod.2015.05.001.

14 Faludi, J., Bayley, C., Bhogal, S., and Iribarne, M. (2015). Comparing environmental impacts of additive manufacturing vs traditional machining via life-cycle assessment. *Rapid Prototyp. J.* 21 (1): 14–33. https://doi.org/10.1108/RPJ-07-2013-0067.

15 Ramesh, M., Gopinath, A., and Deepa, C. (2016). Machining characteristics of fiber reinforced polymer composites: a review. *Indian J. Sci. Technol.* 9 (42): 1–7.

16 Hanon, M.M. and Zsidai, L. (2021). Comprehending the role of process parameters and filament color on the structure and tribological performance of 3D printed PLA. *J. Mater. Res. Technol.* 15: 647–660. https://doi.org/10.1016/j.jmrt.2021.08.061.

17 Alghamdi, S.S., John, S., Choudhury, N.R., and Dutta, N.K. (2021). Additive manufacturing of polymer materials: progress, promise and challenges. *Polymers (Basel).* 13 (5): 1–39. https://doi.org/10.3390/polym13050753.

18 Ngo, T.D., Kashani, A., Imbalzano, G. et al. (2018). Additive manufacturing (3D printing): a review of materials, methods, applications and challenges. *Compos. Part B Eng.* 143: 172–196. https://doi.org/10.1016/j.compositesb.2018.02.012.

19 Tümer, E.H. and Erbil, H.Y. (2021). Extrusion-based 3D printing applications of PLA composites: a review. *Coatings* 11 (4): 1–42. https://doi.org/10.3390/coatings11040390.

20 Ertane, E.G., Dorner-Reisel, A., Baran, O. et al. (2018). Processing and wear behaviour of 3D printed pla reinforced with biogenic carbon. *Adv. Tribol.* 2018: https://doi.org/10.1155/2018/1763182.

21 Hanon, M.M., Alshammas, Y., and Zsidai, L. (2020). Effect of print orientation and bronze existence on tribological and mechanical properties of 3D-printed bronze/PLA composite. *Int. J. Adv. Manuf. Technol.* 108 (1–2): 553–570. https://doi.org/10.1007/s00170-020-05391-x.

22 Krajangsawasdi, N., Blok, L.G., Hamerton, I. et al. (2021). Fused deposition modelling of fiber reinforced polymer composites: a parametric review. *J. Compos. Sci.* 5 (1): https://doi.org/10.3390/jcs5010029.

23 Teraiya, S., Vyavahare, S., and Kumar, S. Experimental investigation on influence of process parameters on mechanical properties of PETG parts made by fused deposition modelling. In: *Advances in Manufacturing Processes. Lecture Notes in Mechanical Engineering* (ed. H.K. Dave and D. Nedelcu). Singapore: Springer https://doi.org/10.1007/978-981-15-9117-4_21.

24 Subramaniam, S.R., Samykano, M., Selvamani, S.K. et al. (2019). 3D printing: overview of PLA progress. *AIP Conf. Proc.* 2059: https://doi.org/10.1063/1.5085958.

25 Roy, R. and Mukhopadhyay, A. (2020). Tribological studies of 3D printed ABS and PLA plastic parts. *Mater. Today Proc.* 41: 856–862. https://doi.org/10.1016/j.matpr.2020.09.235.

26 Conte, M. and Igartua, A. (2012). Study of PTFE composites tribological behavior. *Wear* 296 (1–2): 568–574. https://doi.org/10.1016/j.wear.2012.08.015.

27 Parandoush, P. and Lin, D. (2017). A review on additive manufacturing of polymer-fiber composites. *Compos. Struct.* 182: 36–53. https://doi.org/10.1016/j.compstruct.2017.08.088.

28 Ramesh, M., Rajeshkumar, L., and Balaji, D. (2021). Influence of process parameters on the properties of additively manufactured fiber-reinforced polymer composite materials: a review. *J. Mater. Eng. Perform.* 30: 4792–4807. https://doi.org/10.1007/s11665-021-05832-y.

29 Ilyas, R.A., Sapuan, S.M., Harussani, M.M. et al. (2021). Polylactic acid (PLA) biocomposite: processing, additive manufacturing and advanced applications. *Polymers (Basel)* 13 (8): https://doi.org/10.3390/polym13081326.

30 Bankupalli, N., Srinivasa Rao, D., and Vamsi Krishna, T.S. (2020). Role of butadiene content on tribological properties of polymeric components fabricated by FDM. *Mater. Today Proc.* 38: 3019–3026. https://doi.org/10.1016/j.matpr.2020.09.325.

31 Norani, M.N.M., Abdullah, M.I.H.C., Bin Abdollah, M.F. et al. (2021). Mechanical and tribological properties of FFF 3D-printed polymers: a brief review. *J. Tribol.* 29: 11–30.

32 Singh, R., Singh, N., Amendola, A., and Fraternali, F. (2017). On the wear properties of Nylon6-SiC-Al_2O_3 based fused deposition modelling feed stock filament. *Compos. Part B Eng.* 119: 125–131. https://doi.org/10.1016/j.compositesb.2017.03.042.

33 Mital, G., Gajdoš, I., Spišák, E. et al. (2022). An analysis of selected technological parameters' influences on the tribological properties of products manufactured using the FFF technique. *Appl. Sci.* 12 (8): https://doi.org/10.3390/app12083853.

34 Nirmal, U., Hashim, J., and Megat Ahmad, M.M.H. (2015). A review on tribological performance of natural fiber polymeric composites. *Tribol. Int.* 83: 77–104. https://doi.org/10.1016/j.triboint.2014.11.003.

35 Lin, L., Ecke, N., Huang, M. et al. (2019). Impact of nanosilica on the friction and wear of a PEEK/CF composite coating manufactured by fused deposition modeling (FDM). *Compos. Part B Eng.* 177: 107428. https://doi.org/10.1016/j.compositesb.2019.107428.

36 Jagannatha, T.D. and Harish, G. (2015). Mechanical properties of carbon/glass fiber reinforced epoxy hybrid polymer composites. *Int. J. Mech. Eng. & Rob. Res.* 4 (2).

37 Suresh, A., Jayakumar, L., and Devaraju, A. (2020). Investigation of mechanical and wear characteristic of banana/jute fiber composite. *Mater. Today Proc.* https://doi.org/10.1016/j.matpr.2020.07.426.

38 Sınmazçelik, T., Fidan, S., and Ürgün, S. (2020). Effects of 3D printed surface texture on erosive wear. *Tribol. Int.* 144: 106110. https://doi.org/10.1016/j.triboint.2019.106110.

39 Kennedy, D.M. and Hashmi, M.S.J. (1998). Methods of wear testing for advanced surface coatings and bulk materials. *J. Mater. Process. Technol.* 77: 246–253.

40 Shahar, F.S., Sultan, M.T.H., Safri, S.N.A. et al. (2022). Fatigue and impact properties of 3D printed PLA reinforced with kenaf particles. *J. Mater. Res. Technol.* 16: 461–470. https://doi.org/10.1016/j.jmrt.2021.12.023.

41 Berfenstam, R. (1957). A critical review on different types of accident morbidity studies. *Courrier* 7 (7): 349–361.

42 Conference, I. and Pawlak, W. (2018). Wear and coefficient of friction of PLA-graphite composite in 3D printing technology. *Eng. Mech.* 2018: 649–652. https://doi.org/10.21495/91-8-649.

43 Susanti, D.Y., Sediawan, W.B., Fahrurrozi, M., and Hidayat, M. (2020). Studies on kinetics and optimum agitation of phenolic compound extraction from intact red Sorghum. In: *Proceedings of the 6th International Conference and Exhibition on Sustainable Energy and Advanced Materials* Surakarta, Indonesia (16–17 October 2019). Germany: Springer Nature Singapore.

44 Liu, W., Huang, H., Zhu, L., and Liu, Z. (2021). Integrating carbon fiber reclamation and additive manufacturing for recycling CFRP waste. *Compos. Part B Eng.* 215: 108808. https://doi.org/10.1016/j.compositesb.2021.108808.

45 Yang, C., Jiang, P., Qin, H. et al. (2021). 3D printing of porous polyimide for high-performance oil impregnated self-lubricating. *Tribol. Int.* 160: 107009. https://doi.org/10.1016/j.triboint.2021.107009.

46 Dangnan, F., Espejo, C., Liskiewicz, T. et al. (2020). Friction and wear of additive manufactured polymers in dry contact. *J. Manuf. Processes* 59: 238–247. https://doi.org/10.1016/j.jmapro.2020.09.051.

47 Soundararajan, R., Jayasuriya, N., Vishnu, R.G.G. et al. (2019). Appraisal of mechanical and tribological properties on PA6-TiO_2 composites through fused deposition modelling. *Mater. Today Proc.* 18: 2394–2402. https://doi.org/10.1016/j.matpr.2019.07.084.

48 Ramadan, M.A. and Reda, R. (2017). CNTs, Al_2O_3 and SiO_2 reinforced epoxy: tribological properties of polymer nanocomposites. *Tribol. Ind.* 39 (3): 357–363. https://doi.org/10.24874/ti.2017.39.03.11.

49 Eng, R. and Mater, S. (2020). Investigation on the manufacturing variants influential on the strength of 3D printed products. *Res. Eng. Struct. Mater.* 6 (4): 293–313.

14

Influence of Fiber Treatment on the Wear Properties of Biocomposites

Anthony Chidi Ezika[1], Emmanuel Rotimi Sadiku[1], Raphael Stone Odera[2], Uzoma Ebenezer Enwerem[2], Victor Ugochukwu Okpechi[2], Martin Emeka Ibenta[2], and Shadrack Chukwuebuka Ugwu[2]

[1] *Tshwane University of Technology, Institute of Nanoengineering Research (INER) & Department of Chemical, Metallurgical and Material Engineering, Faculty of Engineering, and the Built Environment, Staatsartillerie Rd, 0183, Pretoria West Campus, Pretoria, South Africa*
[2] *Nnamdi Azikiwe University, Department of Polymer Engineering, Faculty of Engineering, Along Enugu-Onitsha Expressway, Ifite Rd. PMB 5025, Awka, Nigeria*

14.1 Introduction

The need for improved mechanical properties of engineering materials, as well as commodity goods, has initiated the inclusion of fibers for reinforcement during the fabrication of materials [1]. Natural fibers (NFs) which are low-cost are often treated before use. NF treatment may be physical, chemical, thermal, and sometimes biological. The type of treatment given to fiber is dependent on the intended properties, nature of the fiber, and the intended application of the treated fiber [2]. Chemical treatment has been employed to minimize toxic elements on NFs, strengthen the fibers, and more especially create bonding sites on the fiber surface by introducing several functional groups [3, 4]. The bonding sites result in better interfacial bonding when a material is filled with chemically treated NF to form a composite [4]. Hence, the resulted composite will possess a strong adhesion force between the contact surfaces of the constituent materials; thereby resulting to an improved composite with good mechanical features. Chemical and physical treatments of NFs have been reported to improve the mechanical properties of the resulted composites [5–9].

Moreover, the inclusion of treated NF into materials not only changes the mechanical behaviors of the composites but also induces biodegradability of the resulted composites [10–15], hence they are regarded as biocomposites. Environmental concern and cost of synthetic fibers have resulted in the idea of utilizing NFs in the reinforcement of polymer materials [16–25]. Thus, polymer matrices filled with NFs are widely attractive due to their high tensile strength [26], biodegradability [19], low friction coefficient [27], light weight, and many other applications that require mechanical and tribological properties, from automotive to household applications.

Tribological Properties, Performance, and Applications of Biocomposites, First Edition.
Edited by Chandrasekar Muthukumar, Senthilkumar Krishnasamy,
Senthil Muthu Kumar Thiagamani, and Ganesan Chinnachamy.

NF treatment increases the fiber-polymer adhesion of the biocomposites, this influences the wear properties of the biocomposites [20].

Herein, the effect of the NF treatments on the wear behaviors of the biocomposites is discussed. The discussion is centered on the polymeric-based biocomposites, ranging from thermoplastic to thermosetting polymers while elucidating the influence of NF treatments on the wear properties of their composites. In addition, types of fiber treatment and classification of biocomposites are brought to light.

14.2 Fibers

Man-made and NFs are the two main types of textile fibers. Man-made fibers are not found in nature, even though they may be like NFs, which are made up of naturally occurring elements.

NF is any hair-like raw material derived primarily from animals, vegetables, or mineral sources that can be spun into nonwoven fabrics such as papers, diapers, felts, or woven fabrics upon spinning into yarns [28]. NF can also be considered as a mass or pile of cells having a small diameter in contrast to their length. In a nutshell, NFs are fibers that are man-made. They are obtained from either animals or plants [29, 30]. Natural fibers from plant or vegetable sources are more properly referred to as cellulose based and can be classified by plant source, while NFs from animal sources are more popularly known as protein-based fibers. Mineral fibers are those that are mined from the earth. Except for silk, all-natural cellulose and protein-based fibers are obtained in short lengths and are called staple fibers. Silk is a continuous filament fiber. Some fibrous plants can grow in a variety of climates, from the far north to southern Arctic Circle, thus making the recycling of carbon dioxide (CO_2) possible. It is worth noting that lignocellulose, natural fibrous resources, and protein-based fibers are perfectly viable, inexhaustible, and decomposable, and their production poses no harm to the ecosystem. NFs are distinguished by their most important properties, such as air porosity, moisture absorbency, and the ability to give off moisture easily. They equally do not release toxic substances that are harmful to human health nor cause allergic reactions makes them very suitable for use [31]. NFs contain a lot of hemicelluloses, lignin, cellulose, and pectins, and they tend to be active polar hydrophilic materials [18, 32, 33].

NFs have several favorable properties over synthetic fibers, including their renewability, availability and relatively inexpensive, minimal specific density and low mass, high-specific strength, recyclability, and degradability [34]. On the other hand, synthetic fibers are man-made fibers that are not derived from natural sources. Synthetic fibers are mostly made from petroleum products. NFs offer inferior qualities to synthetic fibers. These are primarily made from a variety of compounds, each with its own set of properties. Synthetic fibers are made from chemical products such as acrylics, nylon, polyurethanes, and polyesters. These fibers have excellent mechanical properties; they are tenacious and possess high stability, as well as a lengthy life span [35, 36].

Synthetic fibers are categorized into distinctive groups: regenerated fibers (these are fibers that were developed by transforming natural polymers), those made from nonbiodegradable materials, and those manufactured from synthetic polymers [37].

Polyester, polyamide, acrylic, and polyolefin are the most common synthetic fiber, accounting for 60% of global fiber usage. These polymers made up of free macromolecules are driven through small orifices called spinnerets and subsequently solidified in the production process [38]. After fiber creation, which can take the shape of a continuous filament or a discontinuous staple fiber, the material can be processed into a variety of products, including yarns, braiding, knitting, woven fabrics, as well as nonwovens.

14.2.1 NF Reinforcement

The diameter of NFs is frequently expressed in micrometer units. It reflects the fiber's average width over its whole length. Except for silk, all-natural fabrics have a short length. Staple or continuous filament fibers can be found in silk and synthetic textiles. Fiber length has a significant effect on the processing of fibers and yarns, as well as the conversion of fiber strength to yarn strength. A greater fiber length is desired in general. Textile fibers come in two lengths: staple and filament. Staple fibers range in length from 2 to 46 cm, while filament fibers can be any length.

The mechanical properties of composites are influenced by physical parameters such as NF length and diameter [39, 40]. Smaller-diameter NFs have a greater favorable impact on mechanical characteristics than large-diameter fibers [41]. A reduction in the stress exchange capability between the matrix and fiber is imminent if the length of the fiber is less than the critical length. This results in poor mechanical properties [42]. The length of the fibers controls impact strength and energy dissipation across the length of the composite, which has been shown to be more successful with large fibers [43, 44]. The wear extent reduces because of increase in fiber length and weight percent loading, resulting in improved bonding between fiber and matrix with increasing fiber size in a sand abrasion wear test [43, 45]. The wear rate of sour-weed fiber-based composites diminishes as the fiber length increase up to 15 mm. It is possible that when fiber size gets bigger, the adhesion bond strength between fiber–matrix interfaces gets stronger, providing higher resistance to slide wear [43]. The influence of fiber length on the mechanical strength of short roselle and sisal fiber in advanced polyester composites was investigated, and their findings show that a fiber weight content of 45 wt% and a fiber length of 5 mm are excellent for improving the mechanical properties of the composite. The mechanical behavior of banana fiber-reinforced nanocomposites was investigated for different lengths of banana fiber and differing weight percentages of banana fiber content. The nanocomposites had improved mechanical behavior when the banana fiber length was 30–40 mm and the banana fiber content was 40 wt% [46]. Another investigation of the tensile characteristics of palmyra/hemp hybrid polyester composites for various fiber lengths and fiber weight ratios found that the best fiber length and weight ratio were 30 mm and roughly 55%, respectively [46]. Several studies carried out by researchers have shown that fiber length has a significant role in making sure that the interfacial bonding between the matrix and fiber is achieved. Theoretically, it was proven that short lengths reduce the load- bearing agility of fibers, which accounts for increased wear rate. Excessive short lengths result in fibers pulling out, and this implies higher wear rates. A study on sugar cane fiber with varying lengths proved that fibers with medium lengths provide good wear and frictional properties when compared to fibers that are either too short or too long [47].

According to various studies, the fiber length and fiber content are the most important elements that determine the physical and mechanical properties of NF-reinforced composites [48].

14.2.2 Treatments of NFs

Jute, isora, ferula, sisal, flax, okra, hemp, and coir are a few of the NFs that are employed as fillers in the polymer matrix. Many automobile sectors are employing cellulose fibers to replace traditional fibers such as carbon and glass in polymer matrix reinforcement. NFs are advantageous over man-made fibers as reinforcement, but they also have certain disadvantages. At high temperatures, NFs are less stable. These have a moisture absorption capability that limits their use in damp environments [49, 50]. It also causes dimensional instability in fiber composites [51]. The reason for this is concealed in the NF's structure. The primary components of NF are hemicellulose, cellulose, pectin, and lignin. They have a significant attraction for moisture due to the abundance of hydroxy-1 groups, making them hydrophilic in nature [52]. The modification of NF through specific treatments is unquestionably required. This is typically achieved by using a reagent of desired functional group to affect the fiber structures, which results in a change in their composition. Modification of fiber can reduce the moisture absorption capacity of NFs, resulting in an ambient improvement in incompatibility at the interface of polymer matrix and fiber [53].

14.2.2.1 Types of NF Treatment

Purification and surface modification of NF are important steps toward improving fiber–matrix bonding because it modifies the fiber surface and creates an enhanced structure modification through surface dewaxing and impurity removal. Various methods and chemicals have been reported in the literature as having been applied to date palm fibers (DPF) to assess their impact on biocomposite properties. A study showed that treated fiber-filled matrices have better mechanical and morphological properties than untreated fiber-filled matrices [54]. It was determined that treating palm fibers with various percentages of sodium carbonate results in a substantial improvement of mechanical properties [30, 55]. DPF treatment with a combination of alkaline and silane improved DPF-polyester bonding compared to untreated fibers or fibers treated individually with alkaline or silane. The findings revealed that treatment conditions and parameters have a significant influence on the thermal, mechanical, and physical characteristics of DPF [30]. Some other authors proved the advantageous effect of chemical [56, 57] and physical methods [58–60] on the mechanical performances of NF-filled nanocomposite and their sustainability under wet environments [61–65].

Chemical Treatment Alkaline treatment: Treating a NF with reagents of OH functional group optimizes the interaction between the polymer matrix and NFs. This treatment minimizes the OH functional groups on the surface of the fibers while also increasing surface roughness [57, 66–69]. Alkaline treatment is one of the significant processes used to reduce the hydrophilic properties of the fibers and improve their compatibility with polymers. NFs were subjected to a variety of chemical treatments, including alkaline, silane treatment, acetic acid treatment, benzoyl peroxide treatment, potassium permanganate treatment, stearic acid treatment, polymer coating (PT), graft copolymerization, and isocyanate treatment [70]. When compared to untreated composites (UTCs), chemically treated NF composites have higher impact and tensile strength, bending, interlaminar shear strength, and hardness values [66]. Borassus fine fibers were alkali treated with 5% NaOH for 1, 4, 8, and 12 hours owing to a large amount of hemicellulose dissolution, the fibers' tensile strength, tensile modulus, and percent elongation improved by 41%, 69%, and 40%, respectively, after 8 hours of treatment [71]. Tensile behaviors of Borassus fruit fibers were improved by a 5% NaOH treatment due to increased surface area and improved interfacial properties of the fibers. Cane fibers were treated with alkali for 24 hours with varying concentrations of 2%, 4%, 6%, and 8%, and the highest tensile strength was found to increase by 6% [72]. The impact of alkaline treatment on the flexural modulus and flexural strength of jute fiber-filled vinyl ester composites was studied, and it was discovered that the flexural modulus and flexural strength increased by 403% and 98%, respectively, when compared to UTCs [73]. The influence of alkaline treatment on the flexural and impact strengths of coir fiber-filled polyester nanocomposite with 5% weight NaOH solution for 72 hours revealed that treated fiber nanocomposites outperformed untreated fiber nanocomposites due to improved bonding between matrix and fiber. Alkaline treatment was a better treatment compared to other treatments [66].

Silane Treatment: The interfacial bonding between polymer matrix and glass fibers can be improved using organosilanes as coupling agents [74]. To qualify as an efficient composite coupling agent, an organosilane must contain functional groups capable of forming covalent bonds with phases, namely the reinforcing fibers and polymer matrix [75]. With the right functional group design, polymer matrix and cellulose can be covalently bonded by the silane treatment. This can significantly increase the fiber-matrix adhesion. Again, treatment with alkali prior to silane treatment can enhance the effect of silane treatment because an alkaline treatment activates the reactive sites and clears the fiber surface. In the fabrication of biocomposites, silane treatment of NF has proven to be better than alkaline treatment, with matrix-fiber bonding shown to be stronger than when treated with alkaline. This results in a biocomposite with a stronger tensile strength [76–79].

Esterification: This method of fiber surface treatment is necessary because of abundance of hydroxyl ions present on the surface of NFs. Unlike silane treatments, which aim to form weak covalent interactions at the interface of the matrix and fiber, functionalizing the NF surface is the major impact of esterification treatment. Esterification process tends to convert the hydrophilic hydroxyls to hydrophobic ester bonds, increasing the fibers' hydrophobicity and, thus, their compatibility with the matrix. Acetylation is the most commonly used esterification method and was first used on a commercial scale. Acetylation can be accomplished using a variety of compounds, which include acetic acid, acetic anhydride, and acid chloride [76, 80].

Acetylation treatment: Acetylation treatment involves the use of acid catalyst to graft acetyl groups onto the cellulose network of NFs [81]. This method involves the immersion of fibers in acetic acid, followed by treatment with acetic anhydride for some hours in the presence of heat to catalyze the reaction [18]. In this technique, the reaction occurs among hydroxyls and NF anhydride/carboxylic groups. Acetyl is employed to remove the cuticle and wax on the fiber surface, resulting in improved fiber structure. The treatment of hemp and flax fibers with acetylation lowers wettability, removes noncrystalline parts, produces a smoother fiber surface, and improves stress cushioning. Acetylation altered the bulk and surface properties of flax fibers [81].

Benzoylation treatment: This involves the use of benzoyl chloride to improve the fiber's thermal stability and interfacial adhesion by lowering its hydrophilic nature. First, lignin, waxes, and covering oil are extracted in a pretreatment step, followed by wetting the fiber surface with OH groups. Finally, the fibers are soaked with benzoyl chloride, which replaces the OH groups on the fiber surface with benzoyl groups, by adding the OH groups to the cellulose backbone [81].

Peroxide treatment: Here, peroxide serves as a grafting agent. It induces grafting of the polymer onto the fiber surface and, as such, reacts with the OH groups of both the fiber and matrix. Thus, the fiber adheres to the matrix at the interface. The process also results in a reduction in the wettability of the fiber and tends to increase the thermal stability of the fiber [81].

Maleated coupling agents: Maleated coupling agents interact adequately with the functional groups presence at the surface of the matrix and fibers to establish a carbon–carbon link with the polymer chains. The reaction between the OH groups and the maleic anhydride at the amorphous region of the fiber cellulose structure coats the fiber surfaces with long-chain polymers. The covalent bonding between the maleic anhydride and OH groups excludes the hydroxyl group from the fiber cell; this reduces the hydrophilic propensity of the cellulose structure, leading to significant interlocking between the matrix and the fiber and the formation of a network interface [81].

Treatment with sodium chlorite: The process involves the bleaching of fibers with sodium chlorite in acidic medium. This method increases the hydrophobicity of the fiber, dries off the moisture content of the fiber, and enhances its flexiblity [81].

Physical Treatment Physical treatment of NFs improves the coupling between the matrix and the fiber by improving the interfacial structure without altering the chemical characteristics of the fibers [52]. Investigations have shown the impact of physical treatments on NFs. Plasma, heat treatment, corona, fiber beating, and ultraviolet (UV) are physical treatment methods. These techniques have effects only on the surface of the fibers, and as such, they can only alter the geometry of the surface. When a cellulose fiber is treated with corona, the surface energy is altered to the coupling of the matrix with the hydrophilic fiber [81, 82]. On the other hand, the plasma treatment is achieved through the generation of atmospheric pressure by using a high voltage at a low temperature, to alter the surface geometry and energy of cellulose fibers, as seen in corona treatment [81]. According to reports, the corona treatment stimulates natural surface sites all along the polymeric chains, which may react with oxygen to produce an etching impact [82].

Plasma treatment is a physical procedure that has been used effectively to change the surface of a variety of NFs. The mechanical properties of NFs were proven to greatly improve after plasma treatment [83]. Furthermore, plasma treatment can introduce a specific reactive site on the NF surface, which can constitute strong covalent bonds with the matrix, resulting in a strong matrix/fiber interface. Furthermore, surface etching caused by plasma treatment may enhance surface roughness, resulting in an improved interface with the matrices via coupling [84]. Still on the physical treatment, the UV treatment method increases the polarity difference between the fiber surface and the matrix, resulting in improved fiber moisture absorbing capacity and the tenacity of the composite [85]. In the fiber-beating technique, fiber defibrillation, an increase in surface area and mechanical interlocking can result in a 10% increase in NF strength. During the treatment, the fiber degradation temperature should be considered while heating the fiber. This condition has influence on the fibers' chemical, mechanical, and physical properties, such as their degree of polymerization, cellulose crystallinity, water content, strength, and chemistry [85].

Biological Treatment Biological treatments employ naturally occurring microorganisms, such as bacteria and fungi. These treatments take place in aqueous environments and are relatively inexpensive to undertake, but they are time-consuming and contaminate the environment. Retting is a common biological fiber treatment. The controlled degradation of plant stems to free the bast fibers from their fiber bundles and separate them from the woody core and epidermis is regarded as retting. During the retting process, bacteria (mostly Clostridia species) and fungi release enzymes that degrade pectic and hemicellulosic compounds in the middle lamella between individual fiber cells [86].

Fungal treatment is a biological modification that uses enzymes to remove noncellulosic parts from the fiber surface. White-rot fungi produces extracellular oxidase enzymes, which removes lignin from the fibers when reacted with lignin peroxidase [81]. The use of enzymes to treat NFs was put in place due to the ease of the method. Enzymes are biological catalysts with excellent diversity, structure, and reaction that speed up chemical reactions while undergoing no permanent chemical change. The enzymes' primary function in the treatment process is to degum NFs while improving cleanliness, uniformity, fineness, and softness [87]. The enzyme treatment is faster but more sensitive than the chemical treatment because it must be performed at atmospheric pressure of less than 100 °C and a pH of 4–8, whereas the addition of chelators such as ethylenediaminetetraacetic acid to enzyme mixtures can enhance the efficiency of this treatment type. Nevertheless, the high price of enzymes and the proper selection of enzymatic agents limit the use of this method [82, 87].

14.3 Biocomposites

Biocomposites (NF-filled composites) have proven to be sustainable and eco-efficient, which accounts for their advancement for future products, materials, and applications. Tremendous advancements in the utilization of biocomposites can be found in the domestic sector, automotive industries, building materials, circuit boards, and aerospace industry. Different classes of NFs have been studied in conjunction with organic matrices to create composite materials that are competitive with synthetic fiber composites that require considerable attention. Besides that, biocomposites provide opportunities for environmental benefits, lower energy consumption, insulation, and sound absorption [88, 89].

Biocomposites are made up of one or more biologically derived phases, which can be reinforcements (jute, hemp, flax, paper, cotton, kenaf, or recycled wood and sisal) or organic matrices [90]. Biocomposites are created by simply combining fibers with a polymeric matrix [91]. Particle reinforced composites, structural composites, and fiber-reinforced composites are the three types of composites. Biocomposites are composite materials made up of one or more phase materials derived from nature [92–94].

14.3.1 Classification of Biocomposites

Based on the matrix and the fiber, biocomposite systems can be classified into partial biodegradable and complete biodegradable biocomposites [95]. Bio-fibers are used in complete biodegradable biocomposites with matrices made of biodegradable polymers such as renewable biopolymer matrices (e.g. soy plastic, starch plastic, cellulose plastic) or petro-based biodegradable polymer matrices (e.g. aliphatic co-polyester, polyesteramides). Bio-based matrices are reinforced with synthetic fibers in partial biodegradable biocomposites, whereas nonbiodegradable polymer matrices such as traditional thermoplastic polymers (e.g. polypropylene, polyethylene) and thermoset polymers (e.g. epoxy, polyester) are reinforced with fibers [95–98] (Figure 14.1).

14.3.2 Natural Fiber-Polymer Composites

Natural fiber-polymer composites (NFPCs) are indeed composites with their mechanical efficiency determined by the fiber–matrix interface as well as the stress exchange function, this characterizes their tendency to transfer stress from the matrix to the fiber [53, 99, 100]. NFPCs are progressively employed in a wide range of applications due to their environmentally friendly and cost-effective alternative to conventional petroleum-derived materials. The organic reinforcement fiber and the polymer matrix are the two major components of NFPCs. NFs function as reinforcement in an NFPC, boosting the stiffness and strength of the

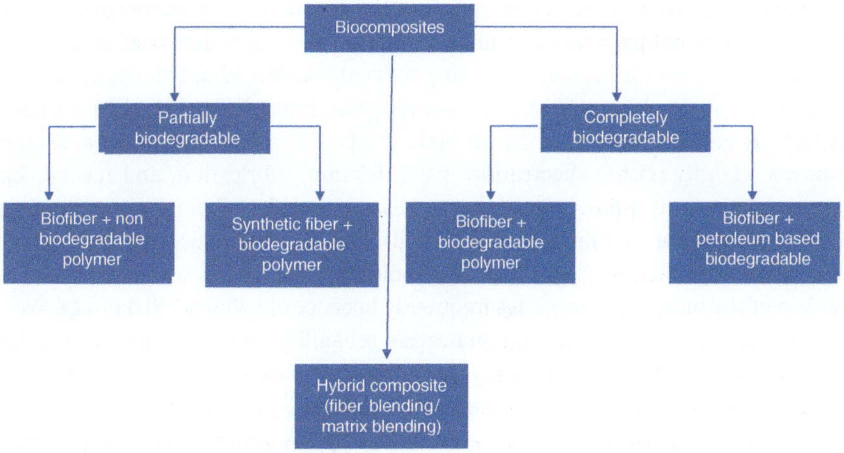

Figure 14.1 Classification of biocomposites.

end composite. Traditional reinforcement fibers, which include carbon, aramid, and glass, can be manufactured with specific performance ranges, whereas NF properties vary greatly depending on the fiber origin, the geometry of the fiber, and the preconditioning process [101]. NFPCs are sustainable and renewable materials made up of NFs contained in a polymer matrix that may also be biological in origin. NFPCs containing nonbiodegradable thermoplastics cannot be biodegraded, but they can be reclaimed easily when compared to thermoset composites. NFPCs, on the other hand, made of a biodegradable polymer matrix, can be broken down into natural degradation products after their desired use. NFPCs have an edge in terms of toxicity, effluent emission, energy consumption, and the enormous amount of disposal options [102]. One of the challenges of using NFs as composite reinforcement is their low thermal stability. The processing temperature of NFs is restricted to 200 °C, limiting the matrix material selection to plastics with low melting temperatures, such as polypropylene (PP), polyethylene (PE), polyvinyl chloride (PVC), and polystyrene (PS). This creates a new challenge because most of these plastics have lower mechanical performance when compared to engineering thermoplastics like polycarbonate, acrylonitrile butadiene styrene, and polyethylene terephthalate (PET) [102, 103]. According to previous research, the properties of NF composites differ due to different types of fibers, moisture conditions, and sources. Mechanical composition, microfibrillar angle [104], structure [26], defects [105], cell dimensions [100], physical properties [18], chemical properties [53], and the matrix–fiber interaction [106], all influence NFPC performance [53]. NF characteristics such as volume fraction, physical properties, impurities, moisture absorption, and orientation are examples of such characteristics that contribute to the mechanical properties of NFPC [53].

14.3.3 Tribological Properties of NF-Reinforced Composites

Fiber-reinforced polymer (FRP) composites play an important role in tribological materials with low noise and self-lubrication abilities. Through various material combinations, FRP provides more opportunities to enhance the tribological performance of frictional parts. As a result, FRP composites are widely used in a variety of applications, and most manufacturing parts are subjected to tribological testing based on their applications, such as clutch pads, brake linings, and brake pads, as well as gears, bearings, and prosthetic joints. Tribo materials have a distinct property identity with a spectrum of wear, friction, lubrication, and mechanical properties that are used to design materials for specific tribological applications [107]. The relative motion of materials with one another causes friction, which results in wear. The material suffers tribological damage because of this. The surface of the material depreciates frequently because of tribological damage. Wear in a material system reduces the material's reliability and durability, which can lead to failure and increased cost. Wear is hugely affected by material properties such as surface characteristics, toughness, hardness, and chemical composition [108]. Tribology performance can be enhanced by the selection of an appropriate reinforcement combination in the polymer matrix [109]. Better wear resistance is provided by friction composites with the optimal weight composition of fiber filler [110]. According to a recent study, increasing the stress reduces the wear rate of the sisal fiber-filled polypropylene nanocomposite [108]. The wear characteristics and friction of kenaf fiber polyurethane composites were investigated. Comparing the pure polyurethane nanocomposite, the fiber filler enhanced the specific wear rate of the composite by approximately 78%. Again, as revealed in a study, the composite of treated woven fiber filler performs well in terms of wear resistance [111].

14.4 Influence of Fiber Treatment on the Wear Properties of NF-Filled Polymer

NFs consist majorly of hemicellulose, cellulose, pectin, lignin, and waxy elements. These organic components are found in the cell wall of the NF with impart on the fiber properties [112–117]. NF in composites presents some challenges, such as high moisture intake, hydrophilic characteristics, and a high variation of mechanical properties [118]. The surface nature of NF is hydrophilic, which is incompatible with the hydrophobic polymer matrix and consequently affects the adhesion of the material and consequently the mechanical and tribological properties of the composite. To improve the interfacial adhesion, chemical modification of the NF is explored [119–121]. Among many fiber treatments, alkalization, acetylation, and salinization are the most common. The alkaline treatment removes impurities, fats, hemicellulose, and lignin, exposing the cellulose and consequently increasing roughness. Acetylation partially removes lignin and hemicellulose and exposes more active hydroxyl groups from cellulose to react with the matrix; this enhances the chemical bonding and salinization processes involving a coupling agent (silane) that improve matrix interfacial adhesion [122–125]. Surface treatment of NF improves the surface roughness and creates more bonding sites by effectively detaching the hemicellulose, impurities, and lignin without degrading the fiber strength [126].

14.4.1 Influence of Fiber Treatment on the Wear Properties of NF-Reinforced Epoxy Composites

The wear behavior of epoxy composite filled with treated NF, as discussed below, shows significant improvement in the composite structure due to an enhanced interfacial relationship between the matrix and fiber. Mohan, Kanny, and coworkers [127] investigated the wear properties and friction of nano-clay treated fiber (NC-BF) reinforced epoxy composite and UTC. Vaccine resin infusion processing technique was used to fabricate short (NC-BF) and long (UT-BF) reinforced composites with fiber concentrations within the range of 20–60 wt%. The results show that treated NC-BF-reinforced composites form transfer layer between wear surface and counterface, hence improving the wear property. Similarly, wear and mechanical behaviors of Kenaf–Aloe Vera–Jute fiber-reinforced epoxy composite were significantly improved upon chemical treatment using sodium hydroxide solution [128]. Different types of NFs (flax, jute, and hemp) were chemically modified and filled with epoxy resin through the hand-layup approach at different hybrid ratios. Experimental results of wear analysis confirmed that the treatment of the plant fibers improved the wear properties of the developed composites as there was enhanced bonding between the resin and fiber [129]. Chemically treated jute fiber-filled epoxy resin showed enhanced tribological behavior and mechanical properties [130]. The effect of surface medication on the mechanical and abrasive wear performances of the composite of coir fiber-filled epoxy resin showed that the bonding between the cellulosic fiber and the hydrophobic resin was improved due to surface modification. Hence improved tensile strength, enhanced wear resistance, greater porosity, and hardness [131]. The effects of silane treatment on the mechanical, tribological, morphological, and water absorption behavior of coir stack fiber and its composites were studied. The result shows improved properties and wears behavior due to better fiber–matrix interaction [132].

The effect of orientation of fiber on wear performance was studied on jute fiber-supported epoxy composite. Pretreatment of jute fiber was done using sodium hydroxide to eliminate impurities from the fiber surface. Results showed improved surface interaction between the fiber and the matrix hence improving the wear performance of the composite [133]. The abrasive wear behavior of chemically treated coir fiber-filled epoxy polymer composite was analyzed using two variables: normal load and sliding speed in a single pin-on-disc wear testing machine by Khan and coworkers [134]. Results from Table 14.1 and graphs below show wear behavior tends to decline by increasing the sliding speed in both pure epoxy and treated coir fiber-filled epoxy composite; more so the rate of abrasion of the composite with treated coir fiber is slightly less when compared to neat epoxy polymer composite (Figure 14.2).

The morphology of untreated/treated coir fiber-reinforced epoxy composite fired up at 1000° can be seen in Figure 14.3. The intercellular gaps shown in the UTC can be marked as cellular units, and the spherical voids in the UTC are extremely reduced in the chemically treated composite. Surface roughness could be attributed

Table 14.1 The values of wear rates for both samples the different sliding speed and load.

			Wear rate
Load in (N)	**Sliding speed in ($r\,m^{-1}$)**	**Wear rate pure epoxy**	**Treated coir filled epoxy**
10	100	0.463×10^{-3}	0.296×10^{-3}
15	200	0.114×10^{-2}	0.817×10^{-3}
20	300	0.147×10^{-2}	0.942×10^{-3}
25	400	0.176×10^{-2}	0.125×10^{-3}

Source: Khan et al. [134]/Open Science Publishers.

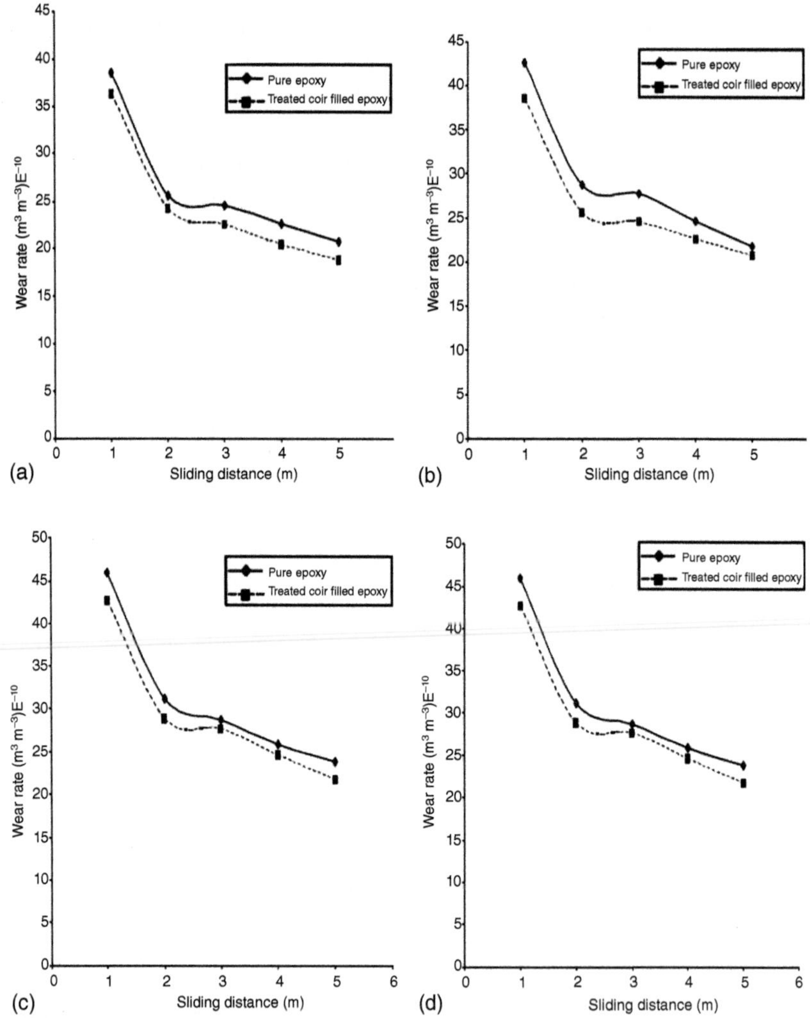

Figure 14.2 Graphical illustration of the effect of sliding distance over wear rate at (a) 100 rpm, (b) 200 rpm, (c) 300 rpm, and (d) 400 rpm. Source: Khan et al. [134]/Open Science Publishers.

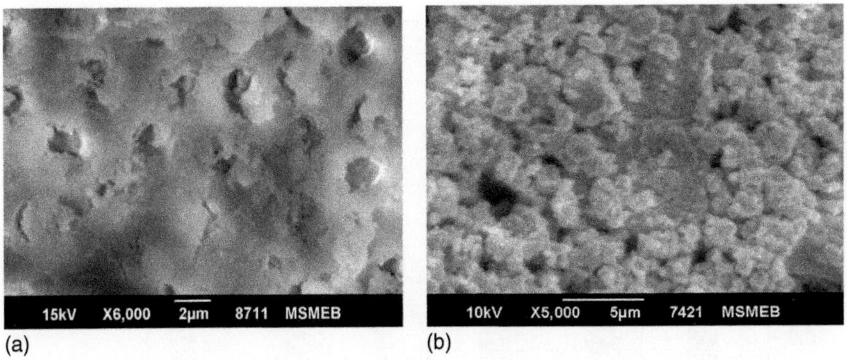

Figure 14.3 SEM analysis of (a) untreated coir fiber and (b) treated coir fiber reinforced epoxy composite fired at a temperature of 1000°. Source: Khan et al. [134]/Academia.

to the effective removal of impurities and the creation of bonding sites for the matrix [134].

The influence of loading red mud on the mechanical properties of bamboo fiber/epoxy composite and equated with the same set of glass/epoxy composite was examined by Biswas and Satapathy [135] treatment of bamboo fiber was done using NaOH to create suitable gaps in the fiber providing sites for bonding with the matrix, hence improving bonding. The results show the wear properties of bamboo fiber composite is superior to glass/epoxy composite. More so, Anu and coworkers [133] studied the influence of alkalization and filler content on the erosion wear and mechanical behavior of cement by-pass dust filled bamboo fiber composite using Taguchi orthogonal array was studied. Results identify significant control factors influencing the erosion wear and also outline significant interaction effects between composite components. Shalwan and Yousif [136] aimed at modifying the surface of DPF used sodium hydroxide before reinforcing epoxy composite wear behavior was significantly improved upon modification when compared to unmodified composite. Behera [137] investigated the influence of fiber surface modification on the structural, thermal, and tribological properties of sisal fibers and their epoxy composite was investigated. Surface treatment was done with alkali (NaOH), glutamic acid, and a combination of both alkaline and glutamic acid. The results from the research show that the wear resistance of chemically treated sisal fiber-reinforced nanocomposite is high when compared to untreated nanocomposite. Sliding wear resistance of jute fiber-reinforced epoxy composites with fiber loadings of changed weight percentages of 10, 20, 30, and 40 wt% and various chemical treatments (alkaline, benzoyl chloride treatment) from the experiment carried out with 30 wt% fiber loading of benzoyl chloride treatment composite showed low sliding wear behavior, followed by alkaline treatment composite and UTC. The combination of 30 wt% fiber loading displays minimum sliding wear resistance at a varied AL and SD, this happens due to the strong interfacial connection among the fiber and matrix material, which raises the wear resistance of the composite Swain and Biswas [138]. The wear behavior and mechanical properties of pulverized poultry eggshell fiber hybrid reinforced epoxy composite were analyzed by Oladele

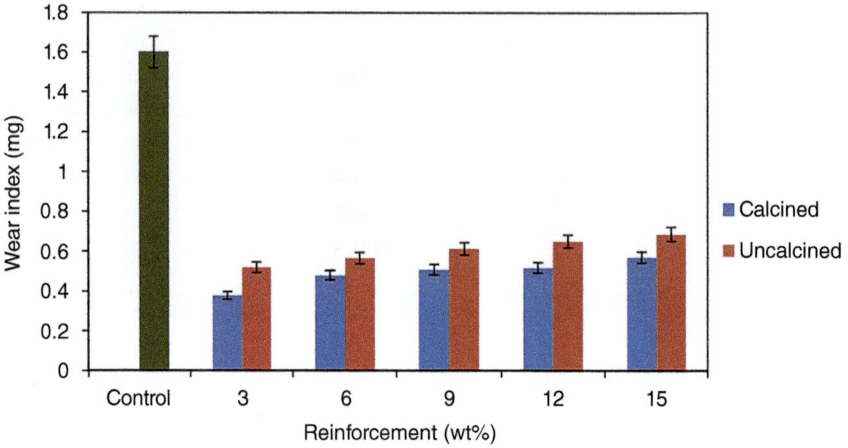

Figure 14.4 Wear index variation of calcined and uncalcined eggshell particles/sisal fiber hybrid reinforced epoxy composites and control. Source: Oladele et al. [139]/IOP Publishing Ltd/Public Domain CC BY 4.0.

and coworkers [139]. Results showed that treatment enhanced the mechanical and wear behavior of the composite (Figure 14.4).

14.4.2 Influence of Fiber Treatment on the Wear Properties of NF- Reinforced Polyester Composites

Dry sliding wear properties of unfilled and organically modified montmorillonite (MMT)-filled coconut sheath fiber reinforced- unsaturated polyester (USP) composites were investigated on a pin-on-disc wear test apparatus. Coconut sheath fiber was treated with silane coupling agent and alkaline, respectively, and hand lay-up fabrication technique and compression molding were used for fabrication of the resulting composite. Experimental results revealed that silane treatment reduced the specific wear rate of the composite, scanning electron microscopy morphology showed silane treated composite had the strongest interfacial adhesion and the smoothest worn surface under given load and sliding velocity when compared to alkaline treated composite Rajini and coworkers [140]. Alkaline treatment improved the frictional and wears behavior of betelnut fiber-reinforced polyester composite under wet contact conditions by 95% and 54%, respectively [136].

Yousif et al. [141] researched the influence of chemical treatment on the tribo-behavior of coir fiber-filled polyester nanocomposite. He reported good interfacial adhesion of the fiber surface to the nanocomposite structure. Yousif et al. [142] also worked on the tribological characteristics of oil palm-based polyester composites and reported a three times increase in wear properties to that of neat polyester. Chand and Dwivedi [143] investigated the frictional performance of sisal fiber-reinforced polyester composite and reported 27 wt% of silane treated sisal fiber-reinforced composite exhibited the lowest wear rate and increased the frictional coefficient due to better interfacial bonding between the matrix and fiber. Similarly, the effect of alkaline treatment and fiber length on the wear performance of Palmyra palm-leaf stalk fiber (PPLSF) filled polyester composite and the usability of the composite in wear resistance was investigated by Shanmugam [144] applications were explored at different speeds and normal loads for constant sliding distance using pin-on-disc wear tester according to ASTM G99 standard. Hand lay-up and compression molding techniques were used to fabricate the composite. The result shows reduced wear loss and coefficient of friction due to reinforcement with PPLSF and a further reduction in wear loss and coefficient of friction for alkaline-treated PPLSF. At high speed and high load conditions, wear loss and coefficient of friction were reduced by 64% and 22%, respectively, for treated and UTC (Figures 14.5 and 14.6).

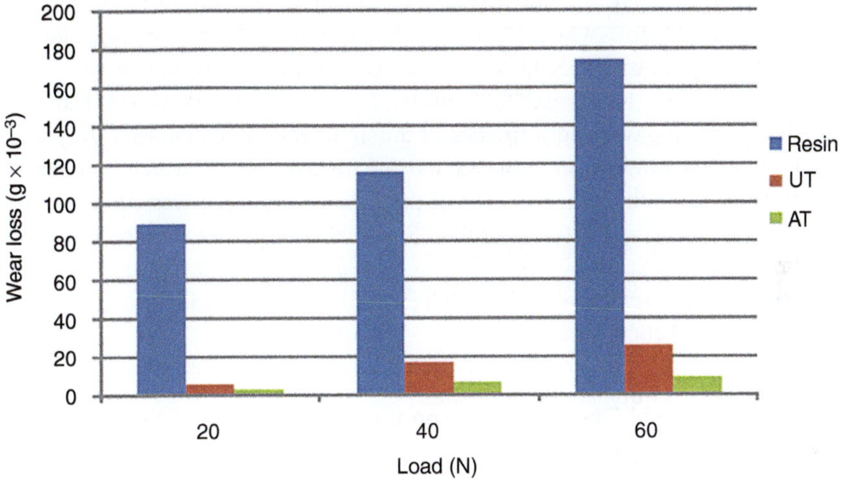

Figure 14.5 Graphical illustration of the effect of alkaline treatment on the wear behavior of PPLSF reinforced polyester composite at 500 rpm. Source: Shanmugam et al. [144]/Taylor & Francis.

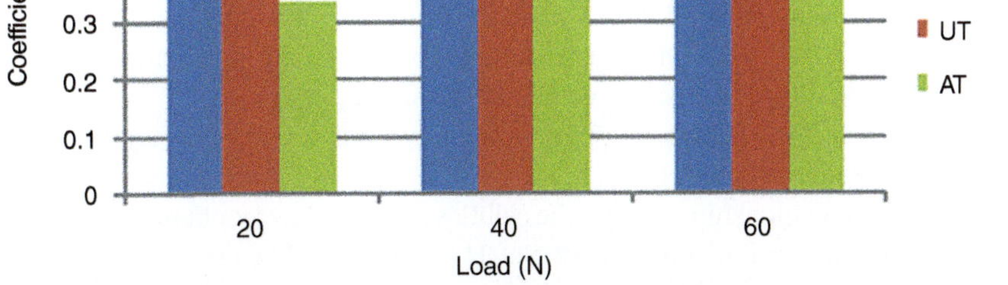

Figure 14.6 Graphical illustration of the effect of alkaline treatment on the coefficient of friction of PPLSF reinforced polyester composite at 700 rpm. Source: Shanmugam et al. [144]/Taylor & Francis.

Similarly, Kumar and coworkers [145] studied the mechanical, physical, and abrasive properties of hemp/nettle NF woven mat filled polyester. The alkaline solution was used to chemically modify the fabric and a hand lay-up and compression-molding method were used to fabricate the composite. From the results, the minimum abrasive wear rate of the hybrid composite is 0.0082 $mm^3\,Nm^{-1}$ found in the 9 wt% hemp/nettle fiber composite at 10 N of normal load and 200 rpm through orthogonal array approach. The influence of various chemical treatments (eco-friendly sodium bicarbonate, potassium permanganate, and potassium hydroxide) on Himalayan agave fiber (HAF) of 7 mm size on its physicomechanical and abrasive wear characteristics of polyester/HAF composite was studied by Kumar and coworkers [146]. Fiber treatment concentration was 5%, 10%, and 15% for 24 hours and reinforced at 10 wt% fiber loading. Results showed that dry abrasive specific wear rate of composite gave the lowest for sodium bicarbonate treated HAF-based composite. Divya and coworkers [147] studied three-body abrasive wear performance of different weight percentage (wt%) filled organo-modified MMT with constant wt% of treated and untreated coir sheath (CS) in USP. The body abrasive wear of the hybrid composite was studied under different filler loading, treatment of the CS, and abrading distance. The results for the abrasive wear test show wear volume increases with an increase in abrading distance, and the specific wear rate is high for the UTC compared to alkaline-treated and silane-treated composite.

14.4.3 Influence of Fiber Treatment on the Wear Behavior of NF Reinforced Vinyl Ester Composite

The mechanical and wear behavior of untreated and treated Roselle fiber-filled vinyl ester nanocomposite was studied as a function of fiber content by Manickam and coworkers [148]. Results revealed that the wear resistance of the nanocomposite fabricated with treated Roselle fibers was better when compared to the UTC (Table 14.2).

Similarly, Kumar and coworkers [149] attempt to explore the effect of hybrid reinforcement on the mechanical and dry sliding wear properties/behavior of vinyl ester composite. Vinyl ester was reinforced with coir fiber, silane aramid fiber, and the inclusion of coconut shell powder in various weight proportions. The coir fiber was treated with an alkali solution to improve its surface roughness. An investigation on the effect of fiber treatment on the mechanical and wear behavior of vinyl ester and its hybrid was evaluated. Results show that treated fiber exhibits a very good interfacial relationship and efficient share transfer between the matrix and the resin, making the resulting composite effective against sliding wear.

Experimental investigation on the evaluation of wear resistance of hybrid polymer composite interlayered petiole date palm fiber (PDPF) and glass fiber (GF). The interfacial relationship between the cellulose fiber and hydrophobic matrix was enhanced by chemical treatment at 5% NaOH for 48 hours. The relevance of the GF is to improve the specific stiffness and dimensional stability of the sample. The wear behavior of four hybrid samples (30PDPF/0GF), (20PDPF/10GF), (10PDPF/20GF),

Table 14.2 The variation of specific wear resistance in Roselle fiber reinforced vinyl ester composite at various applied load conditions.

Fiber content (wt%)	SWR ($cm^3\ Nm^{-1}$) at 5 N		SWR ($cm^3\ Nm^{-1}$) at 15 N		SWR ($cm^3\ Nm^{-1}$) at 20 N	
	Untreated	Treated	Untreated	Treated	Untreated	Treated
10	0.0025	0.0024	0.0027	0.0025	0.0027	0.0026
15	0.0023	0.0022	0.0025	0.0023	0.0026	0.0025
20	0.0021	0.002	0.0023	0.0021	0.0024	0.0023
25	0.0019	0.0018	0.0021	0.002	0.0023	0.0022
30	0.0017	0.0017	0.0019	0.0017	0.0022	0.002
35	0.0018	0.0014	0.002	0.0016	0.0023	0.0022
40	0.002	0.0013	0.0021	0.0015	0.0024	0.0023
45	0.0021	0.0014	0.0022	0.0017	0.0024	0.0024
50	0.0022	0.0015	0.0023	0.0019	0.0025	0.0025

SWR = specific wear resistance.
Source: Manickam et al. [148]/Springer Nature/CC BY 4.0.

and (0PDPF/30GF) was evaluated at a sliding distance of 1000 m, sliding velocity of 1.7 m $second^{-1}$, and applied load of 20 and 40 N. Scanning electron microscopy was performed on both treated alkaline fiber and worn surfaces of the hybrid samples to evaluate the surface modification and wear mechanism, respectively. Wear resistance was improved relatively on the fours samples in different designs and treatment of fiber-enhanced bonding with hydrophilic matrix hence improving the composite characteristics Merzoug and coworkers [150]. Pashaei and Hosseinzadeh [151] had taken walnut shell powder (WSP) as reinforcement material for fabricating green vinyl ester composite and investigated sliding wear properties. The comparison was made for pure vinyl ester green materials and WSP filled composite, and results from the investigation show that the wear volume loss increase with increasing abrading resistance/load. It was found that WSP-filled vinyl ester composites exhibit a higher coefficient of friction and a lower wear rate as compared to vinyl ester green composites due to different interfacial properties of the green fiber.

14.4.4 Influence of Fiber Treatment on the Wear Properties of NF- Reinforced Polypropylene Composites

Comprehending the tribological behavior of NF/polymer composites is just as important as understanding their mechanical properties [142]. In the literature, however, there is little work on the effects of NFs on the tribo-performance of polymeric composites. According to some studies, the tribological behavior of composite polymers based on NF is not intrinsic and is strongly influenced by a variety of processing parameters, including operating parameters, polymer characteristics, physical and interfacial adhesion properties of fiber, additives, and contact condition [152]. One of the key aspects that can alter tribological qualities is fiber treatment [153]. A large level of wear in sliding contact is typically caused by poor bonding between the matrix and the reinforcing fibers [154].

Nijjar and coworkers [155] studied the influence of fiber pretreatment on the wear behaviors of Borassus fruit fine (BFF) fiber-reinforced polypropylene. BFF fiber was treated with alkali to eliminate residual lignin and enhance interfacial coupling between the matrix/fiber contact. Five weight percent maleate polypropylene (MAPP) was employed as a compatibilizer to improve fiber-polymer matrix interface adhesion. Steel counter face was used to analyze the frictional and wear behavior of BFF/PP nanocomposites in relation to the alterations using a pin-on-disc test apparatus in dry-sliding conditions. The addition of BFF to the polypropylene matrix, as well as the treatments, enhanced the wear properties of the virgin polymer matrix. The abrasive wear resistance of polypropylene matrix with MAPP compatibilizer and alkali-treated fiber composite demonstrated improved interfacial bonding, resulting in better tribological behavior under varied loading circumstances. The inclusion of MAPP compatibilizer and alkali-treated fibers resulted in a significant improvement in the coefficient of friction in polypropylene matrix and alkali-treated fiber composites.

Adediran and coworkers [156] investigated the influence of surface modification on the mechanical, temperature, and wear properties of polypropylene matrix filled with coir fiber and yam peel particulate. Coir fibers, both untreated and treated, were used. The treated coir fibers were treated with 5 M NaOH. The yam peel was divided into two groups: treated and untreated. The treated yam peel was altered with a 1 M solution of NaOH and HCl in proportions of 30% and 70%, respectively. The yam peel was mashed and sieved to 45 μm after being sun-dried for 14 days. Treated coir fiber (TCF)/yam peel particulate (YPP) and untreated coir fiber (UCF)/YPP reinforcements were used to create samples with a constant coir fiber content (15%) and varying amounts of yam peel particles (2, 4, 6, and 8 wt%). Mechanical properties, as well as temperature and wear behavior, were investigated in the hybrid composite samples created. The sample was reinforced with treated 4 wt% coir fiber, and yam peel particles were reported to have exhibited the best mechanical qualities. The heat conductivity of composite samples, on the other hand, was enhanced with the addition of fiber. When compared to the control sample, all-composite samples developed were reported to have exhibited better abrasion resistance.

The effect of coupling agents on the abrasive wear characteristics of polypropylene reinforced with chopped-up jute fiber was investigated by Chand and Dwivedi [157]. They concluded that adding an MA-g-PP coupling agent had a significant impact on wear resistance and that coupling agent use resulted in improved wear resistance when compared to the case where coupling agent was not used.

Njoku and coworkers [158] examined the effect of sodium hydroxide treatment on the wear properties of Urena lobata fiber-reinforced polypropylene composites. The composites' physical and abrasion properties were assessed, and alkaline-treated Urena lobata fiber-reinforced polypropylene composites were shown to have improved density and wear resistance.

Ibrahim and coworkers [159] noted a remarkable enhancement of the wear resistance of wood floor- reinforced polypropylene composites with the incorporation of MAPP coupling agent.

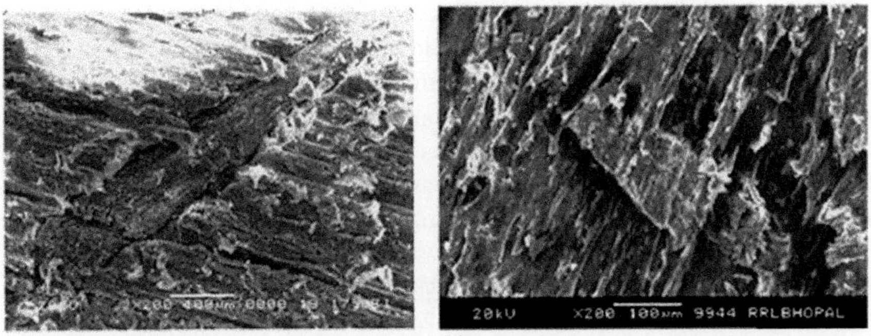

(a) (b)

Figure 14.7 (a) The SEM image of worn surface of untreated sisal fiber-reinforced composites. (b) The SEM image of worn surface of sisal fiber-reinforced composite having 1 wt% MA-g-PP. Source: Dwivedi and Chand [160]/with permission of Elsevier.

Dwivedi and Chand [160] used a Suga abrasion tester to investigate the effects of the MA-g-PP coupling agent on the abrasive wear behavior of sisal fiber-reinforced PP composites. The use of the MA-g-PP coupling agent improved fiber-matrix adhesion, which resulted in the material wearing less. As a result, adding a coupling agent to the composite was reported to improve wear resistance in the sisal fiber-reinforced PP composite. They noted that at first, samples with 1 wt% MA-g-PP displayed maximum wear resistance at low loads (i.e. 1 and 3 N), but at higher applied loads (i.e. 5 and 7 N), samples with 2 wt% MA-g-PP exhibited maximum wear resistance for chopped sisal fiber-reinforced PP composite. The enhanced adhesion between sisal fiber and PP led to improved wear resistance at low loads. At higher loads, the sisal fiber-reinforced composite with 2-wt% MA-g-PP, which is the optimum amount of MA-g PP for the composite that gives the highest wear resistance, provides the best wear resistance. This could be attributed to the right mix of dominant factors, which are modification of PP matrix and improved bonding between sisal fiber and PP. Figure 14.7 shows the scanning electron microscopic (SEM) images of modified and unmodified sisal fiber-reinforced polypropylene composites as observed by Dwivedi and Chand (2009).

14.4.5 Influence of Fiber Treatment on the Wear Properties of NF-Reinforced Polylactic Acid Composites

The influence of surface treatments on the abrasive wear performance of polylactic acid (PLA)/jute fiber composites was investigated by Goriparthi and coworkers [161]. The surface of jute fiber was modified by silane, peroxide, permanganate, and alkali treatments. To evaluate the effect of NFs on tribological properties of biocomposites, researchers used a pin-on-disc apparatus with a sliding speed of 0.418 m second^{-1} and a weight of 9.8 N. The wear resistance of the composite was sensitive to fiber-matrix adhesion, according to the experimental data on abrasive wear. All of the composites were tested with the jute fibers parallel to the normal load and perpendicular to the sliding distance. When compared to untreated jute fiber-PLA composites, surface-modified jute fiber composites showed better wear resistance. As a result, it was determined that the better wear resistance of the treated jute fiber-PLA composites is due to increased fiber-matrix adhesion. The SEM images of the worn surfaces of the untreated and silane-treated jute fiber-PLA composites are shown in Figure 14.8 as obtained by Goriparthi and coworkers [161].

Figure 14.9 depicts the effect of various fiber treatment methods on the wear rate of polylactide/jute nanocomposites as observed by Goriparthi et al. [161]. Because of the strong interfacial adhesion, Silane 2-treated polylactide/jute fiber

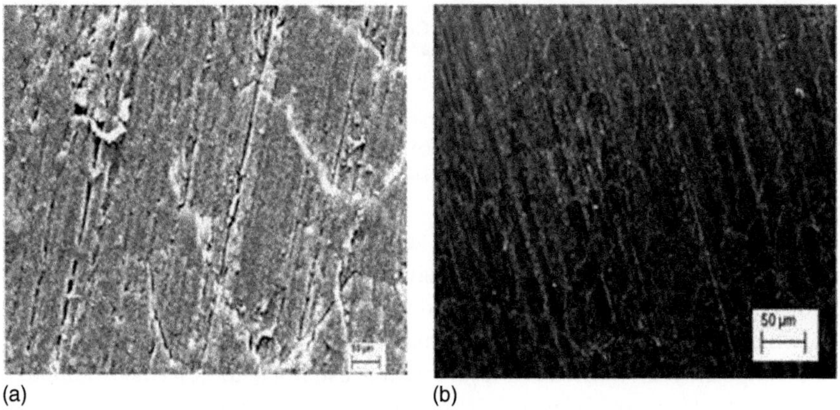

Figure 14.8 SEM images of the worn surfaces of the (a) untreated and (b) silane treated jute fiber-PLA composites. Source: Goriparthi et al. [161]/with permission of Elsevier.

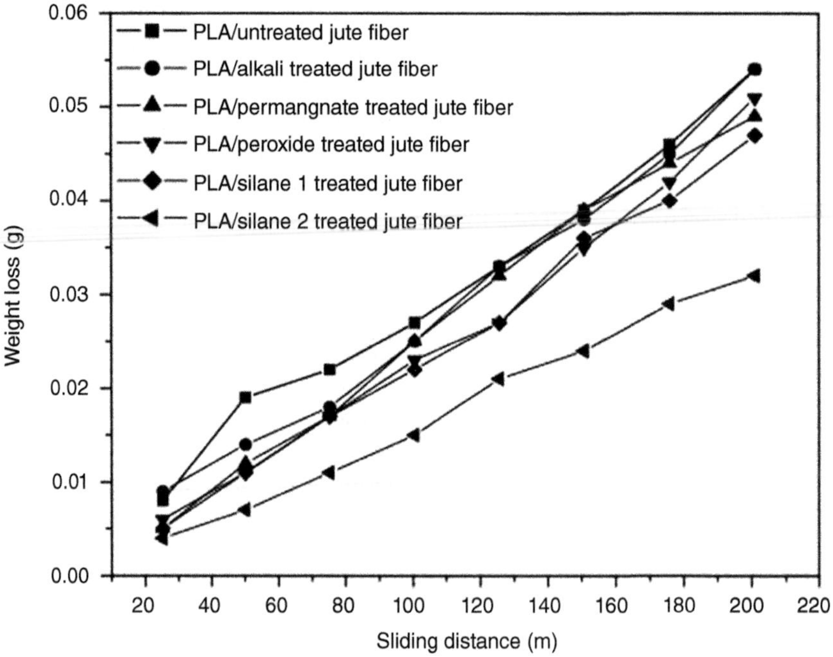

Figure 14.9 Influence of various fiber treatment methods on the wear rate of polylactide/jute composites. Source: Goriparthi et al. [161]/with permission of Elsevier.

nanocomposite showed the best wear resistance among other fiber treatments. Furthermore, weight loss increased when the sliding distance was increased, owing to progressive material removal, including fiber/matrix, as the distance was raised.

Govindan and coworkers [162] treated basalt and sisal chopped fibers with silane and used them in reinforcing PLA. Four distinct weights (10, 20, 30, and 40 N) and four different sliding speeds (100, 150, 200, and 250 rpm) were used to determine wear loss. The resulting composite exhibited improved wear resistance.

14.4.6 Influence of Fiber Treatment on the Wear Properties of NF-Reinforced High-Density Polyethylene Composites

Liu and coworkers [163] used a pin-on-disc wear testing system to study the effect of organosilane treatment on the wear resistance of graphitic nanoplatelets (GNPs) reinforced high-density polyethylene (HDPE) composites at varied sliding velocities. The results showed that silanized-GNP–reinforced HDPE composites had significantly improved wear resistance under various sliding velocities as compared to composites without filler surface modification. Under 1.3 $msecond^{-1}$ sliding velocity, 97% wear resistance improvement was achieved. Furthermore, the organosilane-modified GNPs reduced the effect of sliding velocity on the composite's wear resistance, resulting in good wear resistance over a wide range of sliding velocities, including very high speeds (2.0 ms^{-1}). Superior wear resistance and good wear performance stability at varied sliding velocities show that silanized-GNP/HDPE composites are attractive materials with a lot of potential for tribological applications. Pure HDPE and its nanocomposites were subjected to wear testing at three different sliding velocities. The comparison of their volumetric wear rates obtained by Liu and coworkers [163] is shown in Figure 14.10. As-received graphitic nanoplatelets composites (Comp-asGNP) and high-temperature purified GNPs composites have extremely high wear rates for each sliding velocity, even exceeding pure HDPE. Wear resistance in nanocomposites deteriorates due to two main factors: The presence of GNP aggregates and the residual acid and exposed reactive groups on GNP, which cause poor interfacial interaction with the matrix, resulting in load transfer failure from the polymer matrix to the nanofillers. This was more noticeable in the case of silanized graphitic nanoplatelets composites (Comp-sGNP). For each sliding velocity, however, Comp-sGNP had the lowest specific volumetric wear rate. This indicates that s-GNP can maintain the composite's wear resistance even at high speeds (2.0 ms^{-1}). This implies that the organosilane coating can cover the reactive groups on the GNP surface, and the lengthy silane hydrocarbon tails on the coating may offer enough tangling with the polyethylene chains to result in a much-improved filler–matrix interfacial adhesion as well as wear resistance in the composites.

14.4.7 Influence of Fiber Treatment on the Wear Properties of NF- Reinforced Low-Density Polyethylene Composites

Despite the limited investigation on the influence of fiber treatment on the wear of NF-reinforced low-density polyethylene (LDPE) composites, however, fiber-matrix

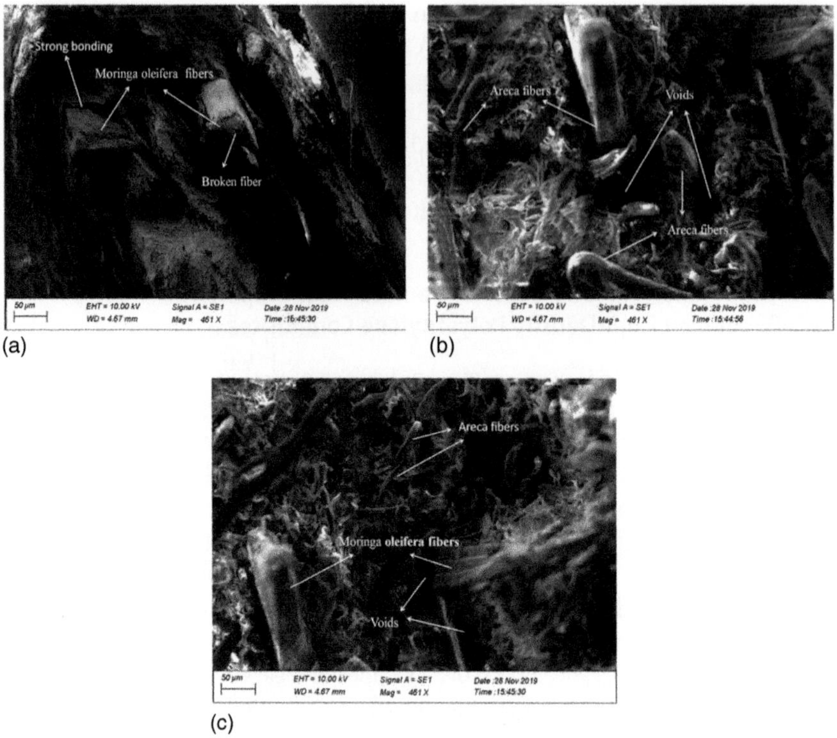

Figure 14.10 SEM image showing fiber interactions in (a) moringa fiber/PET composites, (b) areca fiber/PET composite, and (c) areca/moringa/PET. Source: Nayak et al. [164]/with permission of Taylor and Francis Group.

adhesion has been previously noted as one of the major factors influencing the wear properties of (LDPE) fiber-reinforced composites. Some researchers have studied the influence of fiber treatment on the fiber-matrix adhesion of NF- reinforced LDPE composites. For instance, the influence of alkali and oligomeric siloxane treatment on the characteristics of jute fiber reinforced LDPE composites was investigated by Sever and composites [165]. After alkali treatment, the jute fiber was treated with oligomeric siloxane. The adherence of jute fiber to LDPE is greatly improved by both alkali and oligomeric siloxane treatments. As a result, this treatment is expected to improve the wear qualities of this composite.

Sultana and coworkers [166] studied the effect of oxidation on the properties of cellulosic waste mango saw-dust reinforced LDPE composites. The unbleached raw sawdust fibers were oxidized with sodium hypochlorite. Improved interfacial adhesion was observed between the oxidized sawdust and the LDPE matrix, which has the potential to improve the wear properties of the composite.

Rahman and coworkers [167] improved the compatibility of pineapple leaf fiber (PALF) and LDPE matrix by irradiating PALF–LDPE (50%/50%) composites with gamma rays (Co-60) at doses ranging from 2.5 to 10 kGy, with the 7.5 kGy dose providing the best fiber-matrix compatibility. This is also expected to improve the wear properties of the composite.

14.4.8 Influence of Fiber Treatment on the Wear Properties of NF- Reinforced PET Composites

The effect of different fiber treatments on the wear properties of NF-reinforced PET composites has been investigated, following their application in the automotive sector. Due to reduced wettability as well as the desire to improve mechanical and wear properties, treated NFs are reinforced with PET to achieve the desired wear properties for application in automotive parts. Nayak and coworkers compared mechanical and wear properties of three composites based on PET matrix via areca sheath fiber/PET composite (areca sheath fiber [AF] composite); moringa oleifera fruit fiber/PET (moringa oleifera fruit fiber [MF] composite) and the hybrid of areca/moringa/PET composites. Their result confirmed that fiber treatment with NaOH improved the adhesion of fiber to the matrix; higher adhesion between the fiber and matrix gave rise to improved wear properties [164]. The best mechanical and wear properties were obtained for areca/moringa/PET composite due to better adhesion as seen in the SEM images (Figure 14.10).

The presence of organic fillers has the potential to enhance the wear properties of the composite. Organic date fiber seed nano-fillers were used to improve the wear and friction properties of PET and other polymer composites. The result shows that some level of wear resistance was achieved with an increase in date seed nano-filler content up to the optimum; this improved wear property was observed more in MDPE than PET [168]. For jute fibers, preparatory delignification has been identified to increase adhesion between fibers and resin as well as the degumming process using sodium silicate; this improved adhesion has a direct relationship with the wear and friction properties [169].

The use of lubrication has also been shown by researchers to greatly improve the wear and friction properties of NF-reinforced composites, four types of lubrication via fluid film lubrication, elasto-hydrodynamic lubrication, boundary lubrication, and mixed lubrication. The mostly used lubricants are oils (castor oil, rapeseed oil, etc.), greases, synthetic oils (esters, silicones, etc.), and solid lubricants [170].

14.4.9 Influence of Fiber Treatment on the Wear Properties of NF- Reinforced Polyamide Composites

The need for environmentally friendly engineering materials has birthed the development and application of treated NF-polyamide biomass composite [167].

The effect of alkaline treatment with NaOH and silane treatment on the friction/wear properties of hemp fiber (HF) reinforced plant-derived polyamide 1010 (PA1010). The silane coupling agents used were aminosilane, epoxysilane, and ureidosilane. The research shows NaOH and silane treatment greatly improved the wear properties of HF/PA1010 due to high interfacial adhesion, which brought about change in the mode of friction and wear [171]. This is as shown in Figure 14.11.

In a similar study, Nishitani investigated the effect of pretreating HF with NaOH followed by 3-(2-aminoethylamino) propyltrimethoxy silane; the result shows that a class of biodegradable engineering material with good thermal, mechanical, and wear properties can be formed from HF/polyamide [172]. Also, Nishitani

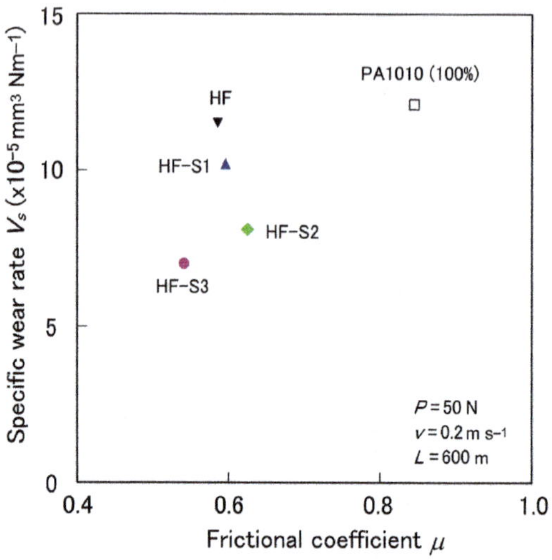

Figure 14.11 Specific wear rate as a function of frictional coefficient for HF treated with aminosilane (HF-S1), epoxysilane (HF-S2), and ureidosilane (HF-S3). Source: Nishitani et al. [171]/MDPI/ Public Domain CC BY 4.0.

produced PA1010 which is plant-originated castor oil by reacting sebacic acid and decamethylenediamine; the result confirmed that incorporation of HFs treated with NaOH and then with aminosilane coupling agent improves thermal, mechanical, and wear properties of HF/PA1010 biomass composites [172].

In another research, two-folds of recycled NF-reinforced plant-derived PA1010 made from recycled jute fiber and recycled sisal fiber were researched. The research revealed that though alkali treatment with NaOH followed by surface treatment with ureidosilane coupling agents showed some level of friction and wear properties, treating with sodium chlorite (NaClO2) followed by ureidosilane coupling agents gave the best surface treatment for friction and wear properties [173]. A similar result shows that by treating wood fibers with a coupling agent (KH-550), the wear properties of polyimide composites were greatly improved [174].

14.4.10 Influence of Fiber Treatment on the Wear Properties of NF-Reinforced Hybrid Biocomposites (Fiber Blending + Polymer Blending) Composites

Recent research is geared toward the elimination of environmental issues posed by the use of nonbiodegradable polymers. This resulted in the advent of natural FRP composites; then comes the problem of desirable properties to meet the requirements for some engineering applications without compromising eco-friendliness. This gave rise to the production of NF hybrid composite. Ramesh and coworkers [175] prepared treated fiber/PLA based hybrid biocomposite by dispersing MMT clay content in a screw extruder after treating kenaf fiber with 6% NaOH. The result shows that composite with good mechanical and wear properties was obtained at 1% MMT clay and 30% treated kenaf fiber. Kenaf fiber, like other NFs, can be modified by chemical treatment for the production of composites with good properties, though more research is needed to improve its use alone in biocomposites.

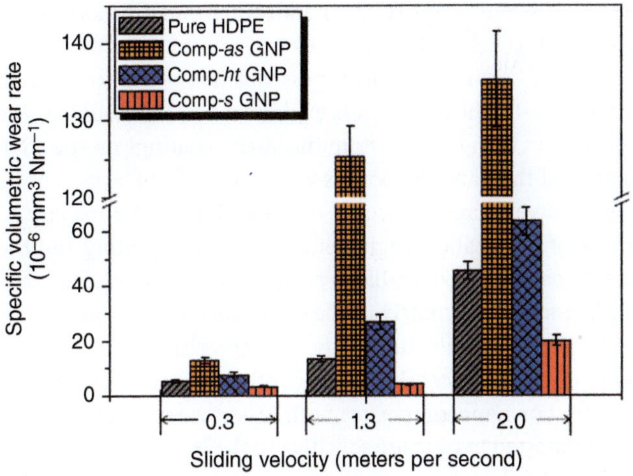

Figure 14.12 The comparison of specific volumetric wear rates of pure HDPE, Comp-asGNP, Comp-htGNP, and Comp-sGNP under pin-on-disc wear tests with three sliding velocities. Source: Liu et al. [163]/with permission of Elsevier.

Blended alkali-treated coir fibers PALFs blend in varied quantity reinforced PLA hybrid composites gave a fiber-matrix adhesion increasing the wear resistance due to alkali treatment of blended coir/pineapple [176]. Hybrid composites from treated NF (*Phaseolus vulgaris*) using USP resin matrix, fly ash filler, and 1% silicon (iv) oxide as secondary filler was investigated. Results show that there is an increase in the wear properties of the composite up to an optimum of 6% fiber content [177] (Figure 14.12).

Oladele and coworkers [178] treated chopped sugarcane bagasse fiber with 0.5 M NaOH solution and used it in reinforcing recycled HDPE. The resulting composite exhibited better wear resistance when compared to the control sample.

Savaş and coworkers [179] investigated the abrasive wear properties of coconut fiber-reinforced HDPE composites using maleic anhydride-grafted polyethylene (PE-g-MA) compatibilization. A pin-on-disc abrasive wear tester was used to examine wear qualities under dry sliding conditions. PE-g-MA had good miscibility between fibers and matrix, which resulted in an increase in hardness and wear resistance of the samples.

Untreated and organosilane-treated carbon nanofibers (CNFs) reinforced HDPE composites were studied by Liu and coworkers [180] to see if wear effects and dielectric characteristics were related. The nanocomposite was subjected to wear testing for up to 120 hours, after which the dielectric permittivity and dielectric loss factor of polymer composites were investigated as wear time increased. The findings show that, for nanocomposites, there are approximate linear connections between permittivity and wear coefficient. Wear resistance was good in composites containing silanized CNFs with a sufficiently thick covering. Figure 14.5, as obtained by Lui and coworkers (2011), compares the wear coefficient and the wear time for oxidized CNFs-reinforced HDPE nanocomposites (Nanocomp-ox) and three different samples of silane treated CNFs-reinforced HDPE nanocomposites (Nanocomp-A, Nanocomp-B, and Nanocomp-C) (Figure 14.13).

The change rate of wear coefficient in Nanocomp-ox was seen to be lower than in Nanocomp-A and Nanocomp-B. The researchers explained that this could be due to the ox-CNFs' surface stiffness being higher than the silane coatings on the CNF surface. The wear resistance of the nanocomposites with thin coating layers on the CNF surface was poor, even lower than Nanocomp-ox, especially for Nanocomp-A. This could be due to a lack of physical entanglement between the coating and the matrix, causing the thinner coatings to be unable to produce strong interfacial interaction between the nanofibers and the matrix in the composites. Nanocomp-C, on the other hand, has the lowest slope, indicating the best wear resistance. As a result, the findings show that the thick silane coating significantly increased the composite system's wear resistance. For nanocomposites with lesser wear resistance, the change in permittivity was reported to be more sensitive to the increased wear coefficient, as shown in Figure 14.14.

The permittivity rose in a linear relationship with the wear coefficient. Nanocomp-ox had a substantially faster rate of permittivity rise than the composites with silane-treated CNFs, based on the slopes of the lines. It was discovered that the silane treatment reduced the presence of polar groups on the carbon fibers' surface, resulting in fewer reactive groups being affected by the external electric

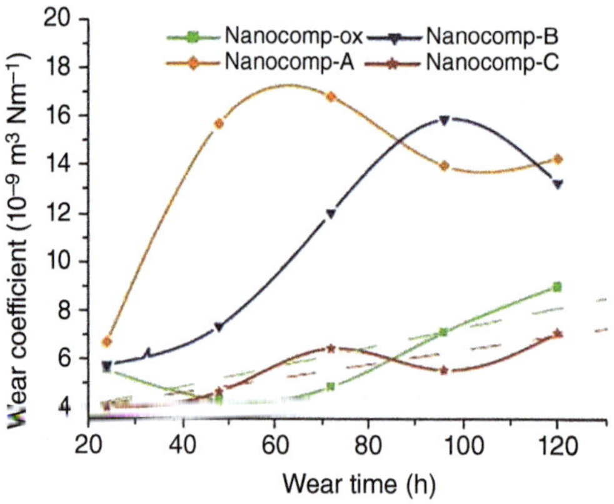

Figure 14.13 Comparison of wear coefficient and wear time for oxidized CNFs-reinforced HDPE nanocomposites (Nanocomp-ox) and three different samples of silane-treated CNFs-reinforced HDPE nanocomposites (Nanocomp-A, Nanocomp-B, and Nanocomp-C). Source: Liu et al. [180]/Springer Nature/Public Domain CC BY 4.0.

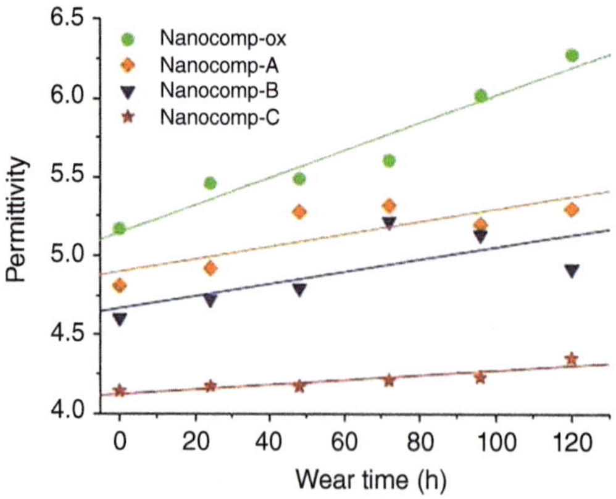

Figure 14.14 Permittivity-changing patterns of the four samples at 1×10^3 Hz with increased wear hours. Source: Liu et al. [180]/Springer Nature/Public Domain CC BY 4.0.

field. For all silane-based materials, wear effects on silanized CNF-reinforced HDPE composites were uniformly less susceptible to the permittivity shift caused by the wear processes. With Nanocomp-C, the thickest coating on the CNFs, and the smallest slope, the trend is evident that increasing the coating thickness caused a loss in sensitivity.

14.5 Conclusion

In this overview, the phenomenon of fiber treatment and the different types of treatments, with their various examples, have been discussed in this chapter. The idea of reinforcing polymers with NF to achieve a biocomposite with desired mechanical and tribological properties was also elaborated. The various research works consulted and practical examples analyzed, show that fiber treatment influences the wear properties of biocomposites, as shown using polymeric biocomposites. These influences on the wear properties can be in the form of wear rate, wear weight loss, sliding distance, wear time, and wear resistance, which are of high significance for the applications of biocomposites, especially in the production of automotive parts. Through the treatment of NF, which increases the interfacial adhesion at the contact surface between the polymer matrix and the filled fiber, a biocomposite with improved wear properties is achieved.

References

1 Yao, W., Li, J., and Wu, K. (2003). *Mechanical properties of hybrid fiber-reinforced concrete at low fiber volume fraction. Cement $ch.* 33 (1): 27–30.

2 Prado, M., Marques, J.N., Pereira, G.D. et al. (2017). *Evaluation of different surface treatments on fiber post cemented with a self-adhesive system. Mater. Sci. Eng. C* 77: 257–262.

3 Brígida, A.I.S., Calado, V.M.A., and Gonçalves, L.R.B. (2010). et al, Effect of chemical treatments on properties of green coconut fiber. *Carbohydr. Polym.* 79 (4): 832–838.

4 Ahad, N.A., Parimin, N., Mahmed, N. et al. (2009). *Effect of chemical treatment on the surface of natural fiber. J. Nucl. Relat. Technol.* 6 (1): 155–158.

5 Rong, M.Z., Zhang, M.Q., Liu, Y. et al. (2001). *The effect of fiber treatment on the mechanical properties of unidirectional sisal-reinforced epoxy composites. Compos. Sci. Technol.* 61 (10): 1437–1447.

6 Yahaya, R., Sapuan, S.M., Jawaid, M. et al. (2015). *Effect of layering sequence and chemical treatment on the mechanical properties of woven kenaf–aramid hybrid laminated composites. Mater. Des.* 67: 173–179.

7 Latif, R., Wakeel, S., Khan, N.Z. et al. (2019). *Surface treatments of plant fibers and their effects on mechanical properties of fiber-reinforced composites: a review. J. Reinf. Plast. Compos.* 38 (1): 15–30.

8 Cisneros-López, E.O., González-López, M.E., Pérez-Fonseca, A.A. et al. (2017). *Effect of fiber content and surface treatment on the mechanical properties of* *natural fiber composites produced by rotomolding. Compos. Interfaces* 24 (1): 35–53.

9 Varley, R.J., Tian, W., Leong, K.H. et al. (2013). *The effect of surface treatments on the mechanical properties of basalt-reinforced epoxy composites. Polym. Compos.* 34 (3): 320–329.

10 Yussuf, A.A., Massoumi, I., and Hassan, A. (2010). *Comparison of polylactic acid/kenaf and polylactic acid/rise husk composites: the influence of the natural fibers on the mechanical, thermal and biodegradability properties. J. Polym. Environ.* 18 (3): 422–429.

11 Huang, Z., Qian, L., Yin, Q. et al. (2018). *Biodegradability studies of poly (butylene succinate) composites filled with sugarcane rind fiber. Polym. Test.* 66: 319–326.

12 Wu, C.-S. (2009). *Renewable resource-based composites of recycled natural fibers and maleated polylactide bioplastic: characterization and biodegradability. Polym. Degrad. Stab.* 94 (7): 1076–1084.

13 Chattopadhyay, S.K., Singh, S., Pramanik, N. et al. (2011). *Biodegradability studies on natural fibers reinforced polypropylene composites. J. Appl. Polym. Sci.* 121 (4): 2226–2232.

14 Takagi, H. (2019). *Review of functional properties of natural fiber-reinforced polymer composites: thermal insulation, biodegradation and vibration damping properties. Adv. Compos. Mater* 28 (5): 525–543.

15 Zhang, M.Q., Rong, M.Z., and Lu, X. (2005). *Fully biodegradable natural fiber composites from renewable resources: all-plant fiber composites. Compos. Sci. Technol.* 65 (15-16): 2514–2525.

16 Kalia, S., Kaith, B.S., and Kaur, I. (2009). *Pretreatments of natural fibers and their application as reinforcing material in polymer composites—a review. Polym. Eng. Sci.* 49 (7): 1253–1272.

17 Bledzki, A.K., Sperber, V.E., and Faruk, O. (2002). *Natural and Wood Fibre Reinforcement in Polymers*, vol. 13. United Kingdom: iSmithers Rapra Publishing.

18 Faruk, O., Bledzki, A.K., Fink, H.P. et al. (2012). *Biocomposites reinforced with natural fibers: 2000–2010. Prog. Polym. Sci.* 37 (11): 1552–1596.

19 Sahari, J. and Sapuan, S.M. (2011). *Natural fibre reinforced biodegradable polymer composites. Rev. Adv. Mater. Sci* 30 (2): 166–174.

20 Omrani, E., Menezes, P.L., and Rohatgi, P.K. (2016). *State of the art on tribological behavior of polymer matrix composites reinforced with natural fibers in the green materials world. Eng. Sci. Technol. Int. J.* 19 (2): 717–736.

21 Sanjay, M.R., Siengchin, S., Parameswaranpillai, J. et al. (2019). *A comprehensive review of techniques for natural fibers as reinforcement in composites: preparation, processing and characterization. Carbohydr. Polym.* 207: 108–121.

22 Rogovina, S.Z., Prut, E.V., and Berlin, A.A. (2019). *Composite materials based on synthetic polymers reinforced with natural fibers. Polym. Sci. Ser. A* 61 (4): 417–438.

23 Nurazzi, N.M., Harussani, M.M., Aisyah, H.A. et al. (2021). *Treatments of natural fiber as reinforcement in polymer composites—a short review. Funct. Compos. Struct.* 3 (2): 024002.

24 Sanjay, M.R., Madhu, P., Jawaid, M. et al. (2018). *Characterization and properties of natural fiber polymer composites: a comprehensive review*. *J. Cleaner Prod.* 172: 566–581.

25 Vigneshwaran, S., Sundarakannan, R., John, K.M. et al. (2020). *Recent advancement in the natural fiber polymer composites: a comprehensive review*. *J. Cleaner Prod.* 277: 124109.

26 Ku, H., Wang, H., and Pattarachaiyakoop, N. (2011). et al, A review on the tensile properties of natural fiber reinforced polymer composites. *Composites Part B* 42 (4): 856–873.

27 Menezes, P.L., Rohatgi, P.K., and Lovell, M.R. (2012). *Studies on the tribological behavior of natural fiber reinforced polymer composite*. In: *Green Tribology*, 329–345. Germany: Springer.

28 Chandramohan, D. and Marimuthu, K. (2011). *A review on natural fibers*. *Int. J. Res. Rev. Appl. Sci.* 8 (2): 194–206.

29 Ticoalu, A., Aravinthan, T., and Cardona, F. (2010). *A review of current development in natural fiber composites for structural and infrastructure applications*. In: *Proceedings of the Southern Region Engineering Conference (SREC 2010)*. Toowoomba, Australia (11–12 November 2010). Australia: Engineers Australia.

30 Alzebdeh, K.I., Nassar, M.M., and Arunachalam, R. (2019). *Effect of fabrication parameters on strength of natural fiber polypropylene composites: Statistical assessment*. *Measurement* 146: 195–207.

31 Kozłowski, R.M. and Mackiewicz-Talarczyk, M. (2020). *Introduction to natural textile fibres*. In: *Handbook of Natural Fibres* (ed. R.M. Kozłowski and M. Mackiewicz-Talarczyk), 1–13. India: Elsevier.

32 Jawaid, M., Khalil, H.P.S., Hassan, A. et al. (2012). Bi-layer hybrid biocomposites: chemical resistant and physical properties. *BioResources* 7 (2): 2344–2355.

33 Gurunathan, T., Mohanty, S., and Nayak, S.K. (2015). *A review of the recent developments in biocomposites based on natural fibres and their application perspectives*. *Composites Part A* 77: 1–25.

34 Mohd Nurazzi, N., Khalina, A., Sapuan, S.M. et al. (2017). *A review: fibres, polymer matrices and composites*. *Pertanika J. Sci. Technol.* 25 (4): 1085–1102.

35 Sanjay, M.R., Siengchin, S., Parameswaranpillai, J. et al. (2019). *A comprehensive review of techniques for natural fibers as reinforcement in composites: preparation, processing and characterization*. *Carbohydr. Polym.* 207: 108–121.

36 Gangil, B., Ranakoti, L., Verma, S. et al. (2020). *Natural and synthetic fibers for hybrid composites*. In: *Hybrid Fiber Composites: Materials, Manufacturing Process Engineering*, 1–15. Germany: Wiley.

37 Desai, A. and Kant, R. (2016). *Geotextiles made from natural fibres*. In: *Geotextiles*, 61–87. Sawston, UK: Elsevier.

38 Cesa, F.S., Turra, A., and Baruque-Ramos, J. (2017). *Synthetic fibers as microplastics in the marine environment: a review from textile perspective with a focus on domestic washings*. *Sci. Total Environ.* 598: 1116–1129.

39 Kerni, L., Singh, S., Patnaik, A. et al. (2020). *A review on natural fiber reinforced composites*. *Mater. Today Proc.* 28: 1616–1621.

40 Duval, A., Bourmaud, A., Augier, L. et al. (2011). *Influence of the sampling area of the stem on the mechanical properties of hemp fibers*. *Mater. Lett.* 65 (4): 797–800.

41 Shalwan, A. and Yousif, B. (2013). *In state of art: mechanical and tribological behaviour of polymeric composites based on natural fibres. Mater. Des.* 48: 14–24.

42 Ranganathan, N., Oksman, K., Nayak, S.K. et al. (2016). *Structure property relation of hybrid biocomposites based on jute, viscose and polypropylene: the effect of the fibre content and the length on the fracture toughness and the fatigue properties. Composites Part A* 83: 169–175.

43 Patel, V.K., Chauhan, S., and Katiyar, J.K. (2018). *Physico-mechanical and wear properties of novel sustainable sour-weed fiber reinforced polyester composites. Mater. Res. Express* 5 (4): 045310.

44 Jayabal, S., Velumani, S., Navaneethakrishnan, P. et al. (2013). *Mechanical and machinability behaviors of woven coir fiber-reinforced polyester composite. Fibers Polym.* 14 (9): 1505–1514.

45 Ramesh, M., Palanikumar, K., and Reddy, K.H. (2013). *Mechanical property evaluation of sisal–jute–glass fiber reinforced polyester composites. Composites Part B* 48: 1–9.

46 Ganeshan, P., Kumaran, S.S., Raja, K. et al. (2018). *An investigation of mechanical properties of madar fiber reinforced polyester composites for various fiber length and fiber content. Mater. Res. Express* 6 (1): 015303.

47 Parikh, H.H. and Gohil, P.P. (2015). Tribology of fiber reinforced polymer matrix composites—a review. *J. Reinf. Plast. Compos.* 34 (16): 1340–1346.

48 Sreenivasan, V., Rajini, N., Alavudeen, A. et al. (2015). *Dynamic mechanical and thermo-gravimetric analysis of Sansevieria cylindrica/polyester composite: effect of fiber length, fiber loading and chemical treatment. Composites Part B* 69: 76–86.

49 Sood, M. and Dwivedi, G. (2018). *Effect of fiber treatment on flexural properties of natural fiber reinforced composites: a review. Egypt. J. Pet.* 27 (4): 775–783.

50 Sawsen, C., Fouzia, K., Mohamed, B. et al. (2015). *Effect of flax fibers treatments on the rheological and the mechanical behavior of a cement composite. Constr. Build. Mater.* 79: 229–235.

51 Punyamurthy, R., Sampathkumar, D., Ranganagowda, R.P.G. et al. (2017). *Mechanical properties of abaca fiber reinforced polypropylene composites: effect of chemical treatment by benzenediazonium chloride. J. King Saud Univ.-Eng. Sci.* 29 (3): 289–294.

52 Faruk, A., Bledzki, A.K., Fink, H.P. et al. (2012). Biocomposites reinforced with natural fibers: 2000–2010. *Prog. Polym. Sci.* 37: 1552–1596.

53 Mohammed, L., Ansari, M.N.M., Pua, G. et al. (2015). *A review on natural fiber reinforced polymer composite and its applications. Int. J. Polym. Sci.* 2015: 15.

54 Eslami-Farsani, R. (2015). *Effect of fiber treatment on the mechanical properties of date palm fiber reinforced PP/EPDM composites. Adv. Compos. Mater* 24 (1): 27–40.

55 Amroune, S., Bezazi, A., Belaadi, A. et al. (2015). Tensile mechanical properties and surface chemical sensitivity of technical fibres from date palm fruit branches (*Phoenix dactylifera* L.). *Composites Part A* 71: 95–106.

56 Fiore, V., Di Bella, G., and Valenza, A. (2015). The effect of alkaline treatment on mechanical properties of kenaf fibers and their epoxy composites. *Composites Part B* 68: 14–21.

57 Kabir, M., Wang, H., Lau, K.T. et al. (2012). Chemical treatments on plant-based natural fibre reinforced polymer composites: an overview. *Composites Part B* 43 (7): 2883–2892.

58 Rao, J., Bao, L., and Wang, B. (2018). et al, Plasma surface modification and bonding enhancement for bamboo composites. *Composites Part B* 138: 157–167.

59 Militký, J. and Jabbar, A. (2015). *Comparative evaluation of fiber treatments on the creep behavior of jute/green epoxy composites. Composites Part B* 80: 361–368.

60 Sarikanat, M., Seki, Y., Sever, K. et al. (2016). *The effect of argon and air plasma treatment of flax fiber on mechanical properties of reinforced polyester composite. J. Ind. Text.* 45 (6): 1252–1267.

61 Fiore, V., Sanfilippo, C., and Calabrese, L. (2019). *Influence of sodium bicarbonate treatment on the aging resistance of natural fiber reinforced polymer composites under marine environment. Polym. Test.* 80: 106100.

62 Jain, D., Kamboj, I., Bera, T.K. et al. (2019). *Experimental and numerical investigations on the effect of alkaline hornification on the hydrothermal ageing of Agave natural fiber composites. Int. J. Heat Mass Transfer* 130: 431–439.

63 Jiang, Y., Han, X., and Gong, J. (2019). et al, Laboratory investigation of epoxy asphalt binder modified by brominated SBS. *Constr. Build. Mater.* 228: 116733.

64 Chen, D., Li, J., and Ren, J. (2011). *Influence of fiber surface-treatment on interfacial property of poly (l-lactic acid)/ramie fabric biocomposites under UV-irradiation hydrothermal aging. Mater. Chem. Phys.* 126 (3): 524–531.

65 El-Abbassi, F.E., Assarar, M., Ayad, R. et al. (2015). *Effect of alkali treatment on Alfa fibre as reinforcement for polypropylene based eco-composites: mechanical behaviour and water ageing. Compos. Struct.* 133: 451–457.

66 Sathish, S., Karthi, N., Prabhu, L. et al. (2021). *A review of natural fiber composites: extraction methods, chemical treatments and applications. Mater. Today Proc.* 45: 8017–8023.

67 Mahjoub, R., Yatim, J.M., Sam, A.R.M. et al. (2014). *Characteristics of continuous unidirectional kenaf fiber reinforced epoxy composites. Mater. Des.* 64: 640–649.

68 Athith, D., Sanjay, T.G., Yashas Gowda, P. et al. (2018). *Effect of tungsten carbide on mechanical and tribological properties of jute/sisal/E-glass fabrics reinforced natural rubber/epoxy composites. J. Ind. Text.* 48 (4): 713–737.

69 Manimaran, P., Senthamaraikannan, P., Sanjay, M.R. et al. (2018). *Study on characterization of Furcraea foetida new natural fiber as composite reinforcement for lightweight applications. Carbohydr. Polym.* 181: 650–658.

70 Jena, P.K., Mohanty, J.R., and Nayak, S. (2022). *Effect of surface modification of vetiver fibers on their physical and thermal properties. J. Nat. Fibers* 19 (1): 25–36.

71 Reddy, K.O., Maheswari, C.U., Shukla, M. et al. (2013). *Tensile and structural characterization of alkali treated Borassus fruit fine fibers. Composites Part B* 44 (1): 433–438.

72 Chikouche, M.D.L. et al. (2015). *Influence of alkali treatment on the mechanical properties of new cane fibre/polyester composites. J. Reinf. Plast. Compos.* 34 (16): 1329–1339.

73 Rashid, B., Leman, Z., Jawaid, M. et al. (2017). *Influence of treatments on the mechanical and thermal properties of sugar palm fibre reinforced phenolic composites. BioResources* 12 (1): 1447–1462.

74 Li, X., Tabil, L.G., and Panigrahi, S. (2007). *Chemical treatments of natural fiber for use in natural fiber-reinforced composites: a review. J. Polym. Environ.* 15 (1): 25–33.

75 Hu, L., Wan, Y., He, F. et al. (2009). *Effect of coupling treatment on mechanical properties of bacterial cellulose nanofibre-reinforced UPR ecocomposites. Mater. Lett.* 63 (22): 1952–1954.

76 Zhang, L., Zhong, J., and Ren, X. (2017). *Natural fiber-based biocomposites.* In: *Green Biocomposites* (ed. M. Jawaid, M.S. Salit, and O.Y. Alothman), 31–70. Springer.

77 Zimniewska, M., Wladyka-Przybylak, M., and Mankowski, J. (2011). *Cellulosic bast fibers, their structure and properties suitable for composite applications.* In: *Cellulose Fibers: Bio- and Nano-Polymer Composites*, 97–119. Springer.

78 Fuqua, M., Huo, S., and Ulven, C. (2012). *Natural fiber reinforced composites. Polym. Rev.* 52 (3): 259–320.

79 Gwon, J.G., Lee, S.Y., Chun, S.J. et al. (2010). *Effect of chemical treatments of wood fibers on the physical strength of polypropylene based composites. Korean J. Chem. Eng.* 27 (2): 651–657.

80 Zhang, W., Vinueza, N.R., Datta, P. et al. (2015). *Functional dye as a comonomer in a water-soluble polymer. J. Polym. Sci., Part A: Polym. Chem.* 53 (13): 1594–1599.

81 Gholampour, A. and Ozbakkaloglu, T. (2020). *A review of natural fiber composites: properties, modification and processing techniques, characterization, applications. J. Mater. Sci.* 55 (3): 829–892.

82 Koohestani, B., Darban, A.K., and Mokhtari, P. (2019). et al, Comparison of different natural fiber treatments: a literature review. *Int. J. Environ. Sci. Technol.* 16 (1): 629–642.

83 Oliveira, F.R., Erkens, L., Fangueiro, R. et al. (2012). *Surface modification of banana fibers by DBD plasma treatment. Plasma Chem. Plasma Process.* 32 (2): 259–273.

84 Cruz, J. and Fangueiro, R. (2016). *Surface modification of natural fibers: a review. Procedia Eng.* 155: 285–288.

85 El-Zayat, M.M., Abdel-Hakim, A., and Mohamed, M.A. (2019). *Effect of gamma radiation on the physico mechanical properties of recycled HDPE/modified sugarcane bagasse composite. J. Macromol. Sci. Part A* 56 (2): 127–135.

86 Ravi, M., Dubey, R.R., Shome, A. et al. (2018). *Effect of surface treatment on natural fibers composite.* In: *IOP Conference Series: Materials Science and Engineering*. Mangalore Institute of Technology & Engineering, Badaga Mijar, Moodbidri, Karnataka, India (2–3 March 2018). IOP Publishing.

87 Konczewicz, W. and Kozłowski, R. (2012). *Enzymatic treatment of natural fibres.* In: *Handbook of Natural Fibres*, 168–184. India: Elsevier.

88 Bharath, K.N. and Basavarajappa, S. (2016). *Applications of biocomposite materials based on natural fibers from renewable resources: a review. Sci. Eng. Compos. Mater.* 23 (2): 123–133.

89 Rodriguez, L.J., Peças, P., Carvalho, H. et al. (2020). *A literature review on life cycle tools fostering holistic sustainability assessment: an application in biocomposite materials. J. Environ. Manage.* 262: 110308.

90 Akampumuza, O., Wambua, P.M., and Ahmed, A. (2017). et al, Review of the applications of biocomposites in the automotive industry. *Polym. Compos.* 38 (11): 2553–2569.

91 Wei, L. and McDonald, A.G. (2016). *A review on grafting of biofibers for biocomposites. Materials* 9 (4): 303.

92 AL-Oqla, F.M. and Omari, M.A. (2017). Sustainable biocomposites: challenges, potential and barriers for development. In: *Green Biocomposites* (ed. M. Jawaid, M.S. Salit, and O.Y. Alothman), 13–29. Springer.

93 Nagalakshmaiah, M., Afrin, S., Malladi, R.P. et al. (2019). *Biocomposites: present trends and challenges for the future.* In: *Green Composites for Automotive Applications* (ed. G. Koronis and A. Silva), 197–215. Elsevier.

94 Ilyas, R. and Sapuan, S. (2020). *Biopolymers and biocomposites: chemistry and technology. Curr. Anal. Chem.* 16 (5): 500–503.

95 Bahrami, M., Abenojar, J., and Martínez, M.Á. (2020). *Recent progress in hybrid biocomposites: mechanical properties, water absorption, and flame retardancy. Materials* 13 (22): 5145.

96 Mohanty, A.K., Misra, M., and Drzal, L.T. (2005). *Natural Fibers, Biopolymers, and Biocomposites.* CRC Press; Taylor and Francis Group.

97 Drzal, L.T., Mohanty, A., and Misra, M. (2001). *Bio-composite materials as alternatives to petroleum-based composites for automotive applications. Magnesium* 40 (60): 80–93.

98 dos Santos Rosa, D. and Lenz, D.M. (2013). *Biocomposites: influence of matrix nature and additives on the properties and biodegradation behaviour. Biodegrad. Eng. Technol. Intech Rij Croat.* 433–475.

99 May-Pat, A., Valadez-González, A., and Herrera-Franco, P.J. (2013). *Effect of fiber surface treatments on the essential work of fracture of HDPE-continuous henequen fiber-reinforced composites. Polym. Test.* 32 (6): 1114–1122.

100 Thakur, V.K. and Thakur, M.K. (2014). *Processing and characterization of natural cellulose fibers/thermoset polymer composites. Carbohydr. Polym.* 109: 102–117.

101 Väisänen, T., Haapala, A., Lappalainen, R. et al. (2016). *Utilization of agricultural and forest industry waste and residues in natural fiber-polymer composites: a review. Waste Manage. (Oxford)* 54: 62–73.

102 Väisänen, T., Das, O., and Tomppo, L. (2017). *A review on new bio-based constituents for natural fiber-polymer composites. J. Cleaner Prod.* 149: 582–596.

103 Lilholt, H. and Lawther, J.M. (2000). Natural organic fibers. In: *Comprehensive Composite Materials*, vol. 1 (ed. A. Kelly and C. Zweben). New York: Pergamon Press.

104 Al-Oqla, F.M. and Sapuan, S. (2014). *Natural fiber reinforced polymer composites in industrial applications: feasibility of date palm fibers for sustainable automotive industry. J. Cleaner Prod.* 66: 347–354.

105 Hänninen, T., Thygesen, A., Mehmood, S. et al. (2012). *Mechanical processing of bast fibres: the occurrence of damage and its effect on fibre structure. Ind. Crops Prod.* 39: 7–11.

106 Dai, D. and Fan, M. (2014). *Wood fibres as reinforcements in natural fibre composites: structure, properties, processing and applications.* In: *Natural Fibre Composites*, 3–65. Elsevier.

107 Mohan, K. and Rajmohan, T. (2020). *Tribological characteristics of natural fiber composite—review.* In: *IOP Conference Series: Materials Science and Engineering.* IOP Publishing.

108 Vigneshwaran, S., Sundarakannan, R., John, K.M. et al. (2020). *Recent advancement in the natural fiber polymer composites: a comprehensive review. J. Cleaner Prod.* 277: 124109.

109 Matějka, V., Fu, Z., Kukutschová, J. et al. (2013). *Jute fibers and powderized hazelnut shells as natural fillers in non-asbestos organic non-metallic friction composites. Mater. Des.* 51: 847–853.

110 Liu, Y., Ma, Y., Yu, J. et al. (2019). *Development and characterization of alkali treated abaca fiber reinforced friction composites. Compos. Interfaces* 26 (1): 67–82.

111 Narish, S., Yousif, B., and Rilling, D. (2011). *Adhesive wear of thermoplastic composite based on kenaf fibres. Proc. Inst. Mech. Eng. Part J: J. Eng. Tribol.* 225 (2): 101–109.

112 Ramesh, P., Durga Prasad, B., Narayana, K.L. et al. (2018). Characterization of kenaf fiber and its composites: a review. *J. Reinf. Plast. Compos.* 37 (11): 731–737.

113 Monteiro, S.N., Calado, V., Rodriguez, R.J.S. et al. (2012). Thermogravimetric stability of polymer composites reinforced with less common lignocellulosic fibers—an overview. *J. Mater. Res. Technol.* 1 (2): 117–126.

114 Zini, E. and Scandola, M.J.P.C. (2011). Green composites: an overview. *Polym. Compos.* 32 (12): 1905–1915.

115 Chandramohan, D. and Marimuthu, K. (2011). A review on natural fibers. *IJRRAS* 8 (2): 194–206.

116 Biagiotti, J., Puglia, D., and Kenny, J.M. (2004). A review on natural fibre-based composites-part I: structure, processing and properties of vegetable fibres. *J. Nat. Fibers* 1 (2): 37–68.

117 Almeida, J.H.S. Jr., Amico, S.C., Botelho, E.C. et al. (2013). Hybridization effect on the mechanical properties of curaua/glass fiber composites. *Composites Part B* 55: 492–497.

118 Shanmugasundaram, N. and Rajendran, I. (2016). Characterization of raw and alkali-treated mulberry fibers as potential reinforcement in polymer composites. *J. Reinf. Plast. Compos.* 35 (7): 601–614.

119 Ahmad, F., Choi, H.S., Park, M.K. et al. (2015). A review: natural fiber composites selection in view of mechanical, light weight, and economic properties. *Macromol. Mater. Eng.* 300 (1): 10–24.

120 Pickering, K.L., Efendy, M.G.A., Le, T.M. et al. (2016). A review of recent developments in natural fibre composites and their mechanical performance. *Composites Part A* 83: 98–112.

121 Azwa, Z. and Yousif, B.F. (2013). *Characteristics of kenaf fibre/epoxy composites subjected to thermal degradation. Polym. Degrad. Stab.* 98 (12): 2752–2759.

122 de Macedo, J.R.N. and dos Santos Rosa, V.N.C.D. (2016). *Differences of alkali and thermal treatment on cotton residues applicable to green composites. Mater. Sci. Eng. Adv. Res.* 1 (4): 13–18.

123 Kalia, S., Kaith, B.S., and Kaur, I. (2009). Pretreatments of natural fibers and their application as reinforcing material in polymer composites—a review. *Polym. Eng. Sci.* 49 (7): 1253–1272.

124 Xie, Y., Hill, C.A.S., Xiao, Z. et al. (2010). Silane coupling agents used for natural fiber/polymer composites: a review. *Composites Part A* 41 (7): 806–819.

125 John, M.J. and Anandjiwala, R.D. (2008). Recent developments in chemical modification and characterization of natural fiber-reinforced composites. *Polym. Compos.* 29 (2): 187–207.

126 Enwerem, U.E., Obasi, H.C., Onuegbu, G.C. et al. (2021). Effect of surface modification of cotton weaves on the mechanical and morphological properties of polyester composites. *J. Text. Inst.* 113: 1–12.

127 Mohan, T. and Kanny, K. (2019). *Tribological properties of nanoclay-infused banana fiber reinforced epoxy composites. J. Tribol.* 141 (5): 051101.

128 Sailesh, A., Arunkumar, R., and Saravanan, S.J.M.T.P. (2018). Mechanical properties and wear properties of Kenaf–aloe vera–jute fiber reinforced natural fiber composites. *Mater. Today Proc.* 5 (2): 7184–7190.

129 Chaudhary, V., Bajpai, P.K., Maheshwari, S. et al. (2018). *An investigation on wear and dynamic mechanical behavior of jute/hemp/flax reinforced composites and its hybrids for tribological applications. Fibers Polym.* 19 (2): 403–415.

130 Valášek, P., D'Amato, R., Müller, M. et al. (2018). *Mechanical properties and abrasive wear of white/brown coir epoxy composites. Composites Part B* 146: 88–97.

131 Liu, Y., Xie, J., Wu, N. et al. (2019). *Influence of silane treatment on the mechanical, tribological and morphological properties of corn stalk fiber reinforced polymer composites. Tribol. Int.* 131: 398–405.

132 Alshammari, F., Saleh, K.H., Yousif, B.F. et al. (2018). *The influence of fibre orientation on tribological performance of jute fibre reinforced epoxy composites considering different mat orientations. Tribol. Industry* 40 (3): 335–348.

133 Anu, G., Ajit, K., Amar, P. et al. (2012). *Effect of filler content and alkalization on mechanical and erosion wear behavior of CBPD filled bamboo fiber composites. J. Surf. Eng. Mater. Adv. Technol.* 2.

134 Khan, A., Ahmad, M.A., Joshi, S. et al. (2014). Abrasive wear behavior of chemically treated coir fibre filled epoxy polymer composites. *Am. J. Mech. Eng. Autom.* 1 (1): 1–5.

135 Biswas, S. and Satapathy, A.J.M. (2010). *A comparative study on erosion characteristics of red mud filled bamboo–epoxy and glass–epoxy composites. Mater. Des.* 31 (4): 1752–1767.

136 Shalwan, A. and Yousif, B.J.M. (2014). *Influence of date palm fibre and graphite filler on mechanical and wear characteristics of epoxy composites. Mater. Des.* 59: 264–273.

137 Behera, S., Gautam, R.K., Mohan, S. et al. (2021). *Dry sliding wear behavior of chemically treated sisal fiber reinforced epoxy composites. J. Nat. Fibers* 1–14.

138 Swain, P.T.R. and Biswas, S.J.M.R. (2017). *Abrasive wear behaviour of surface modified jute fiber reinforced epoxy composites. Mater. Res.* 20: 661–674.

139 Oladele, I.O., Makinde-Isola, B.A., Adediran, A.D.E.O.L.U.A.D.E.S.O.J.I. et al. (2020). *Mechanical and wear behaviour of pulverised poultry eggshell/sisal fiber hybrid reinforced epoxy composites. Mater. Res. Express* 7 (4): 045304.

140 Rajini, N., Mayandi, K., Siengchin, S. et al. (2021). Tribological properties of *Cyperus pangorei* fibre reinforced polyester composites (friction and wear behaviour of *Cyperus pangorei* fibre/polyester composites). *J. Nat. Fibers* 18 (2): 261–273.

141 Yousif, B., Leong, O.B., Ong, L.K. et al. (2009). *The effect of treatment on tribo-performance of CFRP composites. Recent Patents Mater. Sci.* 2 (1): 67–74.

142 Yousif, B., Lau, S.T., and McWilliam, S. (2010). *Polyester composite based on betelnut fibre for tribological applications. Tribol. Int.* 43 (1-2): 503–511.

143 Chand, N. and Dwivedi, U.J.P.C. (2008). *Sliding wear and friction characteristics of sisal fibre reinforced polyester composites: Effect of silane coupling agent and applied load. Polym. Compos.* 29 (3): 280–284.

144 Shanmugam, D., Thiruchitrambalam, M., and Thirumurugan, R.J.C.I. (2016). *Wear behavior of Palmyra palm leaf stalk fiber (PPLSF) reinforced polyester composites. Compos. Interfaces* 23 (2): 89–103.

145 Kumar, S., Prasad, L., Patel, V.K. et al. (2021). *Experimental and numerical study on physico-mechanical properties and Taguchi's designed abrasive wear behavior of hemp/nettle-polyester hybrid composite. Polym. Compos.* 42 (12): 6912–6927.

146 Kumar, S., Prasad, L., Kumar, S. et al. (2020). Physicomechanical and Taguchi optimized abrasive wear behaviour of $KOH/KMnO_4/NaHCO_3$ treated Himalayan Agave fiber reinforced polyester composite. *Mater. Res. Express* 6 (12): 125353.

147 Divya, G., Kakhandaki, A., and Suresha, B. et al, *Wear behaviour of coir reinforced treated and untreated hybrid composites. Int. J. Innovative Res. Dev.* 3 (5): 632–639.

148 Manickam, C., Kumar, J., Athijayamani, A. et al. (2015). *Mechanical and wear behaviors of untreated and alkali treated roselle fiber-reinforced vinyl ester composite. J. Eng. Res.* 3 (3): 1–13.

149 Kumar, P., Suresha, B., and Hemanth, R. (2018). *Mechanical and tribological behavior of vinyl ester hybrid. Composites Part A* 40 (2).

150 Merzoug, A., Akgul, Y., Bouhamida, B. et al. (2020). *Sliding wear properties of palm/glass fiber hybrid reinforced vinylester resin.* In: *Advances in Materials, Mechanics and Manufacturing*, 103–114. Cham: Springer.

151 Pashaei, S. and Hosseinzadeh, S. (2017). *Sliding wear behaviour of walnut shell powder filled vinyl ester/WSP green composites. Iran. Chem. Commun.* 5 (212123615): 138–146.

152 Shalwan, A. and Yousif, B.J.M. (2013). *In state of art: mechanical and tribological behaviour of polymeric composites based on natural fibres. Mater. Des.* 48: 14–24.

153 Omrani, E., Menezes, P.L., Rohatgi, P.K. et al. (2016). *State of the art on tribological behavior of polymer matrix composites reinforced with natural fibers in the green materials world. Eng. Sci. Technol. Int. J.* 19 (2): 717–736.

154 Sahoo, P., Das, S., and Acharya, S.J. (2017). 2. Tribological aspects of natural fiber composites. In: *Green Composites: Materials, Manufacturing and Engineering* (ed. J. Paulo Davim), 1–58. Berlin, Boston: De Gruyter.

155 Nijjar, S., Sudhakara, P., Sharma, S. et al. (2021). *Influence of alkali treatment and maleated polypropylene (MAPP) compatibilizer on the dry-sliding wear and frictional behavior of borassus fruit fine fiber (BFF)/polypropylene (PP). Polym. Compos. Various Eng. Appl.* 2021.

156 Adediran, A.A., Balogun, O.A., Akinwande, A.A. et al. (2021). *Effect of surface modification on the properties of polypropylene matrix reinforced with coir fibre and yam peel particulate. Sci. World J.* 2021: 8891563.

157 Chand, N. and Dwivedi, U.K. (2006). *Effect of coupling agent on abrasive wear behaviour of chopped jute fibre-reinforced polypropylene composites. Compos. Sci. Technol.* 261 (10): 1057–1063.

158 Njoku, C.E., Omotoyinbo, J.A., Alaneme, K.K., and Daramola, M.O. (2021). *Physical and abrasive wear behaviour of Urena lobata fiber-reinforced polymer composites. J. Reinf. Plast. Compos.* 40 (9-10): 341–351.

159 Ibrahim, M.A., Hirayama, T., and Khalafallah, D. (2019). *An investigation into the tribological properties of wood flour reinforced polypropylene composites. Mater. Res. Express* 7 (1): 015313.

160 Dwivedi, U. and Chand, N.J. (2009). *Influence of MA-g-PP on abrasive wear behaviour of chopped sisal fibre reinforced polypropylene composites.* 209 (12-13): 5371–5375.

161 Goriparthi, B.K., Suman, K.N.S., Rao, N.M. et al. (2012). *Effect of fiber surface treatments on mechanical and abrasive wear performance of polylactide/jute composites. J. Mater. Process. Technol.* 43 (10): 1800–1808.

162 Govindan, P., Arul Jeya Kumar, A., Lakshmankumar, A. et al. (2021). *Wear and morphological analysis on basalt/sisal hybrid fiber reinforced poly lactic acid composites.* 236: https://doi.org/10.1177/14644207211067657.

163 Liu, T., Li, B., Lively, B. et al. (2014). *Enhanced wear resistance of high-density polyethylene composites reinforced by organosilane-graphitic nanoplatelets. Wear* 309 (1-2): 43–51.

164 Nayak, S., Khuntia, S.K., Mohanty, S.D. et al. (2020). *An experimental study of physical, mechanical and morphological properties of alkali treated Moringa/Areca based natural fiber hybrid composites. J. Nat. Fibers* 19: 1–12.

165 Sever, K. (2010). *The improvement of mechanical properties of jute fiber/LDPE composites by fiber surface treatment. J. Reinf. Plast. Compos.* 29 (13): 1921–1929.

166 Sultana, S., Nur, H.P., Saha, T. et al. (2012). *Fabrication of raw and oxidized saw dust reinforced low density polyethylene (LDPE) composites and investigation of their physico-mechanical properties. Bangladesh J. Sci. Ind. Res.* 47 (4): 365–372.

167 Rahman, R. and Putra, S.Z.F.S. (2019). *Tensile properties of natural and synthetic fiber-reinforced polymer composites. Mech. Phys. Test. Biocompos. Fibre Reinf. Compos. Hybrid Compos.* 81–102.

168 Elkhouly, H.I. (2021). *Comparison of the effects of nano date seed as reinforcement material for medium-density polyethylene (MDPE) and polyethylene terephthalate (PET) using multilevel factorial design. Polym. Polym. Compos.* 29 (9): 1462–1471.

169 Shireesha, Y. and Nandipati, G. (2019). *State of art review on natural fibers. Mater. Today Proc.* 18: 15–24.

170 Vinayagamoorthy, R.J. (2020). *Friction and wear characteristics of fibre-reinforced plastic composites. J. Thermoplast. Compos. Mater.* 33 (6): 828–850.

171 Nishitani, Y., Kajiyama, T., and Yamanaka, T. (2017). *Effect of silane coupling agent on tribological properties of hemp fiber-reinforced plant-derived polyamide 1010 biomass composites. Materials* 10 (9): 1040.

172 Nishitani, Y., Mukaida, J., Yamanaka, T. et al. (2016). Thermal properties of hemp fiber filled polyamide 1010 biomass composites and the blend of these composites and polyamide 11 elastomer. *AIP Conf. Proc.* 1713: 120007. AIP Publishing LLC.

173 Nishitani, Y., Sugawara, N., Kawasaki, K. et al. (2019). Friction and wear properties of recycled natural fiber reinforced plant-derived polyamide 1010 biomass composites. *AIP Conf. Proc.* 2055: 050001. AIP Publishing LLC.

174 Gang, D.J.S. (2018). *The influence of surface treatment on the tensile and tribological properties of wood fiber-reinforced polyimide composite. Surf. Interface Anal.* 50 (3): 304–310.

175 Ramesh, P., Prasad, B.D., and Narayana, K. (2019). Morphological and mechanical properties of treated kenaf fiber/MMT clay reinforced PLA hybrid biocomposites. *AIP Conf. Proc.* 2057: 020035. AIP Publishing LLC.

176 Siakeng, R., Jawaid, M., Asim, M. et al. (2020). *Alkali treated coir/pineapple leaf fibres reinforced PLA hybrid composites: evaluation of mechanical, morphological, thermal and physical properties. Express Polym. Lett.* 14 (8).

177 Balachandran, G.B., David, P.W., Alexander, A.B. et al. (2021). *Saccharum barberi grass bagasse ash-based silicone rubber composites for electrical insulator applications. Iran. Polym. J.* 30 (12): 1285–1296.

178 Oladele, I.O., Aliu, S.O., Taiwo, A.S. et al. (2022). *Comparative investigation of the influence of stone-dust particles and bagasse fiber on the mechanical and physical properties of reinforced recycled high-density polyethylene bio-composites. Compos. Adv. Mater.* 31: https://doi.org/10.1177/26349833221077701.

179 Savaş, S., Gurbanov, N., and Doğan, M. (2019). *Effect of fiber type, fiber content, and compatibilizer on two-body abrasive wear performance of HDPE matrix composites. J. Compos. Mater.* 53 (19): 2743–2760.

180 Liu, T., Wood, W., and Zhong, W.-H. (2011). *Sensitivity of dielectric properties to wear process on carbon nanofiber/high-density polyethylene composites. Nanoscale Res. Lett.* 6 (1): 1–9.

Index

a

abrasive wear (AW) 10, 28, 52, 59, 90, 109, 147, 225
- adhesive and 147
- dimension 225
- mechanism of 59

acetic acid treatment (AAT) 251
acoustic emission monitoring test 241
acrylonitrile butadiene styrene (ABS) 221, 255
additive manufacturing (AM)
- of biocomposites 220
  - fabrication of 222
- process parameters 220
- specimen deterioration 227–228
- wear and frictional characteristics of 228
- wear and friction test 228
  - acoustic emission monitoring test 241
  - pin-on-disc or tribometer 239
  - pin-on-drum 239–240
  - repeated impact wear test 240
  - Rubbing test 241
- wear behaviour 222
  - abrasive wear 224–225
  - adhesion wear 223–224
  - corrosive oxidative wear 226
  - erosion wear 225–226
  - fatigue wear 226
  - fretting wear 226

adhesion wear 223–224
Amonton's formula 148
animal fiber 135
areca sheath (AS) fiber 69
atomic force microscopy (AFM) 89, 197

b

bamboo fibers 141, 162–163
- based epoxy composites 29

banana fibers 4, 14
benzoyl chloride treatment 259
benzoyl peroxide treatment (BPT) 251
betel nut fibers 7–8
biobased composites
- banana reinforcement 4
- Betel nut fibers 7–8
- *Calotropis gigantea* reinforcement 5–6
- coconut coir reinforcement 4
- flax reinforcement 3
- hemp reinforcement 4–5
- Kenaf reinforcement 6–7
- Ramie reinforcement 5
- tribological characteristics
  - fillers on wear behavior 11
  - morphology analysis 12–14
  - reinforcement orientation on wear behavior 8–9
  - reinforcement volume fraction on wear behavior 9–10
  - surface modification on wear behavior 11–12
- tribological characterization 2

biocompatibility 186, 187, 189, 191, 193, 197, 205
biocomposites 71, 187
- additive manufacturing (AM) 220–222

Tribological Properties, Performance, and Applications of Biocomposites, First Edition.
Edited by Chandrasekar Muthukumar, Senthilkumar Krishnasamy,
Senthil Muthu Kumar Thiagamani, and Ganesan Chinnachamy.

biocomposites (*contd.*)
  classification 254, 255
  natural fiber polymer composites 254–255
  natural fiber-reinforced composites 255–256
  orthopaedic applications 200–203
    friction and wear of biocomposites 203–207
  sliding wear 173–177
biomaterials 185
  classification of
    ceramic biomaterials 193
    composite biomaterials 193–194
    metallic biomaterials 189–193
  dental applications 197
  desired properties of 188
  wear of biomaterials
    friction and technique 196–197
    wear testing methods 195–196
bio-polymer composite
  biofiller incorporated 173–175
  synthetic/inorganic filler incorporated 175–177
block-on-disc (BOD) machine 4, 26
block-on-ring apparatus 84
2-body abrasive wear 68, 225
3-body abrasive wear 67, 225, 262
Boehmeria nivea 5
bone/implant surfaces 201
Borassus fruit fine fiber (BFF) 92, 264

c

Calotropis gigantea
  fiber-reinforced epoxy composites 28
  reinforcement 5–6
cane fibers 251
carbon fiber reinforced carbon (CFRC) 201
carbon footprint 173
carbon nanofibers (CNFs) 271
cellulose 77
cellulose fibers 131, 250, 253
cellulosic natural fibers 21–23, 121, 161
cement by-pass dust (CBPD) 68, 259
ceramic biomaterials 193
ceramic matrix composites (CMC) 185
ceramic palm fiber composite 161
chemical wear 147
cobalt-based alloys 189, 191
coconut coir reinforcement 4
coconut fiber 164, 271
Cocos nucifera 164
coefficient of friction (COF) 3, 27, 69, 86, 108, 144, 170
composite biomaterials 193–194
composite materials 19, 20, 29, 34, 48, 51, 65, 117, 131, 135, 185, 187
compression molding method 40, 262
conventional composites 60, 67, 199, 200, 219
corn stalk fibers 12, 13
corrosive wear 2, 90, 170, 226
cotton fiber 137
cotton plant 137
crystalline polymers 131

d

date palm fibers (DPF) 250, 262
dental applications 197
  friction and wear of
    dental resins composites in vitro 199–200
    dental restorations in vivo 200
  non-metallic implants in 198
dental resins composites in vitro 199–200
dental restorations in vivo 200

e

energy-dispersive X-ray spectroscopy (EDXS) 196
erosion wear 225, 259
erosive wear 70, 90, 169, 225–226
ethylenediaminetetraacetic acid (EDTA) 253
Eulaliopsis binata fiber-reinforced epoxy composites 29

f

fatigue wear 2, 90, 147, 226
Fe-based alloys 191
fiber reinforced plastics (FRP) 91, 201
fiber-reinforced polymer composites (FRP) 149, 255
fiber-reinforced polymers 187
fibers 133, 248
fiber volume fraction 10

fibre cell 77
fibre-reinforced composites 20
fibre-reinforced polymers (FRPs) 19, 21
  tribological properties 25–33
flax/epoxy based composites
  alkali treatment 40, 43
  compression moulding method 40
  method 40
  wear results 41–43
  wear testing 40
flax fiber 139
fluorocarbon base (FA) 50
fretting wear 90, 226
friction 142
frictional resistance 10
friction test method 148

g

glass fiber (GF) 22, 32, 68, 149, 262
glass-vinyl ester (GVE) composites 67
graft copolymerization (GC) 251
graphitic nanoplatelets (GNP) 267

h

Handy lay-up method 52
hardness 225
hemicellulose 13, 21, 49, 77, 118, 161
hemp fiber (HF) 4, 112, 139, 269
high corrosion resistance 170, 189
high-density polyethylene (HDPE) composites 271
high wear resistance 187, 189, 207
himalayan agave fiber (HAF) 262
human body
  different polymers and composite 194
  implant material 188
hybrid polyester–composites 50–51
hydrophilic fibers 161, 253
hydroxyapatite (HA) 193

i

Ipomoea carnea fibre-reinforced epoxy composites 28
isocyanate treatment (IT) 251
Izod method 240

j

jute fiber (JF) 111, 112, 138, 161, 265
jute-reinforced polymer composites 162

k

Kenaf fibers 6, 7, 26, 111, 140, 270
kinetic friction 106, 143

l

lamina 3, 4, 8, 10–15, 51
lignin 5, 23, 77, 80
low-density polyethylene (LDPE) composites 267–268

m

maleated coupling agents 252
maleate polypropylene (MAPP) 264
maleic anhydride-g-polypropylene (MA-g-PP) 111
Malvaceae 6, 137, 139
man-made fibers 22, 75, 248, 250
metallic-based biomaterials 189
metal matrix composites (MMC) 185
metal oxide fillers, biocomposite
  bamboo fibre 162–163
  coconut fiber 164
  jute fiber 161–162
  oil palm fiber 160–161
micro emulsion silicon (MS) 50
milkweed 5
mineral fiber 141
mudar 5
multi walled carbon nanotube (MWCNT) 5, 176
Musaceae fiber (MF) 69–71
Musa fiber 4

n

nano-clay treated fiber (NC-BF) 257
natural fiber-based PLA composite 71
natural fiber-based VE composite 70
natural fiber-filled composites 254
natural fiber-filled polymer
  wear properties of 257–260
natural fiber flax 40
natural fiber-polymer composites (NFPCs) 254
natural fiber-reinforced high-density polyethylene composites 267
natural fiber-reinforced hybrid biocomposites 270–273

natural fiber-reinforced low-density polyethylene composites 267–268
natural fiber-reinforced polyamide composites 269–270
natural fiber reinforced polyester composites 260–262
natural fiber-reinforced polyethylene terephthalate composites 269
natural fiber reinforced polyimide composites
  driving forces 115–117
  hand layup 120–121
  materials and methods 117–118
    natural fibres 118–119
    production of 119
  natural fibers/polyimides composites 121–122
  polyimides 121
  stratification 120
  triblogical applications of
    natural fibers/polyimides composites 122–124
  vacuum bagging technique 121
natural fiber reinforced polylactic acid composites 265–267
natural fiber reinforced polypropylene composites 263–265
natural fiber reinforced thermoplastic composites
  load 107
  motion type 107
  performance metrics from wear test 108–109
  temperature 108
  test duration 108
  velocity 107–108
  wear behaviour of composite 107
  wear test 109–112
  wear testing methods 105–107
natural fiber reinforced vinyl ester composite 262–263
natural fiber reinforcement 249
natural fiber reinforcement polyamide composites
  applications 149–150
  fibers 133–134
  friction 142–145
  friction test methodologies 148
  natural fibers 135–141
  polyamide 134–135
  polymer materials 129
  production 141–142
  thermoplastic composites 131
  thermoplastic polymer matrix 132
  thermoplastic polymers 130
  thermosetting polymers 130
  wear 145–147
  wear test methodologies 148
natural fibers (NFs) 1, 20–22, 47, 65, 70, 75, 109, 247, 248
  advantages 124
  biological treatment 253–254
  categories of 78
  chemical treatment
    acetylation treatment 252
    alkaline treatment 251
    benzoylation treatment 252
    esterification 251
    maleated coupling agents 252
    peroxide treatment 252
    silane treatment 251
  as compared to conventional reinforcing glass fiber 136
  composites 115
  mechanical and chemical characteristics 79
  natural fiber reinforced polyimide composites 118–121
  natural fiber reinforcement polyamide composites 135–141
  natural fibre reinforced polypropylene composites 76–80
  physical treatment 252–253
  polyester based biocomposites 49–50
  treatments of 250
  tribological studies 86
natural fibers/polyimides composites 121
natural fibers polymer composites (NFPCs) 1, 159
natural fibre composites 49
natural fibre PP composites
  wear rate and friction coefficient 87
natural fibre-reinforced epoxy composites
  environment and industry 23–25
  FRP 25
  tribological properties of 25–30
  tribological test 30

natural fibre-reinforced epoxy hybrid composites 30–33
natural fibre reinforced polypropylene composites 80
  dry sand rubber wheel test 84
  friction coefficient of 86–90
  natural fibres 76–80
  pin-on-drum test 84
  polypropylene 76
  tribological properties of 83–86
  wear behaviour of 90–94
natural fibre reinforced vinyl ester composites
  composite materials 67
  natural fiber-based PLA composite 71
  natural fiber-based VE composite 70–71

o
oil palm fiber 56, 160–161
organosilanes 251
orthopaedic applications 200–203
  friction and wear of biocomposites 203–207
oxidative wear 226

p
palm fiber composites (PFC) 161
palmyra palm-leaf stalk fiber (PPLSF) 260
partially crystalline polymers 131
pectin 13, 22, 77, 80, 250, 256
petiole date palm fiber (PDPF) 262
pin-on-disc type test 198
pin on disk (POD) 56, 106, 239
  method 196
  tribometer 84
pin-on-drum 84, 239–240
plant-based bio-resources 66
plant fiber 135
polyalthia longigolia seed (PLS) 68
polyamide
  matrix composites 132
  natural fiber reinforcement polyamide composites 134–141
polycarbonate (PC) 255
polyester based biocomposites
  driving forces 47–48
  handy lay-up method 52
  hybrid polyester–composites 50–51
  natural fibre composites 49–50
  natural fibres 49
  stratification 51
  tribological applications 59–60
  tribological characteristics of 53–59
  tribological tests on natural fibre 52–53
  vacuum bagging technique 52
polyester composites 48
polyester hybrid composites 56
poly ether-ether ketone (PEEK) 197
polyethylene terephthalate (PET) 255
polyethylene terephthalate-glycol (PETG) 221
polyimides 121
polylactic acid (PLA) 3, 109, 110, 221, 265, 270
polylactide/jute composites 266
polymer-based composites 65
polymer coating (PT) 251
polymer composites 185
polymeric resin 10
polymer matrix composites (PMC) 167, 185
polymers
  classification of 130
  wear classification 59
polypropylene (PP) 76, 111
polytetrafluoroethylene (PTFE) 149, 172, 203
polyurethane/KF composite 111
polyvinyl alcohol (PVA) 69
potassium permanganate treatment (PPT) 251
pressure-velocity (PV) 204
pristine polyurethane (PU) 110

r
Ramie fiber (RF) 5, 6, 112
response surface methodology (RSM) 31, 112
restorative dentistry 197
Rubbing test 241

s
scanning electron microscope/light microscope (SEM/LM) 196

scanning electron microscopy (SEM) 4, 198, 241, 260, 263, 266
semi-crystalline polymers 131
short fiber reinforcements 133
silane treatment (ST) 251, 257, 260, 267, 269, 272
silk fibers 135
sisal fiber (SF) 8, 29, 32, 50, 89, 91–93, 112
sliding friction 144
sliding wear 168, 169
  in bio composites 173
  bio-filler incorporated bio-polymer composite 173–175
  measure wear method 170–171
  in polymer composites 171–173
  synthetic/inorganic filler 175–177
  wear general aspects 168–170
specific wear rate (SWR) 40, 41
  vs sliding distance 42
  vs sliding speed 42
specific wear resistance (SWR) 263
stainless alloys 191
static friction 80, 143
static *vs.* kinetic friction 143
stearic acid treatment (SAT) 251
stratification 51, 120
subtractive manufacturing 220
surface-to-surface friction 188
syngonanthus nitens 27
synthetic fibers 19, 22, 25

t

thermoplastic composites 89, 131–132
thermoplastic polymer matrix 132–133
thermoplastic polymers 130–131, 133, 221, 254
thermosetting polymers 65, 67, 130
TiO_2 nanoparticles 163
titanium-based alloys 187, 191
tribological effects 2
tribological tests 29, 30, 52, 53, 256
tribometer 2, 83, 106, 239
tribotester 83
tricalcium phosphate (TCP) 193

u

ultra-high molecular weight polyethylene (UHMWPE) 149, 203–206
untreated Jute fibers (UJF) 92, 111, 265

v

vacuum bagging technique 52, 121
VE composites 68–71
VE/*Polyalthia longifolia* seed (PLS) composites 68
vinyl ester (VE) resin 66

w

wax 13, 22, 77, 80, 252, 256
wear 145–147
  characteristics 224
wear and friction test
  acoustic emission monitoring test 241
  pin-on-disc or tribometer 239
  pin-on-drum 239–240
  repeated impact wear test 240
  rubbing test 241
wear mechanism 7, 25, 58, 71, 90, 94, 123, 162, 176
wear of biomaterials
  friction and technique 196–197
  wear testing methods 195–196
wear parameters 146
wear rate *vs.* friction coefficient 150
wear test 40
  merits and demerits of 196
  methodologies 148
  methods 105–107, 195–196

x

X-ray photoelectron spectroscopy (XPS) 197